International Scientific-Practical Conference "Transformation of Corporate Governance Models under the New Economic Reality" (Conf-Corp 2020)

SHS Web of Conferences Volume 89 (2020)

Ekaterinburg, Russia
20 November 2020

Editor:

Irina Tkachenko

ISBN: 978-1-7138-2411-4

Printed from e-media with permission by:

Curran Associates, Inc.
57 Morehouse Lane
Red Hook, NY 12571

CURRAN ASSOCIATES INC. proceedings.com

Some format issues inherent in the e-media version may also appear in this print version.

For additional information, please contact EDP Sciences – Web of Conferences at the address below.

EDP Sciences – Web of Conferences
17, Avenue du Hoggar
Parc d'Activité de Courtabœuf
BP 112
F-91944 Les Ulis Cedex A
France

Phone: +33 (0) 1 69 18 75 75
Fax: +33 (0) 1 69 28 84 91

contact-edps@webofconferences.org

Additional copies of this publication are available from:

Curran Associates, Inc.
57 Morehouse Lane
Red Hook, NY 12571 USA
Phone: 845-758-0400
Fax: 845-758-2633
Email: curran@proceedings.com
Web: www.proceedings.com

TABLE OF CONTENTS

Preface

We are pleased to present this collection of scholarly articles of the International Scientific-Practical Conference «Transformation of Corporate Governance Models under the New Economic Reality» (CC-2020). The conference was held in Ural State University of Economics, Yekaterinburg, Russia, on November 20, 2020.

The key topics of the Conference were:

- A shift in the corporate governance paradigm in the context of the technological transformation and the coronavirus crisis;
- Boards of Directors as drivers of business transformation in the new post-pandemic reality;
- Modernization of the stakeholder approach in the development of corporate governance models, understanding of the interests of stakeholders and evaluation of their contribution to the formation of the value and social capital of the business;
- The impact of new technologies (big data, artificial intelligence, neural networks) on the development and efficiency of corporate governance systems;
- Coordination of the corporate governance system with new business formats: ecosystem, platform, distributed companies;
- The impact of digitalization on the formation of strategic corporate interests of a company;
- CSR and corporate strategies: problems of transformation to achieve sustainable development.

The conference was attended in an online format by almost 70 participants from leading universities in 10 countries: Russia, Italy, Hungary, Poland, Bulgaria, Belarus, Uzbekistan, Moldova, Kazakhstan, DPR. The geography of Russian participants is also extensive - Moscow, St. Petersburg, Novosibirsk, Kazan, Saratov, Khanty-Mansiysk, Elista, Omsk, Kemerovo, Ryazan, Nizhny Novgorod, Grozny and other cities.

The conference was held with the informational support of the Russian Institute of Directors, NP "Elite Club of Corporate Conduct". In addition to representatives of the scientific community, the conference was also attended by representatives of the Moscow Exchange, investment companies, representatives of enterprises in the real sector of the economy - NPO Automation, UVZ and others.

An exchange of views, an interesting discussion took place, they talked about how to interpret the processes taking place in corporate governance and in corporate business in general.

Under the financial support of the Russian Foundation for Basic Research (RFFI) within the framework of the scientific project no. 20-010-22044.

The Organizing Committees of CC-2020
Ural State University of Economics

Irina Tkachenko

Marina Evseeva

Statement of Peer review

In submitting conference proceedings to *Web of Conferences*, the editors of the proceedings certify to the Publisher that

1. They adhere to its **Policy on Publishing Integrity** in order to safeguard good scientific practice in publishing.
2. All articles have been subjected to peer review administered by the proceedings editors.
3. Reviews have been conducted by expert referees, who have been requested to provide unbiased and constructive comments aimed, whenever possible, at improving the work.
4. Proceedings editors have taken all reasonable steps to ensure the quality of the materials they publish and their decision to accept or reject a paper for publication has been based only on the merits of the work and the relevance to the journal.

Title, date and place of the conference

Conf-Corp 2020 - International Scientific-Practical Conference «Transformationof Corporate Governance Models under the New Economic Reality»
November 20, 2020
Ekaterinburg, Russian Federation

Proceedings editor(s):

Tkachenko Irina

Date and editor's signature

04/12/2020

Management of small agricultural cooperation and corporatization

Alexey Tolmachev[1,*], *Vitaliy* Smirnov[1], *Evgeniy* Grishin[1] , and *Denis* Syromyatnikov[2]

[1]Kuban State Agrarian University named after I.T. Trubilin, Kalinin Str., 13, 350044 Krasnodar, Russia
[2]North-Caucasus Federal University, Pushkin Str., 1, 355017 Stavropol, Russia

Abstract. The possibility of applying the principles of corporate governance in the practice of small agricultural businesses is considered. The authors believe that cooperative associations of small market participants can transform into corporate associations at a certain stage of development. The scientific novelty is represented by an improved economic mechanism of two-level cooperation of small agricultural enterprises with the financial participation of municipalities, regional and Federal authorities and the formation of their own cooperative Bank. The mechanism recommended by the authors is aimed at motivating small business participants to cooperate by the possibility of subsequent corporatization. The implementation of the cooperation development plan will require temporary exemption of its participants from the tax burden, while the state should practice joint construction of processing, logistics and other necessary facilities in the agricultural sector, with the subsequent transfer of its shares to consumer cooperatives of small businesses.

1 Introduction

To date, there is no short complete definition of corporate management that meets all the set goals. Moreover, there is no such precise definition in Russian-language literary sources, and in foreign ones [1, 2, 3]. In our opinion, this definition includes a systematic method of managing a cooperative Association, aimed at a fair distribution of the final result of economic activity in the agricultural sector. In this way, in our opinion, participants in shared equity cooperative activities can also get an opportunity to continue closed or open corporatization within their associations, and to exercise control over the Association's management.

According to the Civil code of the Russian Federation (paragraph 1, article 65), in our country, corporate participants are considered to be founders, who form their own Supreme management body, beginning with a Chief Executive Officer (CEO). This may include all legal entities of a business that bring profits – or losses – to their majority and minority owners, regardless of whether they call themselves corporations or not.

According to Federal law No. 193 of 08.12.1995 "On agricultural cooperation", an agricultural cooperative is an organization formed by farmers of private subsidiary farms of

* Corresponding author: tolmachalex@mail.ru

citizens on the basis of voluntary joint production, as well as other economic activities, on the basis of combining property for the purpose of solving material and other joint problems of cooperative members. More than 15 persons or two justice units which produce products in agriculture may set up cooperative with one vote from one member. In this regard, cooperative associations of small participants in agricultural business at a certain stage of development can be transformed into corporate associations, corporations.

Until recently, research on the development of corporate governance was more focused on operational management, but in the recent period, this started to include a strategic horizon with control of corporate activity [4, 5]. In connection with the above-mentioned provisions, the study of the problems of small scaled cooperative development with subsequent joint-stock transformation appears to be an urgent task.

2 Materials and Methods

The methodology of this study is based on scientific results obtained by foreign and Russian experts. References, normative indicators, published materials of open scientific periodicals, information from the world Internet network, logical techniques, and authors' calculations have become the information base [6, 7]. For the purposes of this study, the authors have used abstract, comparative, monographic, and computational methods and techniques of econometric modeling.

The scientific novelty is represented by an improved economic mechanism of three-level cooperation of small agricultural enterprises united by the efforts of their participants with the financial support of municipalities, regional and Federal government entities with the formation of their own cooperative bank. The mechanism recommended by the authors is aimed at increasing motivation of small business participants in cooperation with the possibility of subsequent corporatization.

3 Results and Discussion

Currently, agricultural cooperative development is experiencing significant difficulties when working in conditions of perfect competition, the absence of market power, confidence of its participants in this form of organization, access to relatively "cheap" credit facilities, and other problems [8, 9]. To compensate for the uneven market conditions, Krasnodar regional and Federal government entities provide some subsidies, which is still insufficient within the framework of the agribusiness development program (Table 1).

Table 1. Subsidies of state support for the regional agriculture, million rubles

Indicators	2013	2017	2018	2019
Federal budget	6,317.6	4,431.7	3,642.6	4,091.6
including subsidies for: legal entities (LE) and sole proprietors (SP)	6,063.6	4,367.9	3,589.5	4,008.2
municipalities	117.0	63.8	53.1	83.4
other participants	136.9	–	–	–
regional budget	3,374.3	3,019.3	4,105.2	3,973.5
including subsidies for: LE and SP	1,264.0	1,492.6	2,038.2	2,330.5
municipalities	117.6	168.7	105.2	135.0
other participants	326.0	1357.9	1,962.0	1,508.0
local budgets	29.3	–	–	–
additional supply	1,637.1	–	–	–
Total:	9,691.9	7,451.0	7,747.8	8,065.1

Since 2014, the amount of state support funding has been significantly reduced due to the well-known sequestration of Federal spending, and so far, even without taking annual inflation into account, which has not reached the level of 2013. This lag has reached 23.1% by 2017, while capital investment has been transferring to legal entities, individual entrepreneurs and other market participants at the expense of regional budgeting funds that were growing significantly.

The program for the year of 2019 provided by Central Federal government and local budgeting in the amount of 8,295.2 million rubles was implemented on 8,065.1 million rubles, or, in other words, 97.2% of the dotation was spent.

The state development program also includes financial support for agricultural cooperation for the implementation of investment development projects in the Krasnodar territory. The 2019 subprogram was planned to provide grants to agricultural cooperatives for development of the material and technical base. 220.0 million rubles were allocated for this purpose, but 214.7 million rubles, or 97.6% of the planned amount of funds, were disbursed. The balance of 5.3 million rubles was formed due to the absence of applicants who meet the conditions and requirements for granting. As a result of the subprogram's implementation, the material and technical base of eight agricultural consumer cooperatives was modernized with the planned four units.

The modern system of agricultural consumer cooperation is formed by first-level cooperatives that operate within certain regional entities. Regional-level cooperatives are forming the second level of cooperation, while Federal districts can also organize the third level. In 2019, there were 160 cooperatives operating in the Krasnodar territory, including 51 credit institutions, 21 processing companies, including supply, sales, and service companies, and 88 of others. The total number of participants (farmers) in these cooperatives was 2,537. Over the past 10 years, the number of cooperatives has decreased by 6.4% and the number of their participants – by 7.9 %, which is a negative factor in agricultural production (Table 2).

Table 2. The dynamics of decline of the number of regional agricultural cooperatives

Number of cooperatives, units	**Years**						
	2010	**2014**	**2015**	**2016**	**2017**	**2018**	**2019**
At the beginning of the year	171	166	165	163	163	161	160
Newly created	4	5	3	1	4	2	3
Liquidated	9	6	5	1	6	3	4

In order to stimulate the development of the system of agricultural consumer cooperation in the Krasnodar territory, regional target programs with the participation of the state had been developed and approved. As a part of these programs, the government of the Krasnodar territory has formed a target guarantee Fund for supporting second-level cooperatives by annual assets in level about 350 million rubles. The main goal of the guarantee Fund is expanding the access of small agricultural producers to financial collateral for agricultural development projects. Consumer cooperatives Garant Kuban and Business Partner are acting as performers of the regional system of the guaranteed financial security created this way.

However, the dynamics of the retrospective state support for the development of the system of agricultural consumer cooperation shows a decrease in its volume in recent years, which is the reason for the above-mentioned decrease in the number of cooperatives and their participants.

The mechanism of the guarantee Fund (GF) is as follows: the funds of the GF cooperatives of the 1st level are obtained and safe money on the accounts in bank institutions, wich perform the role of provision. At the same time, district cooperatives can apply for a loan guarantee to a level 2 coop while simultaneously requesting a loan from a bank, the cash flow is shown in table 3.

Table 3. Performance Indicators of the regional guarantee fund, 2017–2018, thousand rubles

Indicator	**Cooperatives**		**Total**
	Garant Kuban	**Business Partner**	
Regional budget	157,231	113,715	270,946
Returned money to budget	27,450	17,370	44,820
Guarantee Fund assets	196,030	138,264	334,293
Sureties granted to level 1	407,722	217,923	625,645
Quantity, units	117	68	185

Practice shows that the channel of consumer credit cooperatives is often the only way to attract additional financial resources. With an annual decrease in the number of cooperative participants, their lending volumes show a relatively steady increase in financing of small businesses through credit cooperatives against the background of a decrease in bank loans.

We present data on the directions of spending credit funds by small agrarian business participants (Table 4). Their main users were small entities in the fields of crop production, animal husbandry, and greenhouse vegetable growing.

Table 4. Directions for spending credit funds by small farms, 2017–2018

Direction of expenditure	Value, тыс. руб.	Specific weight, %
Purchase of mineral fertilizers	142,034	22.7
Purchase of seeds	86,035	13.8
Plant protection products	83,844	13.4
Purchase of fuel and lubricants	49,430	7.9
Acquisition of land plots	38,168	6.1
Purchase of equipment	31,914	5.1
Purchase of cattle	28,782	4.6
including cows	24,403	3.9
Purchase of vehicles	23,151	3.7
Spare parts	23,151	3.7
Greenhouse construction Costs	21,274	3.4
Purchase of tractors and agricultural machinery	19,397	3.1
Construction of production facilities	19,397	3.1
The cost of building materials	18,771	3.0
Other directions	15,643	2.5
Subtotal	625,645	100.0

Analysis of the economic activity of small agricultural entities shows that increasing the availability and efficiency of obtaining loans in credit agricultural cooperatives for the purchase of fertilizers, plant protection products, seeds, machinery, spare parts, equipment, purchase of livestock and poultry, reconstruction of livestock facilities, organization of procurement, purchase of feed, etc. entails an increase in the gross volume and profitability of the agricultural business. Note that in 2017 through 2018, 34,335 thousand rubles were returned to the regional budget from the guarantee fund.

In the Figure 1 we show the possibility and project improving the existing scheme of vertically integrated cooperation in order to increase efficiency of regulating the development of consumer cooperation of the subjects of small farming with the possibility of subsequent corporatization.

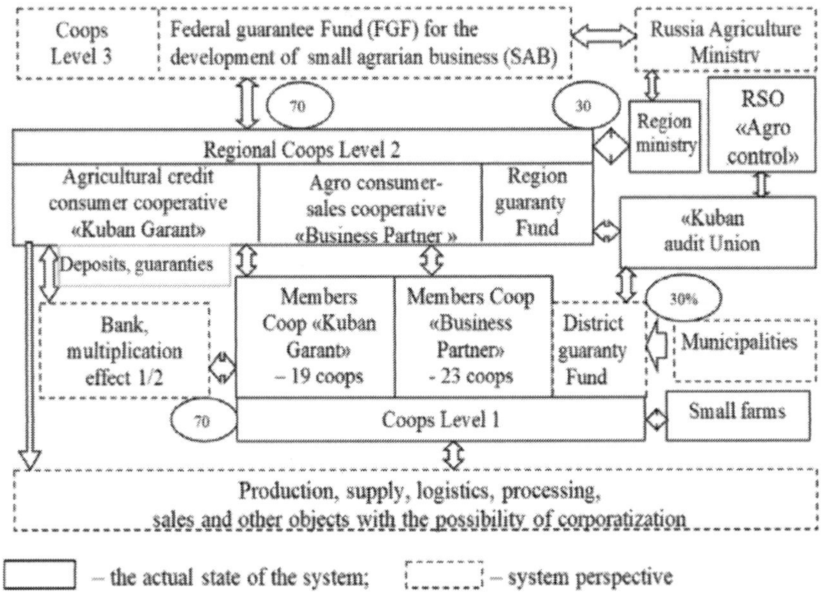

Fig. 1. Vertically integrated system for regulating the development of small agrarian cooperation with the possibility of corporatization (projection)

Given the low current motivation of small business participants for construction and development, we recommend creating infrastructure facilities for joint management with the active involvement of the state. In terms of supply and sales cooperation, it is necessary to actively create an economic infrastructure, using state capabilities to carry out the increasing volumes of production, supply, logistics, processing, sales and other objects that are in demand on practice. In conclusion, we note that the state's participation in the construction of the agricultural cooperation economy is at a low level; it does not correspond to the tasks of import substitution, the development of a small agricultural economy, social factors of employment, and food security.

4 Conclusions

Modern state-subsidized support for small businesses does not compensate for the deformation of market development, perfect competition, and the lack of market power of the small farming in relation to other market participants. In this regard, in the external economic environment of agricultural production, there is an objective need for state regulation of the development of small agricultural entities. At the same time, foreign experience is still underestimated [10], and these factors will continue to hinder the development of small businesses, dooming them to a slow, painful pace.

In the process of the development of the subjects, issues of social protection of participants in small agricultural enterprises should be addressed in a timely manner. We support the initiative of small business participants to introduce small business employees into the unified classifier of professions. We believe in the importance of direction of development of small business by creating a structure of lobbying of interests of small business opportunities regional chamber of commerce and industry, associations of small agrarian business, other instruments to protect the interests of small business entities for land use, resource providing, lending, product sales, corporatization of ownership of cooperatives, and other fields.

We are also proposing an improved scheme of property transformation in the system of vertical cooperation of small agrarian entities with the participation of the state, which differs from the existing approaches by increasing motivation for cooperation by corporatization of new business assets.

The state can create its own production, supply, logistics, processing, marketing and other capacities for the production, processing, storage, warehouse storage of material reserves and other objects that are in demand in practice. Subsequently, based on the willingness of participants created with the participation of the state infrastructure units can be transferred, sold partially, and privatized fully by the members of cooperatives with their further corporatization and the introduction of corporate governance. The shared participation of the regional and Federal governments in the creation and management of assets of agricultural cooperatives of the first two levels will allow to save these capacities from reprofiling until a certain time.

References

1. D. Ahlstrom, D. J. Cumming, S. Vismara, Corporate Governance, **26** (2018)
2. H. A. Shirma, M. Onuk, Jour. of Corporate Governance Res, **4** (2020)
3. Mette Neville, Corporate Governance Intern. Jour. of Busn. In Soc., **11** (2011)
4. E.M. Akhmetshin, R.H. Ilyasov, E.A. Sverdlikova, European Research Studies Journal, **21** (2018)
5. E.G. Pupynina, V.A. Zhukova, A.V. Tolmachev, I.I. Saenko, E.N. Kriulina,

Research Journal of Pharmaceutical, Biological and Chemical Sciences, **10** (2019)

6. D. F. Larcker, G. Ormazabal, D. J. Taylor, Jour. of Finan. Econ., **101** (2011)
7. A. D. Jagolinzer, D. F. Larcker, G. Ormazabal, D. J. Taylor, The Journal of Finance, **75** (2020)
8. A.V. Tolmachev, A.B. Melnikov, Yu.I. Bershitskiy, N.K. Vasilyeva, E. I. Artemova, Journal of Advanced Research in Law and Economics, **8** (2017)
9. J. Yu, A. Smith, D. A. Sumner, American Journal of Agricultural Economics, **100** (2018)
10. S. T. Holden, R. W. Lunduka, American Journal of Agricultural Economics, **96** (2014)

Successful privatization of state property: factors and their impact on efficiency

Mikhail Khamidulin[1,2], *Dilfuza* Rakhimova[1], *Shakhlo* Abdullaeva[2], *Muyassarhon* Ganieva[3], and *Bahramjon* Mamatov[1]

[1]Academy of Public Administration under the President of the Republic of Uzbekistan, st. Karimov 45, 100066 Tashkent, Uzbekistan
[2]Tashkent branch of the Russian University of Economics. G.V. Plekhanov, st. Shahriobod, 3, 100164 Tashkent, Uzbekistan
[3]Tashkent State University of Economics, st. Karimov 49, 100066 Tashkent, Uzbekistan

Abstract. Successful privatization of state property depends on the investment attractiveness of the objects being sold, as well as the general investment climate of the country and the level of development of the financial infrastructure. In the Republic of Uzbekistan, with the adoption of the Decree of the President of Uzbekistan dated October 27, 2020. № UP-6096 "On measures to accelerate the reform of enterprises with the participation of the state and the privatization of state assets", a new stage of privatization of state property began. A significant place in the privatization program is given to the restructuring of state property, improving the efficiency of managing state assets and bringing shares of Uzbek issuers to the world capital markets (IPO). The article examines the degree of influence of an effective corporate governance system and the level of capitalization of the national securities market, contributing to an increase in investment attractiveness and the most effective practical manifestation of the generic advantages of a joint-stock company, ensuring the achievement of the main goal of corporatization - attracting direct investment.

1 Introduction

The formation of a market economy, the effectiveness of which in the socio-economic development of society has been proved by the experience of the world development of mankind, is not possible without the creation of a multi-structured economy, the basis of which is the competitive competition of various forms of ownership. Denationalization and privatization laid the foundation and became a catalyst for the formation and development of the national stock market, the development and application of mechanisms for effective practical implementation of internationally recognized principles of corporate governance of the OECD [1]. With regret, it must be admitted that over the past years, despite the ongoing work to transform state property, no more than 10-15 percent have actually been privatized, mainly in the field of consumer services, trade, local industry, public catering, housing and communal services, etc. P. The enterprises of the basic sectors of the economy of Uzbekistan are de facto, almost 100%, still owned by the state. A new stage of privatization of state property in the Republic of Uzbekistan, the main target parameters for the implementation of which were laid down by the Decree of the President of the country dated 10.27. № UP-6096 "On measures for accelerated reform of enterprises with state participation and privatization of state assets", which provides for the introduction of transparent and market-based principles in the basic industries, "... creation of broader opportunities for participation in the reform process of private capital, reduction of state participation in the economy and further improvement of the competitive environment "[2].Despite the detailed privatization program approved by the aforementioned decree, certain doubts arise that further attempts to form truly market-based property relations in the republic's economic sector may again turn out to be ineffective, primarily in terms of attracting direct investment in privatization objects.At the same time, despite the presence of a formal infrastructure that ensures the attraction of investors to privatization objects through their transformation into joint-stock companies and the sale of issued shares, in the domestic scientific community there is a lack of theoretical developments that reveal the essential characteristics of a joint-stock company, allowing full use of the generic advantages of corporate forms of management in attracting direct, long-term, financial investments. As a result, the adopted practical management decisions do not consider the investment attractiveness of joint-stock companies in the relationship and interdependence of the applied corporate governance system with the level of development of the securities market, especially its secondary segment.

2 Materials and methods

Conceptual foundations in the formation and development of economic science on corporate governance were introduced by leading foreign economists: A. Burleigh, G. Means, B. Black, R. Kreckman, E. Rosenbaum. The theory of corporate finance management was developed in the scientific works of M. Miller, F. Modigliani, M. Gordon, D. Lintner, R. Litzenberger, R. Brailey, S. Myers. Standard neoclassical schemes of a firm's behavior in various environments and organizational features of a corporation are investigated in the works of S. Fischer, R. Dornbusch, R. Schmalenzi, R. Pindike, D. Rabinfeld, R. Friedman, A. Rapensky, M. Hessel, V. Galanov, I.Mazura, and M.Iontseva.The concept of this article, as well as the development of practical skills in managing the processes of attracting investment resources to joint-stock companies formed on the basis of privatized enterprises; require the disclosure of the author's understanding of such methodologically important concepts as corporatism and corporation.

Corporatism is the co-ownership of the property of the corporate community or partnership, contractual relationship in meeting personal and public interests. We are in solidarity with the authors who believe that corporatism is a compromise management in order to ensure a balance of interests [3]. The ability to achieve a relative balance of interests on the basis of consensus and compromises is a distinctive feature of the corporatist model. Corporation - from the English corporation - an association, a set of persons uniting to achieve common goals, carrying out joint activities and forming an independent subject of law - a legal entity [4].

According to professors of economics C. McConnell and S. Brue, a corporation is a legal form of business that is distinct and separate from the specific individuals who own them. These government-recognized "legal entities can acquire resources, own assets, manufacture and sell products, borrow, provide loans, sue and be sued in court, and perform all those functions that are performed by any other type of enterprise. »...

Some researchers present corporate (joint-stock) property as a form of manifestation of private property. The rationale for this position does not go from economics to law, but, on the contrary, from the existing legislative acts and their interpretation in the legal literature to the economic interpretation of legal provisions. However, the legal and not the political and economic interpretation of the category "corporate property" or "collective property" [6] does not reveal its essential content as a historically defined form of appropriation of material and spiritual benefits. As a result, it is impossible to determine the place of this form of ownership in the system of property relations.

In other words, without defining the nature of the appropriation, it is impossible to separate it into a separate, independent form of ownership. The main arguments for considering corporate ownership as one of the varieties of private ownership are:

1. The existence in a market economy of the domination of capital, which determines the entire system of production relations. Consequently, "corporate property" under these conditions does not cease to be capital and the presence of collective property does not cancel the reproduction of the relationship "capital - wage labor", but only gives it certain specificity.It consists in the fact that each worker treats "like his own capitalist to himself as a wage worker" [7]. Therefore, this is the first argument, they believe, that there is no reason to represent joint stock ownership as distinct from private ownership.

2. Each member of an enterprise based on collective and joint-stock ownership acts as a private owner of a certain share of the means of production, expressed in value. In accordance with the above, collective and joint-stock ownership is nothing more than a form of manifestation of private ownership.Some economists, in particular V.P. Patrikeev and I.V.Nekrasov, they propose to classify the shared variety of collective ownership as private and joint ownership as a public form of ownership. So, arguing about two types of collective property, in particular, V.P. Patrikeev notes: "Both of these types of property are completely unsuccessfully combined under the term"collective". Under the public property of V.P. Patrikeev understands "types of property that do not have an owner ..." [8], omitting the important fact that property does not exist without an owner.

3. The most important feature, a concrete expression of the economic isolation of the producer is the presence of his own private or isolated interest. With the formation of collective property, collective interests are also formed (where the interests and aspirations of individual owners are leveled and largely mitigated), and the collective of owners is already an independent subject of production relations, isolated within the economic system [7]. Thus, they believe, collective and joint-stock ownership is also private due to the complete isolation of labor collectives from each other and from society as a whole, as a set of private producers [9].

Also, the authors cannot agree with the extremely opposite point of view of other researchers, who distinguish collective property as a special, independent one, different from both private and public forms of ownership [10]. In our opinion, the specific nature of collective and, therefore, joint-stock ownership determines their intermediate location between private and public forms of ownership in the system of property relations. As follows from the Civil Code of the Republic of Uzbekistan, the property of the corporation is created at the expense of the contributions of the founders, but, on the other hand, the society itself has the right, as a legal entity, to own the property acquired in the course of its activity. The complexity of the issue is aggravated by the fact that the acquired corporations, as a legal entity, in the course of their activities, tangible and intangible assets still belong to the founders and can be distributed by them among themselves in proportion to their contributions. At the same time, corporate managers have the right (albeit somewhat limited) to independently make decisions about the use and ownership of property, including the right of alienation. Different subjects, entering into different kinds of communities, must dialectically combine their rights, interests and duties with the rights, interests and duties of the associations they created - legal entities. First of all, we are talking about a deep discrepancy between formal property rights and real possibilities of control in corporate enterprises, which leads to numerous manifestations of ineffective

https://doi.org/10.1051/shsconf/20208901002

functioning of property relations, violation of contractual rights and property rights, ineffectiveness of bankruptcy procedures, the existence of legal and semi-legal forms of property redistribution in favor of political decision makers, inefficient use of resources, etc. A number of arguments speak in favor of expanding privatization.First, privatization provides conditions for the formation of social groups materially interested in the accelerated development of market relations. Secondly, the long-term parallel existence of public and private companies inevitably gives rise to numerous price "distortions" and especially favorable conditions for counterproductive speculative transactions.Third, as you know, the weighted average price of capital of a firm decreases with an increase in gearing to a 50:50 ratio, and then begins to decline smoothly. This is due to the fact that with an increase in the level of gearing, the risk for investors increases and they, as a rule, agree to remain shareholders of the company only if they guarantee a higher yield than is offered on average in the market, and an increase in the size of dividends means an increase in the price of equity capital. Hence the idea of reducing or completely removing the state from the number of owners of joint-stock companies the sale of shares will make it possible to fully earn the domestic securities market, especially its secondary segment, due to which the performance of the JSC will be assessed using market, objective methods.

3 Results and discussion

Thus, the most important feature of a corporation, confirming the ambiguity of its position in the "real capital - fictitious capital" JSC wish to set up a "double funding". On the one hand, financing is carried out through amortization, retained earnings and bank loans, and on the other, through the issue and sale of securities. The first type actualizes the real financial opportunities, and the second - fictitious -in capital (this distinction is conditional, since only captures the accents).The analysis allowed the authors to identify the characteristic economic, legal and organizational features of modern corporations, the main of which are:

- separation of ownership from management functions;
- dual (dichotomous) nature of ownership of corporation assets;
- a complex property complex formed by the contributions of the founders and the corporation itself in the course of economic activity;
- managers and directors in corporations have legal decision-making power;
- independence as a legal entity;
- limited liability of individual investors;
- the possibility of transferring shares to other persons;
- and, most importantly, the possibility of double financing in the "real capital - fictitious capital" system.

Investment assets of open joint-stock companies are the most mobile and incomparably larger than in the case of corporate enterprises with a different organizational and legal form. and shareholders, as owners, are endowed with rights, but also have interests, which, often, may not only not coincide, but also be directly opposite. This applies not only to participants, whose relationships are built on the basis of subordination and mutual control, but also within groups of participants with the same rights (large shareholders - small shareholders, shareholders, owners of ordinary shares-shareholders, owners of preferred shares, etc.) ... Here lies one of the main problems of corporate governance, which is that, without violating the rights of one of the participants, his interests may be infringed. At the same time, if the rights of a shareholder are protected by both legislation and the corresponding legal infrastructure, then the protection of the interests of the participants becomes fully a function of good corporate governance [12]. In the scientific literature, economic analysis is mainly carried out within the framework of traditionally established trends and views. The Organization for Economic Co-operation and Development (OECD) defines corporate governance as "the internal mechanisms through which companies are governed and controlled, [1] which implies a system of relationships between the company's management board, its board of directors, shareholders and other stakeholders. Corporate governance is the structure used to define the goals of the company and the means to achieve those goals, as well as to control this process. Good corporate governance should provide appropriate incentives for the board and managers to pursue objectives that are in the best interests of the company and shareholders. It should also facilitate effective monitoring, thus encouraging firms to use resources more efficiently "[1].For their part, the authors of the article consider it necessary to take into account the main goal of corporate governance when forming the conceptual framework, namely, to ensure the growth of the long-term value of shareholders' capital, while maintaining the balance of interests of all stakeholders. The concept of "corporate governance" is naturally associated with the concept of "corporation", however, corporate governance should not be interpreted only as the management of a corporation. Nowadays, corporate governance is considered more often in a broad sense, i.e. as management, characterized by a high level of organization and its inherent special properties. Based on the above definition of a corporation, the objects of corporate governance are:

1) the property of the founders, united in a business company or partnership (corporation); 2) tangible and intangible assets owned by the corporation and used by it for the implementation of statutory activities in order to generate income.Liberalization of economic relations and reforming the economy of Uzbekistan, the tasks of restructuring property relations and bringing domestic issuers to the world capital markets, require the development of an adequate corporate governance model for the republic. Analysis of the financial condition of the republic's enterprises shows that investment activity in the real sector of the economy is at a low level, there is no practice of attracting financial resources by issuing and placing securities, joint stock companies do not have an effective dividend policy.

The corporate governance system and the stock market are two sides of the same coin. Corporate governance increases the tradability of shares on the stock market, the stock market, especially its secondary segment, through a wide range of shareholders ensures genuine market control over the activities of the executive body of the JSC and the efficiency of the corporate governance system. The mutual influence of privatization and the stock market can be viewed from two positions. On the one hand, the stock market influences privatization, contributing to the mobilization of internal and external financial resources. Market indicators show the current state of the financial system - the size of the market, the level of liquidity, the possibility of accumulating free savings, i.e. key factors in deciding what, how and when to sell. On the other hand, the deepening of the privatization process increases the number of shares traded on the market, thereby stimulating the further development of the stock market. In countries with developed securities markets, this assessment occurs through a change in the market value of shares: the presence of an effective corporate governance system leads to an increase in the value of JSC shares and vice versa.The ownership structure of enterprises, formed as a result of the first stages of privatization in Uzbekistan, turned out to be unfavorable for the subsequent growth of the stock market and does not contribute to the restructuring of ownership relations through the securities market. This is clearly demonstrated by the analysis of trading volumes in the primary and secondary securities markets of the republic (see Fig. 1).

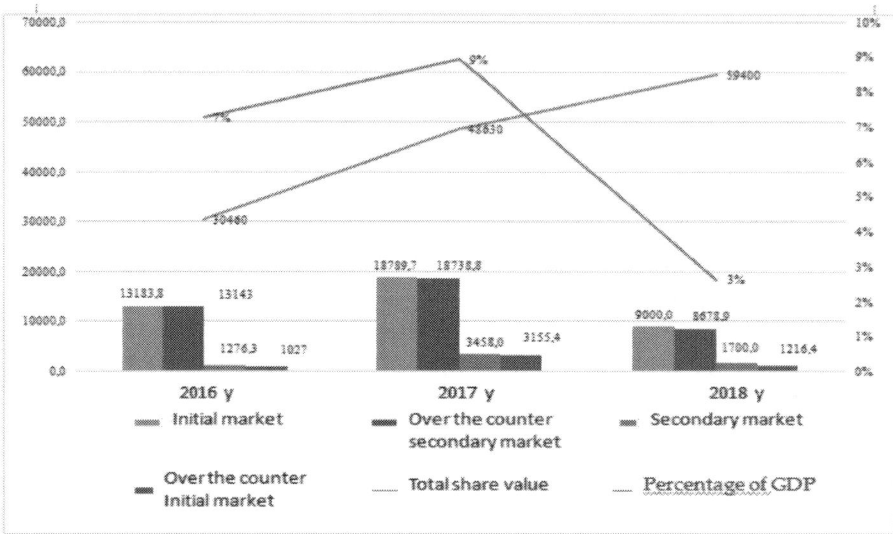

Fig. 1. Indicators of the development of the securities market of the Republic of Uzbekistan (left scale in billion soums, right scale in%) [16]

With the volume of issued shares by 2019 almost 140 trillion soums (at par), the turnover of the primary stock market of the Republic of Uzbekistan in 2018 at market prices did not exceed 9 trillion. sum, or about 7 percent of the available resources. At the same time, the main sales of the primary market occur in the over-the-counter segment, which accounts for up to 85 percent of the turnover. The total turnover of the securities market in Uzbekistan in 2018 amounted to only 3 percent of the country's GDP. The catastrophically low capitalization of the secondary stock market is of particular concern. The problem is that the low capitalization of the secondary market (does not create conditions for truly market pricing in the securities market. Hence, doubts about the fairness of the formed prices for shares, provided for implementation in the process of transformation of state property. Therefore, any delay in the development of the national securities market will lead to a failure of the attempt to conduct an IPO of domestic JSCs, if only because one of the requirements of the world stock markets for an IPO is the regular quotation of at least 20-25% of the issuer's shares on the domestic exchange market. Noting the existence of certain deficiencies in the functioning of the domestic secondary market, we focus on the fact that without the creation of the necessary market mechanisms re allocation of capital in the most efficient production and attract investment for the development of enterprises, leading to the stagnation of the entire stock market. The reason for this is that a feature of the formation of prices in the financial market is a significant impact on this process of the speculative capital circulating on it. Hence, the role of the secondary market is high, the level of development of which is largely determined by the ability of the stock market as a whole to ensure the performance of its functions. Despite the significant legislative and infrastructural development of the stock market in Uzbekistan, it should be noted that there are practically no investment institutions that carry out collective investment, underwriting , market-maker operations that are capable of providing professional assistance to business entities in the rapid placement and quotation of securities. Today, for domestic issuers, the need for infrastructure services is due to the need to increase the investment attractiveness of the securities they issue and to generate demand for them from foreign and domestic investors. A strategic investor feels more confident in a country of which he is a non-resident, when he has the potential to realize his block of shares in the event of a change in market conditions, a change in the business profile, and finally returning to his homeland.

4 Conclusions and proposals

As a result of this structural relationship, the stock market of Uzbekistan does not yet play a full-fledged disciplining and corrective role in the corporate governance system. It is impossible to build a stock market on non-performing assets. The development of the domestic capital market is possible only with a comprehensive solution to all problems. Spot attempts to influence this market can only create the appearance of activity and further discredit (reduce investment attractiveness) the stock market of Uzbekistan, which has a huge investment potential, if only due to the fact that Uzbekistan is one of the ten world leaders in natural resources, at the same time, almost 90 % of assets have not found their true owner. Based on the above, the following conclusions can be drawn:

1. Deepening the privatization processes should provide for the fulfillment of the following conditions:

a shift in emphasis from quantitative indicators of privatization to deep restructuring of industries and improvement of mechanisms for post-privatization support for enterprises; expanding the arsenal of methods and methods of privatization in order to more fully take into account industry and regional characteristics, improve the assessment of privatized property, and its sale on a tender basis; increasing the efficiency of the economy by actively supporting entrepreneurship, implementing the procedures for reorganization and restructuring, as well as liquidating unprofitable enterprises; involvement of the general public in the process of corporatization and further development of the stock market.

2. Privatization can not be considered samodos tatochnoy economic purpose, as rightly noted E. Rosenbaum, F. Bonker and X. Wagener . Instead of "fetishization" of privatization as a means of achieving political consensus it should be emphasized that the ultimate goal of economic transformation is to improve the economic well-being blah Godard increased competition and the effective division of labor. Therefore, according to E. Rosenbaum et al., The question of the cos of the market is the most important, and the role of the privatization should be assessed in the light of its contribution to the development of market institutions.

As the primary necessary steps to rectify the current state of affairs, it is considered necessary:

1. To develop and implement practical mechanisms for attracting funds from the population to the stock market. At the same time, to solve this problem "from above", by initiating separate IPOs and SPOs is to create the appearance of public activity in the capital market ... Moreover, the most difficult and important thing in this process is not underwriting support, market-maker support, road shows, etc., but systematic work carried out with shareholders after the sale of shares. If there is no such work, then, as a rule, this leads to a negative result, since it discredits the IPO and SPO. Suffice it to say that lack of knowledge or unwillingness to use such an effective corporate governance tool as dividend policy leads to a deterioration in the market position of shares. A small shareholder, especially in our country, where the secondary market is not developed and there is no system of quotations for the growth of the share price, sees only one source of return on shares - dividends.

2. A promising direction for attracting financial resources from the securities market should be the active functioning of the initial public offering (IPO) and secondary public offering (SPO) on the domestic market, which will increase liquidity and increase the share of free float shares.The development of mechanisms and procedures for an internal IPO will allow a step-by-step transition to the organization and development of sales of

securities of domestic issuers in foreign countries. In this regard, it is advisable to develop and adopt a special concept for the output of domestic securities to international capital markets, which in the long and short term would provide for:1) measures for the selection of issuers of Uzbekistan, with a well-oiled system of corporate governance, whose shares could be quoted on international stock markets;2) government support for selected domestic issuers, whose securities, ultimately, would circulate on foreign stock markets;3) measures to improve domestic legislation, ensuring the creation of conditions for the issue and circulation of national securities in the securities markets of near and far abroad;4) adaptation of domestic standards for accounting and financial reporting of joint stock companies to international requirements;5) organization of work through foreign economic departments and with the use of representative offices of international organizations for the preparation and conclusion of interstate (bilateral or multilateral) agreements on the possibility of placing domestic securities in the relevant countries and securities of foreign states in our country;6) consultations and negotiations with foreign banks regarding the possibility of placing domestic shares through depositary receipts. At the same time, it is necessary to analyze the mistakes made in the preparation of previous documents, the implementation of which ended in failure, draw the appropriate conclusions, outline a new action plan, taking into account the mistakes of past years and the experience of neighbors from neighboring countries.

References

1. OECD Principles of Corporate Governance, https ://www.oecd.org/
2. A. Begmatov , M. Khamidulin , M. Butaboev, *Corporate culture and corporate governance*, 224 (2007)
3. F. M. Rusinov, E. V. Popova, *The theory of corporate governance in an unstable state of the economy*, 60 (1999)
4. I. I. Mazur , V. D. Shapiro, N. G. Olderogge, *Corporate Management: Handbook for Professionals*, 36 (2003)
5. K. R. Mc Connell , S.L. Brue, Economics: principles, problems and politics, 74 (2002)
6. R. G. Emtsov, M. Yu. Lukin, Microeconomics, 172 (1997)
7. A. Zolotayev, Russian economic journal, **7**, 53 (1997)
8. V. P. Patrikeev, Economics of Construction, **2**, 5 (1993)
9. A. Melentiev, Russian economic journal, **3**, 73 (1993)
10. H. Kurbanov, Corporate finance, Study guide (2005)
11. D. M. Skirdko, *Corporate governance in the real estate market*: Dis. Cand. economy. Sciences, 13 (2004)
12. Law of the Republic of Uzbekistan "On property in the Republic of Uzbekistan", 8
13. Manual on corporate governance, **1(1)**, 28 (2004)
14. S. P. Kukura, The theory of corporate governance, 121 (2004)
15. S.Lyapunov, Problems of theory and practice of management, **6**, 24 (2001)
16. Data of the State Statistics Committee of the Republic of Uzbekistan

Impact of the Covid-19 pandemic on the requirements for accounting and analytical support of business

Roman Chugumbaev[1,*], and *Nina* Chugumbaeva[2]

[1]Moscow state University of Humanities and Economics, Losinoostrovskaya Str., 49, 107150 Moscow, Russia
[2]MIREA - Russian technological University, Vernadsky Avenue, 78, 119454 Moscow, Russia

Abstract. This article proposes to get acquainted with a study that, as a first approximation, assesses the impact of the Covid-19 pandemic on a shift in emphasis in accounting and analytical work. Economic actors once again found themselves in the conditions of a "new normalcy", now it is a global coronavirus crisis. Now society and business are faced with new risks, threats, disasters, opportunities, unforeseen consequences of the COVID-19 pandemic. Of course, the management of the companies is expected to provide the creative solutions, an economically sound strategy or a policy of working in a new environment. The new complex context of decision-making caused by the pandemic needs to keep information and analytical support up to date. Therefore, this article proposes to pay attention to some aspects of information and analytical work that are of particular importance during the period of the impact of the pandemic and its consequences. The list of such aspects includes the mechanisms for ensuring business transparency, the application of stakeholder theory and change management theory, the organization of a systematic business analysis, the concept of social responsibility and ESG criteria. This approach will make it possible to “reveal” the problems caused by the pandemic and its consequences, find the best solutions therefor, implement them correctly and update the business development strategy. The research results based on the analysis of current business needs will help build an adequate system of accounting and analytical support.

1 Introduction

The impact of the pandemic on the costs of human capital, on which economic success largely depends [9], is catastrophic. At the end of March 2020, the State Duma of the Russian Federation adopted a package of laws to combat the consequences of the coronavirus. This time was the beginning of another “new reality” for Russian society and business, when a significant reassessment of priorities and values took place. For now, the COVID-19 pandemic is not receding, raising the level of uncertainty about the economic

* Corresponding author: romanry@ya.ru

impact of the crisis. COVID-19 continues to spread across the world, with countries facing a general economic downturn, rising unemployment rates and declining business confidence. Since these difficult conditions of functioning can persist for a long period and have a long-term negative impact on the economy of organizations, a high-quality system of information and analytical support is needed, which allows to initiate changes in a timely manner and make an economically sound choice of decisions. The success of overcoming the economic crisis caused by the consequences of a protracted pandemic depends thereon.

Given the above, the purpose of the study is to analyze the impact of the COVID-19 pandemic on the needs of information and analytical support of business, namely, its content, priority principles and approaches.

2 Methods

This study offers an initial scientific look at the information and analytical needs of the company managers in a new complex context full of uncertainty and risks. Based on rather limited and recent studies published in the works of domestic and foreign authors and presented at scientific conferences in 2020, as well as on the expert opinion of the specialists who are faced with the problems of making managerial decisions in a pandemic, the most obvious features of the functioning of a business in a new context have been revealed. This study does not claim to provide a sufficiently complete coverage of all issues of accounting and information support. Rather, it is one of the initiatives, which will be followed by further research, both theoretical and empirical, in the field of accounting, analysis and audit of the activities of economic entities in the new conditions.

Currently, high uncertainty affects the need for high-quality information and analytical support for company management. Within the framework of this study, it is proposed to focus attention on some elements of information and analytical work, which, in our opinion, are of particular value for management during a pandemic and the subsequent period of combating its consequences. The list of these elements included as follows:

- application of the mechanisms for the transparent functioning of business;
- analysis and monitoring of the stakeholders' requirements;
- information and analytical support for change management;
- systematic business analysis;
- focus on social responsibility and ESG criteria.

3 Mechanisms for transparent business functioning

The transparent functioning of economic agents [8] in a pandemic is more important than ever as it provides a link with the key business stakeholders. Trust between participants in economic processes is declining under the new conditions. The acute phase of the pandemic in Russia has shown that business and the state in such times need a significant "rapprochement" in order to unite the efforts to ensure the socio-economic sustainability. Mutual transparency is a prerequisite for such convergence. Equally important is the transparency of business intentions and the financial condition of economic entities in relations with banks, which are also committed to a more cautious policy and are more attentive to the borrowers. In a period of high uncertainty and turbulence in the business environment, all the participants in social and economic processes need the reliable, understandable and timely information.

The transparent operation of the company during such a period significantly reduces the likelihood that supervisory authorities, representatives of state authorities, without notification, initiate unscheduled control and audit activities. During a pandemic, such measures are taken much more often than usual, which, of course, can be burdensome for organizations, when many other complex problems that require high attention already arise. Transparency also reduces or even completely eliminates actions that are aimed at abuse and gaining vested interests due to the chaos and uncertainty caused by pandemic consequences and panic in a society. Transparency ensures the authenticity of information about an organization, and during a pandemic, there is a high proportion of fake information in the media space [7]. The openness and transparency of the organization's functioning reduce the consequences of an "information attack" on the part of competitors and other actors in the dissemination of negative information about the company. A more accelerated transition to remote management, remote work during and after a pandemic also requires ensuring the transparency of online business processes. Such internal transparency becomes a prerequisite for the coordinated and efficient functioning of the business. To ensure business transparency on a systematic basis, work should be organized in several directions, in which there is a significant amount of application of analytical procedures [1].

All this indicates a rather high need for the formation and development of systematic processes in economic entities aimed at the formation of transparent information, incl. reporting. It is necessary to provide analytical processing of disclosed information so that key stakeholders receive conveniently organized and understandable information in which you can quickly and easily highlight the relevant part for making decisions in such a difficult period.

4 Consideration of stakeholders' requirements

During a pandemic, "harsh" decisions by the stakeholders do not always take into account the interests of the business, and sometimes create problems that also require a prompt response. At the same time, it is necessary to be able to monitor both internal and external elements of the business environment online using modern information and communication tools.

Thus, some studies confirm that the pandemic has led to a change in consumer behavior [3], which has intensified the search for a differentiated offer, aiming for a greater variety of the products. The very nature of the organization of interaction with customers has changed. Consideration should be given to the changing needs of the customers during the periods of special treatment and restrictions. During the period of the pandemic, the company must not miss the opportunity to reach a new level of trusting relationships with the personnel, the requirements of which are contradictory due to the general uncertainty of the present and the future. Suppliers and contractors in the new environment can, like everyone else, change requirements, priorities, etc., and therefore, you should very carefully monitor the supply chain and be in contact with them.

During a pandemic, various company transformations are required, focused on highly responsible functioning, taking into account the requirements of all key stakeholders. At the same time, the aforementioned requirements for business transparency are also due to the expansion of the circle of stakeholders, who are increasingly actively involved in the management decision-making processes. Since the functioning of a modern company is focused on meeting the requirements of a wide range of stakeholders, it is very important to provide them with information that allows them to give an adequate assessment of alternative solutions. Therefore, one should carefully consider the composition of key stakeholders, and ensure that their requirements are systematically monitored [2]. Consistency is necessary due to the fact that the requirements of stakeholders are subject to change, as well as the influence of external conditions. The consequences of restrictive measures from a pandemic significantly change the context of the operation and development of a business and require timely consideration of changes in the requirements of stakeholders.

5 Managing changes during COVID-19

During a pandemic, we all are faced with challenges, changing the way we do business. In the context of such a massive transition to remote forms of interaction, the transfer to the Internet of some business processes, restrictions on the part of government bodies, many problems and obstacles are obvious in all areas of business functioning: financial, investment and operational ones. But along with the negative aspects that lie on the surface, it is important to see new opportunities, the advantages of the resources available to economic agents. All organizations, including business entities, are trying to initiate and implement changes of varying complexity in order to find a way out of the problems of the negative impact of the pandemic. As you know, it is customary to distinguish transactional organizational changes from transformational ones. At the same time, most of the changes in a pandemic are transformational in nature. If transactional changes focus on individual organizational problems, then transformational ones are more systemic and cover the company as a whole. The need to implement the significant transformational decisions predetermines the need for accountants and analysts to know the provisions of the theory of change management. In a pandemic environment, transformational changes should be based on an analytically grounded and responsible strategy that takes into account the most relevant trends in the organization's functioning environment. To build and implement a strategy that can effectively manage the innovative changes of an enterprise in conditions of overcoming the impact and the consequences of a pandemic, it is necessary to solve the problems of the quality of analytical support:

1. At the stage of initiating changes, the company's management must see the prerequisites for certain directions of organizational changes and take the initiative in a timely manner. It is important for economic agents to recognize the incoming signals for the implementation of changes, to understand their role at this stage of business development and its functioning environment. During a pandemic, there are many different quite clear signals for change. New conditions and restrictions pose challenges for economic entities to master new methods of work, remote methods of personnel control, interaction with partners, methods of delivering goods to customers, etc. Many spheres of entrepreneurial activity have to completely change the business model of their functioning, as well as other strategic elements of the business. The management system of Russian enterprises should be provided with information and analytical tools aimed at identifying the prerequisites for changes in the current regime;
2. At the stage of planning changes, the company's management is faced with the problems of criteria-based assessment of the necessary changes, as well as with the problem of justifying the selection of decisions during the design of changes. Most of the changes during a pandemic are aimed at business survival, which implies a particular importance of sustainability indicators in various aspects, incl. financial and operational ones. We are witnessing when companies, based on taking into account the new operating context, choose to implement the initiatives related to a temporary change in their business focus and produce, on a temporary basis, those products and services that are not inherent in their area of competence. So some companies have decided to produce personal protective equipment, masks, hand sanitizers, etc.;
3. At the stage of implementing the changes, problems arise of how to assess the effectiveness and efficiency of implementation, when a decision should be made on the necessary adjustments;

4. At the stage of assessing the results of the implementation of an innovative solution, it is necessary to assess whether the goals of changes have been achieved and what solutions should be implemented in order to continue the improvement strategy.

6 Business analysis on a systematic basis

A significant transformation of the conditions for the functioning of a business requires the development at the enterprise of a mechanism for initiating and implementing changes, which is possible if the enterprise carries out a systematic business analysis [6]. Business analysis is a new direction of analytical science and a new type of professional activity, a key element of which is monitoring and analysis of the level of necessary and actual fulfillment of the requirements of stakeholders, as well as the possibilities of increasing it through innovative solutions. It appeared as a natural stage in the development of domestic complex economic analysis for the conditions of a turbulent socio-economic environment, characteristic of a modern market economy. The sustainability of a company during a pandemic depends on the number of its key stakeholders, whose requirements become the basis for making decisions on changes in the company. An in-depth study of stakeholder requirements is a strategic objective and is closely related to the categories of strategic management. Business analysis allows you to stay focused in favor of speculative decisions with short-term benefits during a pandemic. On the contrary, by studying the strategic preferences and vision, it identifies the requirements of key stakeholders as the main guideline for implementing changes that would allow to acquire long-term benefits, to recognize the true values for key stakeholders. By monitoring and in-depth analysis of stakeholders' requirements, business needs and identifying business problems, strategic guidelines and vision of the company contribute to its real development in difficult times, and not succumb to chaos and make "short-sighted" decisions.

7 Social responsibility concept and ESG criteria

Several decades ago, when the crisis hit, not only medium-sized, but also large companies and corporations could afford to stop activity and even take a "step back" in areas of activity reflecting the social responsibility. However, now that the role of business in society has evolved significantly, companies must take care of preserving social and environmental capital. During a pandemic, the attention of the public and government regulators to business decisions emphasizes the special role of the concept of social responsibility of business [5] and control of ESG criteria (Environmental, Social, Governance,), especially highlighting the S-factors [4]. During a pandemic, organizations can take advantage of opportunities to be socially responsible and “closer” to society. The conditions in which economic agents find themselves become a test of their real commitment to the principles of sustainability and social responsibility. The attitude of the management to the safety of personnel in the workplaces of their own personnel, the transition to remote work, the provision of PPE and more can reflect a lot about the real concern of the company towards the employees. During a pandemic, many large Russian companies initiate various measures aimed at overcoming the consequences of the pandemic. Some companies carry out the initiatives aimed at supporting healthcare institutions, modernizing intensive care units, purchasing additional equipment and consumables, and personal protective equipment for these institutions. Dignified responsible attention and care of the company in relation to personnel, consumers, suppliers, society and all key stakeholders can remain in the memory of society for a long time and become a reliable social capital that ensures long-term business sustainability. It is clear that the pandemic, in addition to ample opportunities to prove its commitment to responsible, ethical and transparent operation, also poses the threats to business survival. These problems often reach the point where companies are forced to focus on their core operations rather than investing in corporate social responsibility projects. In this regard, there are the prerequisites for an even greater shadowing of the activities of some companies, as well as the use of panic in a society and an increased degree of uncertainty, turbulence of the economic context in speculation and other abuses for the sake of short-term success. These circumstances cannot be ignored by the company's accounting specialists.

8 Conclusion

Thus, in this study, some aspects of accounting work have been noted, which have acquired new relevance for company management in a pandemic. However, there is no doubt that the current experience and the subsequent consolidation of business cases will allow further research on this issue. Concluding this article, we ask the academic community to take part in a thorough study of the problems of information and analytical support of the business. In our opinion, the problem of developing mechanisms for ensuring business transparency during a pandemic remains difficult. In applying the stakeholder approach, the problem of standardizing measurements of various stakeholder requirements has not been solved, the solution of which will open up broad opportunities for comparative analysis of innovations. In addition, it is difficult to determine the real, not exaggerated, stakeholder requirements. For the proper dissemination and application of the principles of sustainable development and social responsibility in Russian enterprises, research is needed to address the role of institutional elements of the functioning environment, such as analytical and rating the agencies, and business associations.

References

1. E. Alloa, Transparency, Society and Subjectivity. – Palgrave Macmillan, Cham, 21 (2018)
2. R. E. Freeman, *Strategic Management: A Stakeholder Approach*, 292 (1984)
3. J. Kim, Australasian Marketing Journal (AMJ), **28(3)**, 108 (2020)
4. E. Ortas, I. Álvarez, A. Garayar, Sustainability, **7(2)**, 1932 (2015)
5. J. M. Palma-Ruiz, J. Castillo-Apraiz, R. Gómez Martínez, International Journal of Financial Studies, **8(3)**, 41 (2020)
6. V.I. Barilenko, M.V. Melnik, R.P. Bulyga, E.B. Gerasimova, O.V. Efimova, V.P. Nevezhin, V.V. Berdnikov, *Fundamentals of business analysis* , 270 (2018)
7. O.S. Deineka, L.N. Dukhanina, A.A. Maksimenko, European Scientific Conference, 326 (2020)
8. O.V. Rozhnova, In the collection: IFRS in Russia as a new intellectual environment, The view of the professional community, 58 (2018)
9. P.D. Shishmanova, Russian technological journal, **8(4)**, 141 (2020)

Theoretical and methodological approaches to identifying models of interaction between government and business in the context of digital transformation

Galina Vorontsova*, *Oksana* Momotova, *Olga* Boris, *Malvina* Ustaeva, and *Rustam* Ustaev

North Caucasus Federal University, Pushkin Str., 1, 355017 Stavropol, Russia

Abstract. The existing approaches to the interaction of business structures and government bodies have been investigated. The factors influencing the formation of the interaction process and the achievement of a balance between the interests of government and business have been determined. A matrix of interaction models is proposed based on the ratio of the strength of the state and the level of integration of business structures, as well as the degree of involvement of subjects in the interaction process. The corporatist model of interaction is defined as widespread in Russia, within the framework of which the factors that worsen the relationship between power and business structures in the Russian Federation have been identified as follows: lobbyism and corruption. It is shown that in modern Russian economic realities, the partnership model is more in line with the tasks of effective interaction between the state and business. The trends in the development of models of interaction between the state and business in Russia and the feasibility of transition to a network partnership model based on breakthrough development in a strategic perspective have been determined. The advantages of the transition to the specified model of interaction between the state and business are have been substantiated, which make it possible to achieve a breakthrough in the digitalization of the real sector and the active introduction of domestic innovative developments, the formation of a special innovative culture, which involves the activation and inclusion of an individual as a full-fledged participant in the innovation process and the growth of entrepreneurial activity of broad strata of the Russian population.

1 Introduction

An important indicator of the level of socio-economic development of the state is the mutually directed desire of the state and business to search for the most effective forms of interaction for solving acute social and environmental problems, achieving high living standards of the population and increasing its spiritual development, equal access of citizens to the benefits of civilization, carried out within the framework of models of social

* Corresponding author: vgv14@mail.ru

responsibility.

Researchers focus on various aspects of the interaction between government and business structures. Despite a fairly large number of theoretical and methodological studies in the field of content, construction and functioning of models of interaction between the state and business, the identification of models and tools for interaction between government and business structures, preferred for the Russian Federation and its individual regions, is a complex process. Among the many factors influencing this process, we consider it necessary to highlight the need to achieve a balance of interests of both parties when justifying the use of a particular interaction model [1].

In modern Russian economic realities, in our opinion, it is the partnership model that meets the objectives of such interaction to a greater extent. Within the framework of the partnership model, significant innovative projects are especially successfully implemented when using such a form of partnership between the state and business, such as, for example, public-private partnership (PPP).

2 Materials and methods

An empirical study of existing approaches to the interaction of business structures and public authorities made it possible to identify the models of such interaction that have formed over the period of development of relations between government and business.

A comparative analysis of the works of Russian and foreign researchers indicates the existence of a variety of approaches to the classification of models of interaction between business structures and government authorities, which use various criteria and grounds for identifying certain models.

3 Results and Discussion

The selection of a model of interaction between the state and business on the part of the representatives of the business community is determined by their interests [2,3]. At the moment, in the Russian Federation, it is big business, due to the high concentration of resources, that has great opportunities in representing its interests in relations with government structures and presents great opportunities for solving social problems of regional development [4]. In general, the degree of participation of the state and business in interaction, their readiness to build relationships based on partnership, in our opinion, can be illustrated graphically (Fig. 1).

This reflects the level of business integration in terms of the creation of large associations and groups, associations of special interest, accumulating and expressing the most common interests of their representatives, or the enlargement of individual business units to the level of monopolists or big business, whose capital impact cannot be ignored. The strength of the state's influence in this case implies the administrative, financial and legislative and regulatory resources that determine the degree of control and regulatory influence.

weak strong state	Suppression and coercion (instruments - confrontation)	Affiliate model based on social responsibility (instruments - collaboration)
	Non-intervention of power (instruments - ignoring)	The corporatist model (neocorporatism/patronage) instruments - collaboration
	low high business integration level	

Fig. 1. Matrix of interaction models based on the "Strength/weakness of the state and the level of integration of the business community" continuum

The analysis of information sources indicates the presence of numerous classifications of models of interaction between the authorities and business structures in the context of digitalization, taking into account a large number of factors and variables.

It is impossible to single out one model of interaction that characterizes the relationship between the state and business throughout the country, we can only talk about the predominance of one or another model [5].

As a result of the study of the points of view of famous scientists, it can be concluded that now a corporatist model (neocorporatist or patronage) of interaction between government and business has taken root in Russian practice. This model of relations between government and business made it possible to overcome the chaos that arises when trying to take into account the multidirectional interests of individual representatives of the business community in various Russian regions. The use of this model is accompanied by the maximum influence of representatives of big business (in Russia, these are mainly natural monopolies, producers of raw materials), as well as associations and unions. These are large representatives expressing the interests of mainly large and medium-sized businesses [6].

Lobbying and corruption continue to be factors that worsen the relationship between government and business structures in the Russian Federation, regardless of the established model of interaction between them. Therefore, the fight against lobbying and corruption is one of the ways to improve the interaction between the state and business. The effect of this factor is largely leveled out within the framework of the use of the partnership model of interaction [7].

Determining the trends in the development of models of interaction between the state and business, in the future, we distinguish two segments on the path towards an optimal model.

The first one has a medium-term planning horizon. This is a partnership model based on social responsibility, which involves joint determination of social policy priorities and those areas in which both the state and business are interested, can and is ready to take an active part [8]. In terms of digitalization development within the framework of this model, e-government services are being developed in Russia.

The second segment of the path towards an optimal model of interaction between the state and business can be presented in the framework of long-term strategic planning. It assumes, as a result, the formation of the most effective model in terms of meeting the interests of the business community and representatives of power structures. We consider the model of network partnership based on breakthrough development to be such a model of interaction between business and government authorities. This type of interaction is based on e-government technology, which implies a deeper digitalization of all branches of government, and not just the executive branch within the framework of e-government. This is due to the delegation of powers by the state to make the most important socio-economic decisions to the representatives of the civil society and business.

The transition to this model makes it possible to achieve a breakthrough in the digitalization of the real sector and the active implementation of domestic innovative developments. The goal for the implementation of interaction between the state and business is the formation of a special innovation culture, which, on the one hand, involves the activation and inclusion of an individual as a full-fledged participant in the innovation process on the part of the demand for innovation, the growth of entrepreneurial activity of broad strata of the Russian population, a critical mass of innovators, and on the other hand, the creation of a wide stratum of Russian entrepreneurs with an innovative focus, ensuring the breakthrough development of the Russian Federation in a strategic perspective.

The very logic of the existence of a networked society determines the direct access of all elements of the network to each other and, accordingly, a fundamentally different non-hierarchical management model. Some experts have introduced the term "digital government" into scientific circulation, or, to be more precise, "electronic government", which implies a new model of management in an information-network society. "Electronic government" can be understood as a way of carrying out information activities in various spheres and aspects of public administration based on information and communication technologies, as well as a new type of state based on the priority use of digital technologies [8].

The creation of an information and communication system of interaction and dialogue is designed to strengthen the role and participation of citizens in public administration and decision-making through digital technologies, which allows us to speak about the development of electronic democracy within the framework of e-government. In fact, this is the most advanced stage, and it must be stated that most often states limit themselves to informing and consulting: providing citizens and businesses with electronic services and introducing the principle of a single window. In this sense, the key role of e-government is reduced to the service function of providing the services, which allows a number of researchers to talk about the phenomenon of a "service state" [9].

The transition to a network partnership model based on breakthrough development will be associated with those global changes that are rapidly growing and are associated with globalization, digitalization, intellectualization of the environment of interaction between the state and business, a trend towards resource conservation and the development of a "green economy", innovative growth. In our opinion, this will lead to a transition from flat to volumetric models, which presuppose the introduction of additional vectors, such as, for example, the readiness of the state to delegate, and business to accept the powers for the balanced development of the economy and society. And also to revise the meaningful characteristics of individual vectors. As indicators of the strength of the state, we propose to use the level of digitalization of the public services sector, the willingness to provide information services to businesses in the format of a "single window", initiatives in the development of state target programs and national projects aimed at supporting business (primarily innovation-oriented) and stimulation of entrepreneurial activity of the population. As a result, the matrix of models of interaction between the state and business will take the following form (Fig. 2).

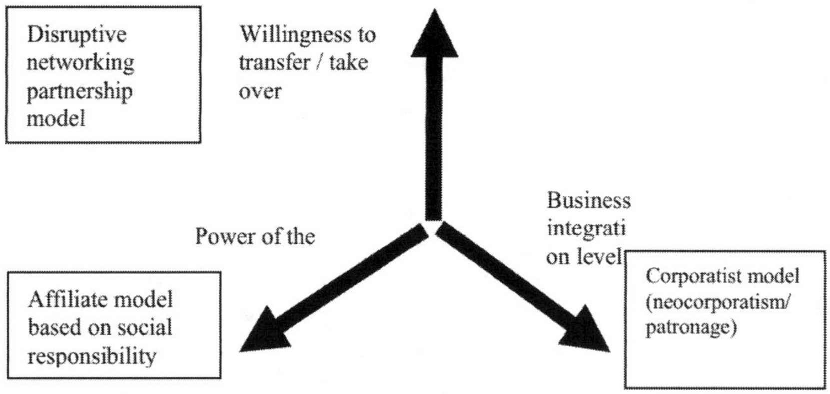

Fig.2. Matrix of interaction models according to the degree of government/business involvement in the delegation/acceptance of powers for the balanced development of the economy and society.

It is problematic to draw a conclusion about the priority of using certain instruments of interaction between the state and business. It is more correct to say that within the framework of certain types of interaction between the authorities and business, the most appropriate tools for such interaction can be identified.

Moiseev V.V., Nitsevich V.V., Nitsevich V.F., Shurtukhina E.A. identify the spheres (areas) of interaction between government and business, each is characterized by the use of specific tools for such interaction [10].

We believe that public and municipal-private partnerships are the predominant instrument for interaction between government and business in the near future.

4 Conclusions

The results of the scientific generalization and empirical research of the practical experience of interaction between government and business allow us to conclude that a theoretical and methodological substantiation of its basic models has been developed.

There are numerous classifications of models of interaction between government and business structures in the context of digitalization, taking into account a large number of factors and variables.

To identify the models of interaction between the state and business, it is advisable to use a matrix of interaction models based on the "Strength/weakness of the state and the level of integration of the business community" continuum.

For the development of ideas about the interaction between the state and business, the expediency of switching from flat to volumetric models has been shown, which involves the introduction of additional vectors, such as the readiness of the state to delegate, and business to take powers for the balanced development of the economy and society.

In real practice, it is necessary to talk about a combination of several types of models, and not about the prevalence of one or another model of interaction between business and government in a particular region and country.

The factors that worsen the relationship between the government and business structures in the Russian Federation have been revealed, first of all, these are lobbying and corruption, regardless of the formed model of interaction between them.

In Russia, a corporatist model of interaction between government and business has developed and is currently dominant.

On the way towards an optimal model of interaction between the state and business, it is advisable to single out two segments:

- in the medium term, it is a partnership model based on social responsibility.
- in the long term - a model of network partnership based on breakthrough development, which makes it possible to achieve a breakthrough in the digitalization of the real sector and the active introduction of domestic innovative developments.

In general, despite the presence of positive shifts in the process of transition from the dominant and competitive models of interaction between the state authorities of the Russian Federation and business entities to the model of mutually beneficial cooperation, this process cannot be called as complete one.

The main driving factor in achieving partnership between government and business is the establishment of an effective public dialogue between the parties.

Despite the presence of positive shifts in the process of transition from the dominant and competitive models of interaction between the state authorities of the Russian Federation and business entities to the model of mutually beneficial cooperation, this process cannot be called as complete one. The main driving factor in achieving social partnership between government and business is the establishment of an open and free dialogue between the parties on an ongoing basis, as a result of which the interests of the subjects of interaction are harmonized.

Acknowledgements

The study has been carried out with the financial support of the RFBR within the framework of the RFBR research project "Development of the concept of forming the effective models of interaction between the authorities and business structures at the regional level within the framework of digital integration of the Republic of Belarus and the Russian Federation", project No. 20-510-00025

References

1. A.E. Chirikova, Bulletin of the Institute of Sociology, **9**, 81 (2014)
2. E. Dorina, O. Meshcheryakova, O. Boris, O. Momotova, G. Vorontsova, Modern Tools for Sustainable Development of Territories, Project Management in the Regions of Russia (MTSDT 2019), 245 (2019)
3. V.Ya. Gelman, World of Russia, Sociology, Ethnology, **2**, 38 (2017)
4. K.Yu. Kisel, *Modern models of interaction between business structures and public authorities: Candidate thesis (politicial sciences)*, 24 (2013).
5. A.A. Yegorov, News of Saratov University, Economy, Management, Law, **15(2)**, 151 (2015)
6. Valentina N. Parakhina, Galina V. Vorontsova, Oksana N. Momotova, Olga A. Boris, Rustam M. Ustaev, Innovational Projects of Technological Growth on the Platform of Public-Private Partnership 2019 Article in a bulletin (WoS)
7. Yu.V. Ragulina, T.V. Butova, Organization of interaction between power structures and business. Monograph, 180 (2014)
8. D.V. Ivanov, A.S. Sokolitsyn, Economy and National Economy Management, Discourse, **5**, 43 (2018)
9. Interaction with public authorities, or government relations, **86** (2015)
10. B.S. Bataeva, *Strategic priorities of socio-economic development of Russia and the concept of corporate social responsibility*: Doctoral thesis (economic sciences): 08.00.05, 368 (2010)

Transformation of corporate governance models in investment funds

Aleksandr Metzger*

CJSC «Management Company», Tramvainyi Ave, 15, 620041 Yekaterinburg, Russian Federation

Abstract. The article is devoted to the phenomenon of corporate governance in investment funds, as one of the key elements ensuring the effective functioning of this institution of financial intermediation. A narrow and formal understanding of the term "corporate governance" in mutual funds is reduced to the relationship between owners and leaders of the management companies. This problem is devoid of practical sense in the context of highly concentrated ownership in these companies. The author proposes an expanded interpretation of corporate governance based on considering an investment fund as an "investment corporation", in which the participants are various categories of investors, fund management, as well as a number of other participants and stakeholders, depending on the nature of the investment process. Considering corporate governance as a system of control and management in Russian investment funds, the author comes to a number of conclusions. The current model considers the protection of the interests of fund investors from the opportunism of managers as a priority and meets the specifics of a certain market segment. The same model, but under different conditions, becomes redundant or even destructive. The analysis of the specifics of the activities of the investment funds for qualified investors makes it possible to formulate adequate changes in the corporate governance model. These include: 1) the transition to a model of coordinating the interests of the investors (controlling and minority), as well as other participants in the investment process, 2) reducing external regulation and moving to internal mechanisms for assessing and making decisions, taking into account the qualifications of the participants themselves, 3) limiting or compensation for the excessive influence of the controlling investor with a concentrated structure of investors.

1 Introduction

The investment fund industry is currently one of the most important segments of the financial market in almost any economically developed country. Like other institutions of financial intermediation (commercial banks, insurance companies, pension funds), this industry ensures, on the one hand, the satisfaction of the interests of various categories of investors, and on the other hand, transforms financial resources into capital for the real sector of the economy. Despite the relevance of such tasks for modern Russia, the

* Corresponding author: metzger@azkapital.ru

development of the investment funds lags far behind the commercial banks dominating in the Russian financial system. The volume of funds accumulated in investment funds has recently slightly exceeded the same amount in pension funds, but rather "thanks" to the extremely unsuccessful state policy towards the latter. In other words, it can be stated that Russian investment funds are unable to unleash their potential and become an attractive instrument for both institutional and ordinary investors.

The overwhelming part of the numerous theoretical studies devoted to the problems of investment funds, for the most part, consider only partial fragmentary aspects of activities, such as the formation of an optimal portfolio of securities, the peculiarities of investment strategies in local markets, etc. At the same time, there are practically no analytical studies devoted to the processes of interaction between the participants in the investment process in funds (investors, managers, stakeholders). Out of the field of view of the researchers are the differences of interests of certain categories of investors, the ability of investment funds to meet these interests. The efforts of the regulator represented by the Bank of Russia and the professional community represented by NAUFOR (National Association of Stock Market Participants) look much more systemic. The system of regulatory documents, standards and rules developed and being improved by them, updating the best practices of behavior of market participants, is aimed at ensuring the stability of both an individual investment fund and this industry as a whole. If we consider an investment fund as a kind of "investment corporation", then all of the above is a model of corporate governance - an integral combination of decision-making and coordination mechanisms, taking into account the interests of both the main participants of the corporation and the stakeholders. The analysis of this model shows that today its priority is to protect fund investors from the opportunistic behavior of the management company (fund managers).

Without denying the existence of this task as applied to investment funds, it should be noted that its relevance in certain market segments has decreased, up to its complete disappearance. As a consequence, the management mechanisms aimed at protecting the interests of investors, at least become excessively costly, i.e. start to work against the interests of investors.

In other cases, it is not so much to protect the interests of investors as to protect them from their excessive influence on the decision-making process (for example, with a concentrated structure of the investors). This influence, amplified through the mechanisms of control over fund management, becomes destructive - it destroys corporate value, formed not only due to the contribution of investors, but also due to the contribution of other participants and stakeholders.

As a result, a unified and seemingly consistent system of control and management decision-making in investment funds leads to serious distortions under certain combinations of conditions and formats for the functioning of the funds. These distortions are manifested in excessive actions (expenses) when solving an urgent problem, or even in actions (costs) devoid of meaning, since they are aimed at achieving a false and irrelevant goal under the given conditions.

Thus, an attempt is made in the work to differentiate the conditions for the functioning of investment funds and to offer models of control and management decision-making (models of corporate governance) adequate to these conditions. Along with a systematic analysis of the conditions for the implementation of the activities of the investment funds, the work uses the methodology for the formation and substantiation of corporate governance models developed for classical entrepreneurial corporations.

2 Materials and methods

In the most general form, *corporate governance* is a set of mechanisms for making and controlling managerial decisions, which reduces the negative effects of the association of various participants and maximizes the positive effects of this association for them. For all the lightness of this definition ("for all good, against all bad"), it allows you to operate with the real interests of various participants, taking into account which ensures not only the stability of the "corporation", but also its development. In this case, it becomes extremely important to single out significant participants, as well as to identify the basic problem typical for this composition of participants. The theory and practice of corporate governance in "entrepreneurial" corporations provides an extensive methodological basis that, with some adaptation, can be used in the case of investment funds.

The issues of corporate governance in investment funds (*fund governance*) have long been the subject of scientific and practical research abroad [1-7]. Similar issues are considered in relation to the industry of Russian investment funds. However, the interpretation of this concept is only one of the concepts of corporate governance in entrepreneurial corporations (resolution of the conflict between the owners and managers). Moreover, the lack of taking into account the specifics of the organization and functioning of the institution of an investment fund leads to additional costs, but does not allow minimizing the specific manifestations of the same opportunism of managers (fund management companies).

For example, in accordance with the Federal Law "On Investment Funds" [8], it is possible to create either a joint stock (JIF) or a mutual investment fund (MIF). If in the first case we are talking about a joint-stock company, then a mutual investment fund is "a separate property complex consisting of property transferred to the trust management of a management company by the founder(s) of trust management ... A mutual investment fund is not a legal entity."

In accordance with the Civil Code of the Russian Federation [9], corporate legal entities (corporations) are legal entities whose founders (participants) have the right to participate (being members) in them and form their supreme body. Thus, the interpretation of a corporation as a business entity is quite appropriate in the case of JIF and, for example, is easily translated into corporate governance bodies familiar to entrepreneurial corporations: the general meeting of shareholders (investors), the board of directors. The role of the management company is also quite understandable - it performs the functions of fund management.

A completely different situation in the case of a mutual fund - the lack of legal entity status formally excludes this fund from the number of "corporations"! However, corporate governance needs to be implemented and the answer is easy to find - within the framework of the fund's management company, which can only be a business entity, i.e. is a "corporation".

The further logic is clear - it is on this corporation that the entire arsenal of mechanisms and practices of corporate governance is projected. The formal spread of effective practices to an object identical in its legal status leads to the absence of the expected result or a negative result, expressed in excess costs and restrictions, which do not lead to an increase in efficiency (the effect of "cargo cult"). Let us explain this statement. Allocation of protection of the interests of owners as a priority goal of corporate governance means that they are understood not as the fund's investors, but as the owners of the management company. It is for them that corporate governance bodies are formed, and corporate governance standards are developed. However, in the context of a super concentrated ownership structure of Russian management companies (the vast majority of them have a single beneficiary), the activities of all corporate governance bodies provide only formal, but costly compliance with the rules and regulations borrowed from the relevant practice of entrepreneurial corporations [10, 11].

The interests of other participants in the process of creation and functioning of investment funds, including investors, is, if not ignored, then considered by this interpretation in passing, indirectly, in the context of the "clients" of the management company.

Nevertheless, it can be argued that in the industry of Russian investment funds, a fairly stable model of real corporate governance has developed over the past two decades, although not without a number of restrictions. In contrast to the formal interpretation of corporate governance, in this case, the focus is not on the insignificant problem along the "management - owners of the management company" axis, but on the real problem of relationships along the "investors - management (management company) of the investment fund" axis. Quite rightly, this problem is identified as the "principal - agent" problem [12, 13], and mechanisms for accounting and protecting the interests of principals are being built in this direction, ie. fund investors. Why has this system of quite effective mechanisms that allow solving the "agency problem" not officially received the status of "corporate governance in investment funds"? It can be assumed that the reason is the fear of going beyond the narrow understanding of corporate governance as the relationship in a corporation between its owners and management. Since the vast majority of Russian investment funds are created in the form of mutual funds (not corporations!), and their investors enter into a trust management agreement (i.e., they are not the owners of the fund!), the whole range of issues related to the interests of investors and other interested parties can be called whatever you like, but not corporate governance!

However, the task of this work does not include the desire to restore terminological justice and call it as a corporate governance, what looks like this management solves the problems of this management and uses the methodology of this management. In our opinion, it is much more important to assess the current model of regulation and control in investment funds ("investment corporations") based on the theory and practice of corporate governance in "entrepreneurial corporations".

As noted earlier, at present, through the efforts of the regulator (the Bank of Russia) and the professional community, an integral system of mechanisms for monitoring and regulating the activities of investment funds has developed. The most significant elements of this system are briefly discussed below.

1. Organizations that make up the so-called "infrastructure of the fund", which, in addition to the management company itself, includes a mandatory specialized depository. In some cases, the infrastructure of the fund is supplemented by an auditor (funds for unqualified investors), a registrar (if the register of shareholders is not maintained by a specialized depository), and an independent appraiser (to assess the assets of the fund for which there is no recognized market quotation). A special role among these organizations is assigned to a specialized depository, which provides not only storage of the fund's property, but also control of the management company for compliance with legislation and in the interests of the fund's investors.

2. Normative documents of the Bank of Russia and self-regulatory organizations of market participants, which regulate the activities of management companies or set guidelines for the development of internal regulations of companies. The most significant areas of regulation include as follows:

- control of requirements for the composition and structure of fund assets;
- risk management of the management company;
- compliance of employees of the management company with qualification requirements and requirements for business reputation;
- control over the business reputation and financial position of the owners of the management company;
- conflict of interests of the management company and its employees in relation to the fund's investors.

3. Special bodies of management and control of the investment fund. These bodies include the general meeting of the shareholders (investors) and the investment committee. In accordance with the current legislation, the general meeting of the shareholders is a mandatory element in closed mutual funds, and the competence of the meeting may include a wide list of significant issues, decisions on which are mandatory for the management company, up to the change of the management company! In turn, an investment committee can be created in investment funds for qualified investors. The powers of this committee, formulated in the Fund's Rules, allow almost complete control over the activities of the management company through the introduction of a permitting procedure for its transactions.

Even a simple listing and brief description of the elements of the system for monitoring and managing the activities of investment funds leaves no doubt about the general focus - protecting the interests of investors from the opportunism of the management company. This conclusion is also confirmed by the introduction into the practice of regulating the investment fund industry of fiduciary responsibility - an attributive element of Anglo-Saxon law and an outsider (market) model of corporate governance. For example, the rules related to fiduciary responsibility for a management company are already spelled out in the Federal Law "On Investment Funds". According to him, the Bank of Russia may order the management company to compensate the client for real damage if it was received due to the fact that the management company did not show due diligence in the interests of the founder of the management.

3 Results and discussion

The evidence and justification of the existing model of "informal" corporate governance does not deny its further development and transformation. This hypothesis is confirmed by the multivariance of the corporate governance model in entrepreneurial corporations. Recall that in this case, theory, and most importantly, practice, distinguish two so-called "basic models" of corporate governance: the already mentioned outsider and insider (stakeholder) models. The dividing line between these models runs along the line, which is set by such parameters as the composition and characteristics of significant participants in the corporation, as well as the key conflict (problem), the resolution of which (solution of which) ensures the sustainable development of the corporation. The correct diagnosis of one version of the model or another largely determines the list of control mechanisms - internal and external institutions, regulatory practices, etc.

For example, the outsider model is characterized by an attributive conflict between ordinary corporation owners and professional management. Dispersed ownership structure, lack of professional competencies among investors, all this determines the priority of the external mechanisms for control and conflict resolution. These mechanisms include, for example, a simple assessment of the effectiveness of management through market indicators, free entry and exit of owners ("voting with their feet"), fiduciary responsibility of management.

On the contrary, the stakeholder model is characterized by a more complex composition of significant corporation participants and stakeholders, as well as a concentrated ownership structure. Accordingly, the basic problem is the coordination of the interests of the participants and stakeholders of the corporation, as a key factor in its sustainable development [14-16]. The set of mechanisms that ensure the solution of the problem naturally changes - first of all, these are internal mechanisms of control and coordination, implemented through the construction of an internal "hierarchy" of management bodies, procedures for their interaction, evaluation of results, and decision-making. Note that building a largely unique "hierarchy" for each corporation is associated with significant internal costs, but ensures the functioning of the corporation in a complex institutional environment [17] or the formation of a more complex result, in the form of the so-called "corporate value".

We can now assess the feasibility of applying similar models to the investment fund industry. The task is greatly simplified if we take into account the attributive property of an investment fund as an institution of financial intermediation - the reflection (adequacy) of the nature of liabilities (fund investors) to the characteristics of its assets (target investment market). In this regard, Russian regulatory practice (as in most developed countries) differentiates investment funds as follows [18]:

- **funds for unqualified investors**, for which target markets (financial instruments or real estate) are limited to markets with high information efficiency and liquidity;
- **funds for qualified investors**, on the other hand, allow investments in poorly efficient and illiquid real estate and financial assets markets (venture capital fund or private equity format), form complex combinations of assets and liabilities (hedge fund format), and even implement entrepreneurial projects (combined fund format).

Analysis of the real Russian practice of the formation and functioning of investment funds allows us to make a number of important comments for the analysis.

The first point concerns the structure of the fund's investors. Almost all funds targeted at unqualified investors have a dispersed structure (the so-called "retail funds"). It can be assumed that these funds are not of interest to large investors. The latter, due to the lack of control mechanisms in this case, and, moreover, pressure on the management company (for example, the general meeting of investors and the investment committee) in order to implement decisions that ensure a profitable transfer of the fund's assets. In addition, the target asset market for these funds (efficient and liquid) significantly reduces the opportunities for transfer pricing transactions. In contrast, qualified investor funds tend to have a concentrated investor structure. The controlling, "anchor" investor gets the opportunity to control the management company through the management bodies [19], and the weakly efficient asset market allows using this opportunity in his own interests.

The second remark characterizes the initiator of the fund creation and further reinforces the situation caused by the structure of the investors. The fact is that the real initiative to create an investment fund is only in theory always the prerogative of the management company. In Russian practice, this option is usually common among "retail" funds, and created in the form of open, interval, and recently in the form of exchange-traded funds. A completely different picture is typical for funds for qualified investors. It can be argued that almost all funds for qualified investors, or in any case created in the form of a closed-end fund, are created at the initiative of the future controlling investor, and the management company is involved to ensure the administration of the fund - formal compliance with the necessary procedures, thereby formalizing the will and interests of the controlling investor

Thus, the whole variety of investment funds is not homogeneous from the point of view of corporate governance issues and objectively there are prerequisites for the formation of not one, but at least two basic models (Table 1).

Table 1. The main approaches to corporate governance and the corresponding basic approaches to corporate governance in investment funds

Outsider (market) model	Insider (stakeholder) model
Corporate governance	
a system of accountability, control and motivation of managers, ensuring that the interests of the fund's investors are met.	a process by which a balance of interests is established, both of the fund's investors and other participants (subjects) of the investment process.

Continuation of Table 1. The main approaches to corporate governance and the corresponding basic approaches to corporate governance in investment funds

The main problem of corporate governance	
violation of investors' interests by fund managers, as one of the options for manifesting an agency problem.	formation of a stable and effective system of interaction between participants, with different, often opposite interests.
Limits of applicability	
The fund's investments are carried out in efficient liquid markets, and its formation takes place at the expense of ordinary unqualified investors. The dispersed structure of the investors prevails.	The fund's investments are carried out in poorly efficient and non-liquid markets, which requires attracting funds from qualified investors and/or involving other groups of stakeholders in the investment process. In Russia, as a rule, there is a concentrated structure of the investors.
Nature of corporate governance mechanisms	
Market - based on the free receipt of objective information and the implementation of market transactions with minimal costs	Hierarchical - based on the coordinated exchange of information between the participants and its assessment in accordance with internal regulations and procedures.

The scale of the potential distribution of the considered models can be estimated in Figure 2.

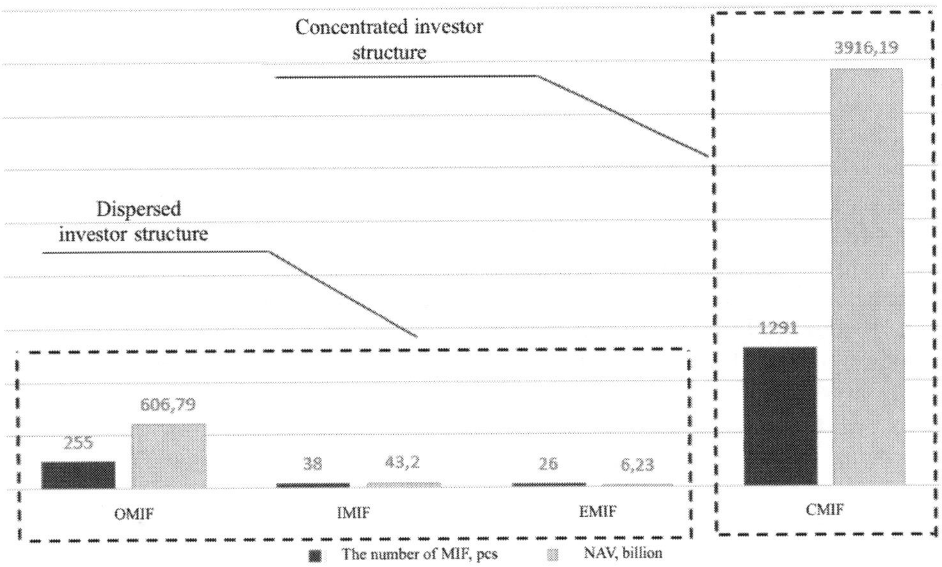

Fig. 1. The number of Russian open (OMIF), interval (IMIF), exchange-traded (EMIF) and closed (CMIF) mutual investment funds and the net asset value (NAV) corresponding to these funds [26,27]

As we can see, the validity of the outsider or market model remains only for open, interval and exchange-traded funds. The absence of large, controlling investors, characteristic of these types, highlights the classic problem of controlling management companies in the interests of ordinary unqualified investors. As noted above, it is this ideology that is embedded in the existing system of control over the activities of investment funds.

The only but rather important flaw of the existing model for this segment of funds is the lack of control over the management style of investment portfolios of funds. The fact is that the dominant (at least in Russian practice) active approach to management, that is, the search for undervalued and overvalued assets and the corresponding restructuring of the portfolio in liquid markets, including exchange markets, is recognized in world practice not only as ineffective, but also one of the most dangerous forms of opportunism of managers of management companies [20-23]. This is the reason for the cardinal shift of the global investment fund industry towards passive management strategies, including in the form of index funds and ETFs (exchange-traded fund, an index fund, whose shares are traded on the stock exchange). A gradual transition in this direction is also observed in Russia, where exchange-traded funds are analogous to *ETFs*.

The introduction into widespread practice of forms of passive investment based on a wide diversification of assets is most in the interests of ordinary inexperienced investors and from the point of view of corporate governance is no less important than control of the composition and structure of assets, transactions and business reputation. Another thing is that, in this case, priority should be given not to tough administrative pressure from the regulator. More justified is the introduction of the obligation to disclose the composition and level of total costs of the fund. It is on this characteristic, and not on the historical profitability of the fund, that ordinary investors' attention should be shifted when choosing a fund. It can be assumed that the management companies placed in these conditions of competition will be forced not only to reduce their remuneration levels, but also to abandon the costly but ineffective style of asset management.

In this case, the already mentioned exchange-traded funds, in which the threat of the emergence of an agency problem is practically excluded, are a kind of target reference point. Therefore, it seems somewhat redundant to extend the entire spectrum of control mechanisms to management companies focused on the management of exchange-traded funds.

The analysis shows that the current model of regulation is quite adequate to the systemic conditions for the functioning of investment funds focused on unqualified investors and on liquid, highly efficient markets. However, as has already been shown (see Figure 1), a much larger volume of funds operates under completely different conditions. Weak information efficiency of markets, a significantly larger range of participants and a more complex combination of their interests, all this makes it difficult to assess the "correctness" of the decisions made. The situation with the private equity fund can be cited as an illustration. In this case, we are already dealing not with an abstract investment portfolio formed on the exchange market, but with the investments in a specific enterprise with a clearly expressed and multidimensional subjectivity (managers of the recipient enterprise, its staff, local authorities, etc.). What is the algorithm for preventing or overcoming the conflicts of interest? Among which subjects? Neither general market regulation nor the markets themselves can answer these and other questions. Moreover, the preservation of priority for external formal mechanisms, or the hope for internal control mechanisms (for example, a meeting of investors and an investment committee), but in the context of a concentrated structure of investors not only does not contribute to the growth of corporate value, but also works in the opposite direction - to destroy it!

Overcoming the current situation (“two specific segments of the investment fund market - one regulation model”) seems to be in the direction of deeper changes in the regulation of funds focused on inefficient markets, ensuring the formation of unique objects and investment schemes, and therefore involving subjects (groups participants) with different, often diverse interests and expectations. The priority in this case is the development of the internal mechanisms of control and motivation. On the one hand, legal administration is difficult in low-efficient markets, and on the other hand, the competence of investors is quite sufficient to both evaluate and even implement internal corporate control procedures. Examples are boards of directors, supervisory boards and investment committees set up in funds for the qualified investors. The search and implementation of best management practices, the development of codes and standards, taking into account the specifics of a particular fund - this is the direction in which the industry of these investment funds is advancing in developed countries [7, 24].

Of course, the concentrated structure of the investors, characteristic of the same category of funds, presents a certain problem. However, the very awareness of this problem allows us to suggest possible solutions to the problem. This can be an analogue of limited partnerships, in which the securing of special rights to manage the fund by the controlling investor is inevitably accompanied by the acceptance of additional (unlimited) liability [25]. The option of limiting the representation of a large investor in control and management bodies may also be quite applicable. At the same time, it is possible to provide for the representation in the same bodies of other groups of participants and stakeholders.

4 Conclusions

An analysis of the conditions for the functioning of the investment funds allows us to single out several significant conditions that determine the difference in the goals to achieve which the system of regulation, control and decision-making should be oriented on. According to the author, these conditions should include 1) the nature of the investment market (effective or ineffective), as well as 2) the characteristics of the fund's investors (not qualified and qualified). In practice, due to legislative restrictions and economic feasibility, the following combinations of these conditions are possible:

1. **Non-qualified investors - a highly efficient (liquid) market.**

Taking into account the traditional dispersed structure of the investors for this option, the goal of maximum protection of the investors from possible manifestations of opportunism of fund managers (management companies) is quite justified. In other words, in this case, the existing system of control and management of investment funds for unqualified investors is quite justified. As practical recommendations, the following can still be attributed

- strengthening the responsibility of management companies in disclosing the aggregate level of the fund's costs and shifting the attention of ordinary investors to this very indicator when choosing an investment fund;
- reducing the regulatory burden on the companies that provide management of exchange-traded funds - the specifics of the functioning of these funds practically excludes the manifestation of the most dangerous forms of management opportunism.

2. **Qualified investors - a low efficient (non-liquid) market.**

This option, on the contrary, is characterized by a concentrated structure of investors, and it is they who, in most cases, initiated the creation of a fund with the involvement of an external or affiliated management company in the administration of the fund. As a result, the problem of the conflict of interests like "controlling investor - fund management" loses its relevance, but a whole range of other potential conflicts of the controlling investor with the minority investors and other participants in the investment process arises. It is important that the composition of these participants, and hence the content of conflicts, changes depending on the format and investment orientation of the fund (private equity or venture capital fund, hedge fund, etc.). In these conditions, the most expedient is to change the model of control and management of the fund, its transformation into a stakeholder model of corporate governance:

- highlighting the coordination of significant interests of all the participants in the investment process and the stakeholders as a priority goal;
- providing the regulator with greater freedom in developing internal procedures, regulations, rules and codes that reflect the specifics of the activities of a particular fund, deliberately adopted or even developed by the participants in the investment process, who in this case inevitably have sufficient qualifications.

In addition, measures aimed at offsetting the excessive influence of the controlling investor may be quite reasonable. For example, limiting the representation of a given investor in an investment committee with a simultaneous expansion of the presence of other participants therein, or increasing the responsibility of the controlling investor, up to imposing full responsibility on him, following the example of limited partnerships.

References

1. S. Bhattacharyya, V. Nanda, Review of finance, **3**, 885 (2013)
2. D Del Guercio, E Genc, H Tran, Journal of financial economics, **3**, 535 (2018)
3. T. Houge, J. Wellman J. Journal of business ethics, **2**, 129 (2005)
4. L. Ma., Y. Tang, J-P. Gomez, Journal of finance, **2**, 587 (2019)
5. M. Qian, journal of corporate finance, **1**, 45(2011)
6. M.Qian, B.Yu B, Applied economics letters, **12**, 967(2015)
7. A.A. Metzger, Bulletin of the Humanitarian University, **2 (2)**, 36 (2013)
8. Federal Law dated November 29, 2001 No. 156-FL «On Investment Funds»
9. The Civil Code of the Russian Federation (Part One): Federal Law No. 51-FL dated 30.11.1994.
10. The basic corporate governance standard for financial institutions that are brokers, dealers, managers, depositories, registrars (draft), http://www.naufor.ru/
11. Guidelines for members of the board of directors (supervisory board) of a financial institution. Appendix to the letter of the Bank of Russia dated 28.02.2019 No. IN-06-28/18, https://cbr.ru/
12. O.I. Shalina, Economics and Management, **6**, 99 (2009)
13. A.E. Abramov et al, *Economics of investment funds: monograph* (2015)
14. I.N. Tkachenko, *Analysis of the concepts of assessing the contribution of stakeholders, or On the issue of stakeholder value*, New trends in the development of corporate governance and business: materials of the Intern. scientific-practical conf. (Yekaterinburg, November 9, 2017), 93(2017).
15. I.N. Tkachenko, A.A. Zlygostev, Manager, **4**,40 (2018)
16. Yu.M. Emohonova, Bulletin of NSUEU, **2**, 193 (2017)
17. A.A. Metzger, Bulletin of the Humanitarian University, **1 (1)**, 64 (2013)
18. Bank of Russia Ordinance No. 4129-У dated September 5, 2016 "On the composition and structure of assets of joint-stock investment funds and assets of mutual investment funds."
19. L.P. Kharchenko, Problems of modern economics, **3 (39)** (2011)
20. D. Bogle, *The Guide to a Reasonable Investor. A reliable way to make a profit in the stock market* (2013).
21. A.A. Metzger, Bulletin of the Humanitarian University, **1**, 15(2018)

22. G. Kommer *Souverän investieren mit Indexfonds & ETF. Wie Privatanleger das Spiel gegen die Finanzbranche gewinnen* (2018)
23. M. Ritzau, *Die große Fondslüge. Falsch beraten von Finanztest, Sparkasse, Banken und Co* (2016).
24. Yu.M. Ermokhina *Interaction of the participants in the investment process with a private equity fund: interests,* Bulletin of NSUEU (2017)
25. Hedge-Fonds Marktentwicklung, Risiken und Regulierung, Positionspapier der Eidg. Bankenkommission. September 2007, https://www.finma.ch/
26. The official website of the Bank of Russia, http://cbr.ru
27. The Cbonds Group of Companies, https://investfunds.ru/about/

The basic algorithm for the design and construction of a monocultural innovation organization

Yuri Polyakov[3], *Andrey* Savchenko[1], *Natalia* Alekseeva[4], *Elena* Panteleeva[2], *Mikhail* Savelyev[1]

[1]Mitra Association, Pushkinskaya Str., 241, 60, 426008 Izhevsk, Russia
[2]M.T. Kalashnikov IzhSTU, Studencheskaya Str., 7, 426069 Izhevsk, Russia
[3]Udmurt State University, Universitetskaya Str., 1, 426034 Izhevsk, Russia
[4]Izhevsk State Agricultural Academy, Studencheskaya Str., 11, 426069 Izhevsk, Russia

Abstract. Algorithmization of management activities in the transition to the digital economy is one of the most important tasks of modern management science. Sketch design is defined by the algorithm creation tool. This algorithmization method is applied to the processes of choosing the type of a homogeneous product, its quality and other consumer properties, utility for the consumer and manufacturer. Products are classified according to their purpose and the resources for their production. The quality of the product is defined as the ratio of the actual properties of the product to the normative ones, while it is necessary to take into account the influence of manufacturing defects, which entails a transition from linear to non-linear quality functions. The cumulative assessment of the product for the consumer is determined taking into account his subjective assessment of the properties of the product, expressed in the weight of the quality of individual utilities. For a manufacturer, the aggregate assessment of product quality directly depends on the main properties of the product, taking into account its other properties, and inversely depends on the cost of purchasing it, taking into account their functional elasticity. The algorithm is recommended for use by specialists dealing with algorithmic management activities.

1 Introduction

In the management of an organization, management decisions play a critical role. So, the development and production of innovative products is the result of the adopted management decisions. The process of developing the management decisions must be designed, i.e. developed in the form of a specific algorithm [1]. Algorithmization of management activities during the transition to the digital economy is one of the most important tasks of modern management science, increasing the competitiveness of the companies. According to M. Porter, the development of the national economy depends on the achievements of individual national companies in the foreign market [2-9].

The development of an algorithm for the design and construction of a monocultural innovation organization (hereinafter referred to as the creation of an organization) is one of the most difficult elements of this task. In our opinion, it is possible to construct a basic algorithm by simplifying the concept of an innovative organization and introducing several concepts:

- a single product produced by an organization.
- unit operation, manufacture of a product.

2 Materials and methods

The main proposed approach to algorithmic design and construction of management activities is the preliminary design. This is an analogue of the conceptual design of a product, the first conceptual stage of organizing the organization's activities.

The organization is created in the following sequence through selection and justification:

- type of homogeneous product for the organization of production activities;
- elements of technological equipment of a single product-operation;
- production capability of the product-operation;
- economic characteristics of the product-operation;
- indicators of the effectiveness and efficiency of the product-operation.
- types of resources for a single product-operation.
- technological elements of a single operation of supplying the homogeneous resources.
- parameters of the production capability of the delivery operation
- indicators of the effectiveness and efficiency of the delivery operation.

3 Results and Discussion

1 stage. Selection and justification of the type of homogeneous product for the organization of production activities.

Let us define a parametric range as a set of products that are similar in design and technology, with the same or similar functional purpose and differing in the quantitative level of the main consumer property.

The basis for choosing and justifying the type of a homogeneous product for organizing production activities is the classification of types of products.

The classification of types of homogeneous products is shown in Figure 1.

- Types of homogeneous products produced on a single product-

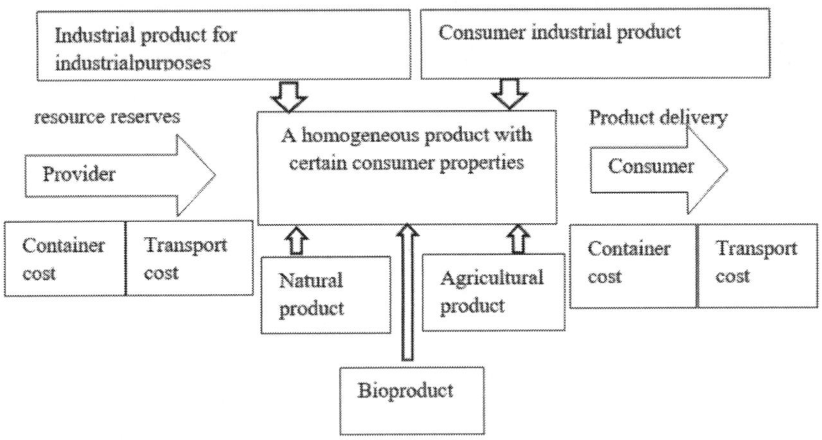

Fig. 1. Classification of product types

To select and justify the types of homogeneous products for production, a selection criterion is required. Such a criterion is the usefulness of the product for the consumer. The usefulness of the product depends primarily on the consumer properties of the product, on the consumer himself, his subjective attitude to individual properties and their totality. The utility of a homogeneous product for a particular consumer may be different, both, relatively high or relatively low. It is necessary to assess the usefulness of the product for the consumer.

Let us present the analysis of the usefulness of a homogeneous product for the consumer in the form of the following sequence:

Usefulness of a product (Cv) is a function of an indicator of product properties (P) useful for a consumer, consumer requirements (U), conditions of consumption (hc), etc.

$$Cv = f(P,U,hc) \quad (1)$$

Useful properties of the products (Q) - a combination of organoleptic, physicochemical and other properties of a product that determine its quality.

2 stage. In most of the quality assessment methods, the property assessment is the ratio of the values of its indicator and the corresponding base indicator:

$$K_{ij} = \frac{P_{ij}}{P_{ij}^{\text{base}}} \quad (2)$$

The quality of the product property function (P) is quantified by metrological measurement i.e.:

$$P = \frac{\oint(X_k)}{\oint(X_o)} \quad (3)$$

Where: P - assessment of the quality indicator of the measured property of the product;

f(Xo) - reference characteristic of the property (base);

f(Xk) - actual characteristic of the property (measured property)

The objective feature, the properties of the product that are part of its quality, the achieved level of properties in relation to the standard are quantitatively measured. The property scoring model establishes a certain relationship between the property index value Pij and its Kij score.

Other methods for assessing the consumer properties of a product.

If we introduce an additional parameter - the defective index Pijdefect, which is taken as the lower permissible value of the indicator, the quality assessment formula takes the following form:

$$K_{ij} = \frac{P_{ij} - P_{ijdefect}}{P_{ij}^{\text{base}} - P_{ijdefect}} \quad (4)$$

This formula differs somewhat from the formula for linear dependence, but describes a different dependence - nonlinear one.

In Harrington's method [11], each property is estimated using a dimensionless coefficient and the mathematical dependence of the estimate on the property indicator is determined by an exponential function:

$$K_{ij} = e^{-(/P^0/)^{mj}} \tag{5}$$

where mj is a positive number within 0<mj<∞;

P0 - a linear function of Pij

$$P^0 = \frac{2P_{ij} - (P_{ij}^{\max} + P_{ij}^{\min})}{(P_{ij}^{\max} - P_{ij}^{\min})} \tag{6}$$

where $P_{ij}^{\max}$ Џ $P_{ij}^{\min}$ - the lower and upper limits of the value of the indicator of the j-th property, provided by the technical specifications.

From here:

P0=-1 when $P_{ij} = P_{ij}^{\min}$

P0=+1 when $P_{ij} = P_{ij}^{\max}$

If the restrictions imposed by the technical specifications concern only the lower limit of the indicator value, then the formula is applied:

$$K_{ij} = e^{-(e^{-P^0})} \tag{7}$$

The described group of techniques has a disadvantage of using the same type of dependence for all indicators. However, the formulas for linear and nonlinear dependence are quite suitable for use. In most of domestic methods, linear dependences are used to assess properties, and in most foreign ones, nonlinear ones.

Different formulas are used to determine a comprehensive quality assessment. As a rule, the following are used to calculate complex quality assessments without taking into account the weight of individual properties:

$$K_0 = \sqrt[n]{\prod_{j=1}^{n} K_{ij}} \tag{8}$$

where n is the number of quality indicators of properties that characterize the quality of the product taken into account.

Due to the simplicity of the calculation, most of the methods for the comprehensive assessment of quality, taking into account the weight of properties, give preference to the arithmetic mean of individual properties.

For techniques using the geometric mean of estimates of individual properties, the basic calculation formula is as follows:

$$K_0 = \sqrt[\sum M_j]{\prod_{j=1}^{n} K_j^{M_j}} \quad (9)$$

Where: Ko - an objective comprehensive assessment of product quality indicators based on metrological methods.

Mj - the weight of individual properties, with 0 ≤ Mj ≤ 1, a 0 ≤ Kj ≤ 1

3 stage. The properties of the function (U) are quantitatively expressed by expert measurements, i.e.

$$U = \frac{f(X_k)}{f(Xstage)} \quad (10)$$

Where: U is an indicator of the usefulness of a product property;

f(Xstage) - subjective attitude towards a product property (score from 0 to 1);

f(Xk) - actual characteristic of the property (measured property)

Consumer requirements (U) - a set of subjective relations of the consumer to the product and its individual properties.

4 stage. Assessment of the usefulness of the product for the consumer

The basic calculation formula of methods using the geometric mean of the consumer's assessments of individual properties is expressed by the formula:

$$Kp = \sqrt[\sum M_j]{\prod_{j=1}^{n} K_j^{M_j}} \quad (11)$$

Where: Kp- subjective comprehensive assessment of product quality indicators;

Mj - the weight of individual properties in the opinion of the consumer, and

≤ Mj ≤ 1, and 0 ≤ Kj ≤ 1

Stage 5.Evaluating the overall utility of a product

The calculated formula for the cumulative indicator of the usefulness of a product is expressed by the following formula:

Ки=Ко*Кр

where: Ки- cumulative assessment of the usefulness of the product for the consumer

Ko –an objective comprehensive assessment of product quality indicators based on metrological methods.

Kp- subjective comprehensive assessment of product quality indicators;

6 stage. Evaluation of the integral quality of the product for the manufacturer

Integral quality can be expressed in the following form:

$$kq = A \times D^a \times Q^b \times P^{-y} \quad (12)$$

Where: Kq -integral product quality;

D is the main parameter of the products determined by the Metrology;

Q-useful properties of products determined
based on the methods and principles of qualimetry;
p - production expenses, quantified by costs
of live and past labor;
a, b, y-elasticity coefficients, show the degree of influence
consumer properties and production costs by the amount of
integral quality;
A-proportionality coefficient, shows the dimension
of integral quality for D=1, Q=1, P=1;
kq-numerically characterizing the level of integral quality qualimetric indicator of this type of product.

4 Conclusions

The results of the algorithmization of the design and construction of a monocultural innovative organization allow in one automatic process to select the type of a homogeneous product, its quality and other consumer properties, usefulness for the consumer and manufacturer. This algorithm is recommended for use by specialists dealing with algorithmic management activities.

Acknowledgements

The reported study was funded by RFBR, project number 20-010-00869

References

1. Yu.N. Polyakov, M.Yu. Savelyev, A.I. Savchenko, Problems of regional economy, **1-2**, 186 (2020)
2. M. Porter, International competition: Competitive advantages of countries, 947 (2020)
3. M. E. Porter, *Competition.: Translated from English*: Textbook, 495 (2000)
4. Michael Porter, *Competitive strategy: Methods for analyzing industries and competitors*, 453 (2015)
5. M. E. Porter, Harvard Business Review, May/June, 43 (1987)
6. M. E. Porter, Strategic Management Journal, **12**, 95 (1991)
7. M. E. Porter, Harvard Business Review, Nov/Dec (1996)
8. A. M. McGahan, M. E. Porter, Strategic Management Journal, **18**, 15 (1997)
9. M. E. Porter, Harvard Business Review, 62 (2001)
10. M. E. Porter, M. R. Kramer, Harvard Business Review, 78 (2009)
11. G.G. Azgaldov, E.P. Reichman, On qualimetry, 172 (1973)

Development of a descriptive monocultural model for creating an innovative organization

Yuri Polyakov[3], *Mikhail* Savelyev[1], *Denis* Perevedentsev[1], *Olga* Titova[2], and *Marina* Galiakhmetova[2]

[1]Mitra Association, Pushkinskaya Str., 241, 60, 426008 Izhevsk, Russia
[2]M.T. Kalashnikov IzhSTU, Studencheskaya Str., 7, 426069 Izhevsk, Russia
[3]Udmurt State University, Universitetskaya Str., 1, 426034 Izhevsk, Russia

Abstract. A methodology for designing an innovative organization has been developed within the framework of a descriptive approach to draft design in monocultural environments of the organization as a whole and of individual managers. The definition of single design elements is proposed as a design tool. The design basis is the conceptual sketch of the organization's activities, which includes the elements of resources, added value, product, its value and demand for it, defined as a single production operation. A classification of types of homogeneous products according to their purpose and resources used has been given. The classification of types of homogeneous products for industrial purposes includes tools, fixtures, machine and equipment, areas, structures and infrastructure. The problems of choosing a product and using the resource have been solved. This methodology is recommended for use by managers and professionals involved in the design of organizations.

1 Introduction

An innovative organization is a pillar of sustainable economic growth, increasing the competitiveness of the national economy. Namely, through the prism of the competitiveness of individual national companies, M. Porter explained that the general level and dynamics of development of the country's economy depends on the achievements of specific firms representing the country in the foreign market [1-12].

The model is a kind of "cognitive map of the area" based on the cognitive theory of learning by Edward Tolman, developed in 1932 [13]. It prescribes what needs to be done to increase the likelihood of a favorable outcome. The development of a descriptive monocultural model of creating an innovative organization (hereinafter referred to as the model) allows you to manifest, visualize the process of designing and constructing an organization.

Designing and constructing a monocultural innovative organization (hereinafter referred to as the organization) is a complex multistage process. In our opinion, it is possible to construct a basic algorithm by simplifying the concept of an innovative organization by introducing several concepts:

A single production operation is a complete technological process or a part of a technological process, in which, over a standardized period of time, one complex of production purposes (tool-device-machine) is used to manufacture a single product, with the participation of a person with a set of technical competencies.

A single product is a finished product manufactured in a single operation, intended for final or intermediate consumption, as well as an element used in a more complex design.

Homogeneous product - products that are identical in design and manufacturing technology, have the same or similar functional purpose and differ in the quantitative level of the main consumer property.

A single homogeneous resource is a type of resource used in a single operation for the manufacture of a homogeneous single product,

A single product for industrial purposes is a type of industrial complex for equipping a single operation.

In our opinion, these are basic elements, constructive bricks, from which, like a mosaic, the activities of an organization are formed.

2 Materials and Methods

The main proposed approach to the design and construction of a single product-operation is a sketch design of the activities of the organization as a whole and the activities of individual managers. Sketch design is an analogue of the conceptual design of a product, the first conceptual stage of organizing the organization's activities.

Conceptually, with the help of sketch design, the design of the following elements of the organization's activities can be shown:

- single operation (production activities of the organization),
- single homogeneous product, the result of a single operation (production activity);
- single homogeneous resource required for a single operation (innovation and procurement activity),
- single product for industrial purposes to equip a single operation (innovation and investment activity);
- The functionality of the organization consists in the following actions:
- formation of the value of a single product,
- formation of a value chain,
- formation of demand for a resource and a product of industrial purpose,
- formulation of equilibrium conditions under which the entire product produced by the organization is consumed.

3 Results and Discussion

An initial conceptual sketch of the organization's activities is shown in Figure 1.

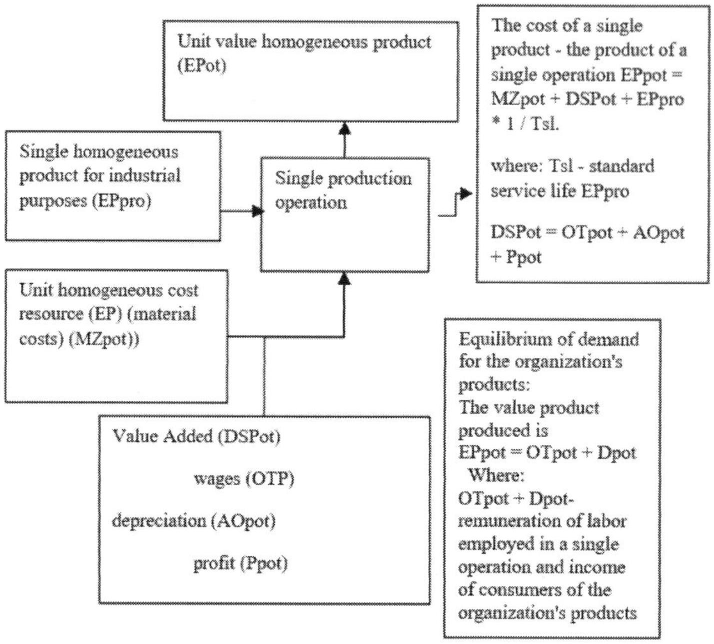

Fig. 1. Conceptual sketch of the organization's activities

It turns out that the simplest, elementary organization has been designed, producing one homogeneous product, consuming one homogeneous resource and a homogeneous product for industrial purposes. The variant of arrangement of the organization's activities from the minimum number of elements has been chosen.

Further design of the organization is carried out in the following sequence in the direction of concretizing the selection of the type of the following elements of activity:

homogeneous product for the organization of production activities;

homogeneous resource for a single product-operation;

production product to equip a single operation.

single operation (parameters of the production capability of the organization).

The choice of the type of a homogeneous product of the organization is as follows: will the organization produce a certain type of agricultural products or a product for industrial purposes or a final product for consumption by the population. The basis for choosing the type of a homogeneous product for organizing production activities is the classification of types of products. The classification of types of homogeneous products is shown in Figure 2. The choice of the organization's product is carried out either on the basis of ready-made or purchased product innovations (industrial design, recipe, utility model, etc.). For each of these choices, the rationing of the value structure of a single homogeneous product is carried out either by the experimental statistical method or by the normative method based on the norms of the value structure or by the analog method based on the structure of the value of analog products. The price of the product is calculated by the method of calculating the cost elements and standard profit. The equilibrium condition is specified, the price of the organization's product is equal to the individual income of an employee in the organization or the individual income of the consumer of the organization's product.

- Types of homogeneous products produced for a single operation

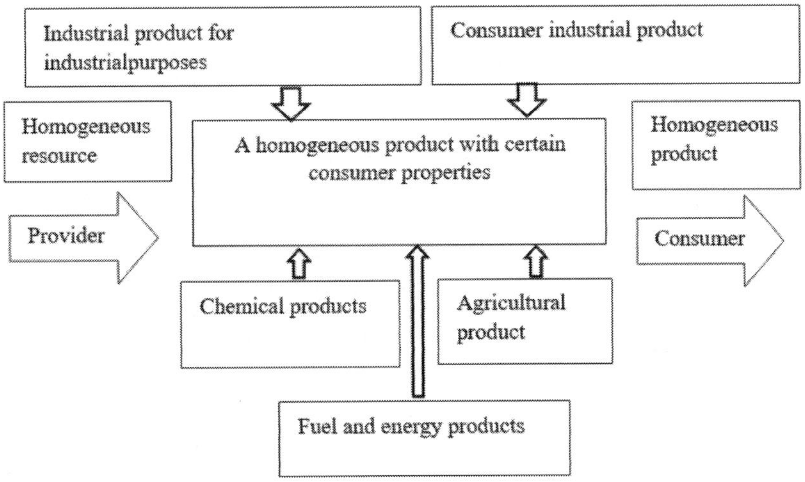

Fig. 2. Classification of types of homogeneous products

The selection of a homogeneous resource is as follows: whether the organization will use a biological resource or agricultural product or an industrial product to produce its product. The basis for choosing the type of a homogeneous resource for the production of an organization's product is the classification of types of the resources. The classification of types of homogeneous resources is shown in Figure 3. The selection and justification of the types of resources for a single product-operation is carried out either on the basis of innovative resources produced by a specialized manufacturer or on the basis of recoverable natural resources with unique properties or extracted biological resources with unique properties. For each of these choices, the rationing of the consumption of resources for the production of the organization's product is carried out either by the experimental statistical method or by the standard method or by the analog method of resource consumption in the manufacture of analog products. An acceptable price of resources is calculated, which ensures the maximum marginal profit, the consumption of a natural resource is rationed and the specific consumption of this resource is calculated per unit of the finished product.

- Resource equipping of a single product-operation with types of a homogeneous resource

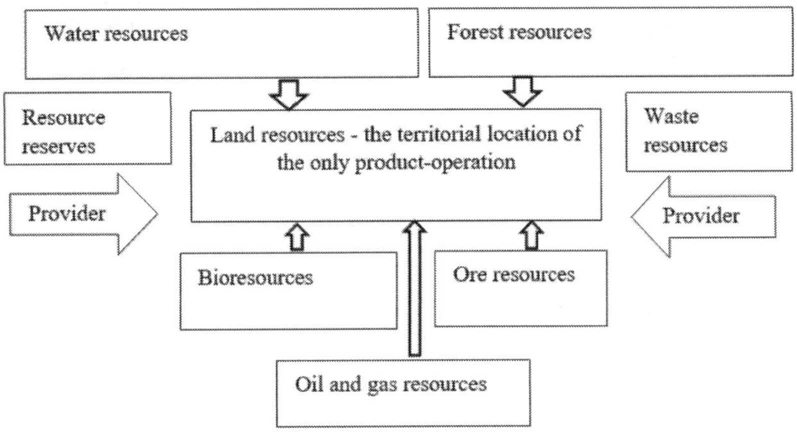

Fig. 3. The types of resources required for a single product-operation

The selection of the type of industrial product for equipping a single operation is as follows: equipping is carried out either on the basis of innovative products produced by specialized manufacturers or on the basis of our own innovative proposals for specialized manufacturers or on the basis of acquired innovative proposals for specialized manufacturers of industrial products.

- Technological equipment of a single product-operation with types of a

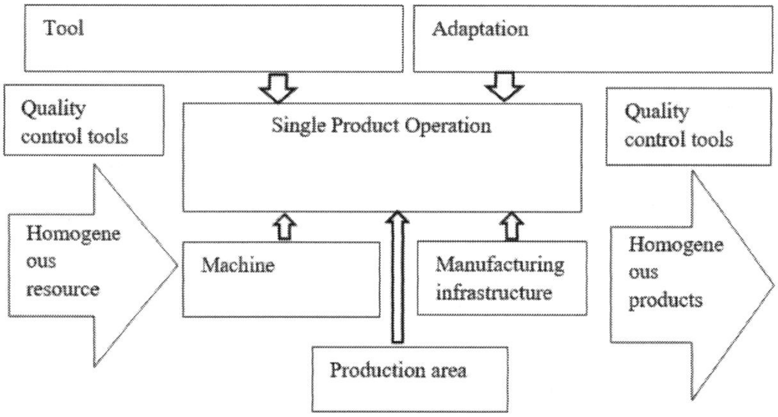

Fig. 4. Classification of types of homogeneous industrial products

For each of these choices, the number of units of a production product for the production of an organization's products is rationed either by an experimental statistical method or a regulatory method or by an analog method of using production products in the manufacture of analog products. The standard service life of a production product is determined, depreciation charges are calculated either on a straight-line basis or by an accelerated depreciation method. An acceptable price for a production product is calculated, which ensures the minimum payback period or the maximum interest income on the funds invested in the production product.

4 Conclusions

The resulting methodological tools are applicable to the creation of an innovative organization in a monocultural environment with regard to product design and resource use. This approach allows to ensure the complexity and integrity of the creation of innovative organizations.

Acknowledgements

The reported study was funded by RFBR, project number 20-010-00869

References

1. M. E. Porter, *Competitive Advantage: Creating and Sustaining Superior Performance*, 592 (1985)
2. M. E. Porter, *Competitive Advantage of Nations*, 896 (1998)
3. M. E. Porter, *On Competition* (1998)

4. M. E. Porter, Takeuchi, H., Sakakibara M., *The Japanese economic model, Can Japan compete?,* 262 (2005)
5. M. E. Porter, Elizabeth Olmsted Teisberg, E. O. Redefining, *Health Care: Creating Value-Based Competition on Results*, 506 (2006)
6. M. E. Porter, Harvard Business Review, May/June, 43 (1987)
7. M. E. Porter, Strategic Management Journal, **12**, 95 (1991)
8. M. E. Porter, Harvard Business Review, Nov/Dec (1996)
9. A. M. McGahan, M. E. Porter, Strategic Management Journal, 18, 15 (1997)
10. M. E. Porter, Harvard Business Review, 62 (2001)
11. M. E. Porter, M. R. Kramer, Harvard Business Review, 78 (2006)
12. Joan Magretta. M. E. Porter, Mann, Ivanov and Ferber, 272 (2013)
13. E. Ch. Tolman, *General psychology*, 2, 479 (2013)

Corporate governance in connection with performance and digital

Igor Dimitriev[1], and *Elena* Krasilnikova[2,*]

[1]GPB, Vvedenskogo Str.12.1, 117342 Moscow, Russia
[2]Central Economics and Mathematics Institute of the Russian Academy of Sciences, Nahimovsky Pr., 47, 117418 Moscow, Russia

Abstract. The article concerned issues, related to the characteristics of Corporate governance and performance. The characteristics of Corporate governance are pervasive: ownership structure, separation from control, Board of directors, protection of rights. For Russian companies traditional organizational characteristics of life cycle stages are not decisive. Currently there is a tendency for owners to get out from management and focus on social, environmental results. The ownership of private insiders is the highest at the stage of formation, and the foreign ownership rises at the stage of growth. Based on the model was estimated significant positive impact of companies' age on an increase in property rights, ownership and control, protecting minority investors encourages R&D. Results of estimation of 62 countries in 2020 represented significant positive relation Technology factors: adaptive attitudes, business agility, IT integration with performance and Corporate governance. An increase in digital is associated with a higher quality of Corporate governance.

1 Introduction: the Corporate Governance Features

The quality of Corporate governance (CG), first of all, is based on taking into account the interests of the company's stakeholders both in the short-term and long-term period. So it could be an internal central, like a system, stream of company's development. CG represents not only control and procedure measures, but also an institute, that provides a framework defines the rights, roles and responsibilities of various stakeholder groups. One of the challenges of CG is related to maintain balance and incentives, required to minimize and manage conflicts of interest. OECD Principles of Corporate governance are constantly improved, and the impact is referred not only on company performance. GC has a significant influence financial stability and economic growth. The resent tasks are associated with digital impact on stakeholder's relations.

Crucial principles of CG are protection rights, especially protecting minority investors. Positive trends are noted for Russian companies according to the Doing business study and Competitiveness report, conducted by the World Bank. Until 2018 year Russia's position in terms of Corporate governance was improving. But in 2019 Russia has worsened position: remains low «property rights» and «conflict of interest regulation» estimation. However,

*Corresponding author: krasilnikova_lena@list.ru

the last measure focuses on director's opportunism and behavior to personally use of firm assets. Traditionally, principal-agent relations have been studied under the owner-manager concept and the positive agency theory in the context of the ownership separation from the control. This is a natural step in the company's development, when the founders delegate their governance functions to professional management, as discussed by A.A. Berle and G.C. Means [1]. Most agent-principal models are aimed at investigating conflicts of interest between the management and the owners. This conflict results from opportunistic behavior in spending the company's resources for the managers' own needs, excessive growth of remunerations, misuse of the company's funds, unreasonably high benefits, payments and preferences to the managers. Such agency problems encourage an increase in agency costs of equity capital for monitoring, control, contribute to lose profit, reduce the effectiveness of strategic projects. Management opportunistic behavior could be solved due to increase in debt financing, which would impose obligations on managers. However, debt attraction as a disciplining factor leads to an increase in agency costs of debt, causes underinvestment, a decreases in free cash flow for shareholders [2-4].

However, there are two controversial cases for Russian companies.Agency models of the capital structure take into account the availability of debt resources, but for more, than 35% of companies in 2009 year, 14% in 2012, and 19,6% in 201limit in attraction finance is the main obstacle to doing business.

Studying Russian companies, researchers have found significant differences in specifics of principal-agent relations, a high concentration of ownership and its impact on corporate activities compared foreign firm in developed capital market. In particular, the classical manager-shareholder conflict (when the management demonstrates opportunistic behavior) is not typical for Russian companies. Many foreign studies conclude that ownership concentration has a negative impact on the company's financial and economic performance and value. But for Russian companies, accumulation of ownership has a different meaning. A high ownership concentration in Russia is a defensive measure due to the lack of protection both for the owners and the company itself. A major shareholder, as a rule, performs functions of the management that controls the company and makes strategic decisions, and is not considered only as an opportunistic hired manager. In companies with a low concentration, the goals of the owners differ mostly and agency contradictions are growing, the presence of a strong owner, who has a strong and predictable strategy, contradiction are mitigated, but an extremely high concentration can lead to discrimination.

2 Methods: Corporate Governance Factors relation with Performance

In the study of the problems of agency relations, it was traditionally assumed that the organization was at the stage of maturity and corporate characteristics at the previous stages were overlooked from the survey. The configuration of ownership and management is associated with the stage of life cycle in developed capital markets, when the founder-entrepreneur is the main initiator of the formation of the company. Moreover, the life cycle stage is mainly defined with management features and characters at various stages.The age and size of the company, its organizational structure, formalization, crises, and preferences of economic agents, agency relation and leadership are characteristics of the life cycle stages according to traditional approaches [5]. I.K. Adizes and L.E. Greiner investigated the stages of the life cycle and development of a company based on the modification of management types and management structure, the presence of entrepreneurial initiative in the founder [6-8]. The dominance of management styles and forms of interaction is primarily focused on solving internal developmental problems, in particular, the study of slowing development and quickly aging. They do not study the strong pressure of the external environment, the weak protection of owners, the uncertain influence of state participation – such important problems typical for Russian companies [9]. For Russian companies, such combinations andtypes of management are not successful in the face of strong pressure from external threats and state participation. I.K. Adizes notes the condition of "silence" within a Russian companies (possibly due to the fear of public leadership qualities, autocracy and desires to control). For Russian companies, the use of the classical theory of the life cycle is complicated, since many companies were formed through privatization. According to the author, the weak protection of the rights and the inability to use market signals for better orientation led to the majority owners in management and a slower owner' delegation functions to management as compared to foreign companies with dispersed ownership. The owners prefer to control flows and holds the post of CEO [10].

2.1 Ownership Separation from Control, link with Board of Directors, Life Cycle and Performance

Although recently there has been a tendency for owners to decrease involvement in management and attracting professional managers. Such changes according to early stages of life cycle in a developed capital market, are characteristic at the maturity stage in Russia. Now the management's interest in the long-term development of the enterprise is increasing.For example, the retail company Ltd. Lenta was established in 1993, and the founding owner was a director. As the company developed and foreign investors were attracted to the Board of directors (BD), the owner headed the BD. With the entry of foreign investors, the owner has stopped management functions, and after that, got out from the equity capital. The BD was chaired by a domestic minority shareholder, and a foreign representative of foreign funds was appointed as CEO.

For growing companies, the role and involvement of the BD is one of important force to sustainable development. At the early stages, in the BD includes those, who are most familiar with the industry and the market. Independent directors with significant knowledge could enforce environmental practices and mitigate agency conflicts [11]. Also Independent directors have a positive correlation with capital structure, owner rights protection, but a negative correlation with avoidance of unpredictability [12]. The Board diversity has a positive impact on sustainability in terms of ecological, social and governance performance. And the number of Independent directors doesn't encourage performance [13]. Not internal drivers stimulate Corporate governance changes, but also external factors like the level of stock exchange listing [14].

Controversial is the issue of development drivers. Within the life cycle concept, research is mainly aimed at changing the company's organizational characteristics related to management. The existing theoretical works are applicable to large companies with dispersed ownership, which makes it difficult to verify their applicability to Russian companies.

Recently, research has focused on fast-growing companies in emerging capital markets and identified a significant discrepancy between the actual situation and the conclusions from the agency theories, which caused the need for more detailed studies of behavioral characteristics of agents.

The preferences of interests affect the corporate strategies. Managers sometimes overestimate their characteristics, consider shares to be undervalued. Optimistic managers overestimate the likelihood of success and underestimate the risk, which leads to an increase in debt financing. From the one hand, in case of the opportunistic management behavior debt attracts to prevent ownership dispersion. From the other hand, debt capital is represented as internal governance, encourages performance and social responsibility [12]. The Corporate governance Index is higher at the growth stage. And debt-to-equity ratio has a positive relation with CG at the slow growth stage. Companies with low CG have a tendency to decrease debt-to-equity, share of liquid assets is higher and they ate more sensitive to market risk.

2.2 Digital relation with Corporate Governance

Digital technologies increase productivity, but change internal interactions. Responsibility of actors will increase through the introduction of Artificial Intelligence (AI), they will transfer from decision makers to agile leaders. And AI implementation will deepen Corporate culture in case of strong relations of human and machines [15]. One of the principles will be trust, openness and transparency. Moreover AI introduction will cause the crucial changes in the profit-maximizing concept. Sometimes less profitable strategies refer to an improvement in ecological, social performance [16].

3 Analysis: Key Factors and Some Resultsand Discussion

The author suggests that the characteristics and features of Corporate governance are related to the stages of the life cycle.

Hypothesis1. Concentration of domestic ownership is the greatest at the stage of formation, and as the company transfers along life cycle, it decreases, and foreign owners get into the ownership structure.Based on regression model, it was revealed that the ownership of private insiders is the highest at the stage of formation, and the ownership of foreign owners rises at the stage of growth.

Further improvement of the model identifies the link between age, ownership structure and protection of rights. Due to the age increase company becomes more mature and acquires reputation.

$$CG_i = c + b_i * X_i + \varepsilon \quad (1)$$

CG_t – corporate governance factors, X_i – age of companies.

The results presented significant positive impact of age on an increase in property rights (2,75 – estimated coefficient in the model) and ownership and control (2,08 – estimated coefficient in the model). Also protecting minority investors encourages R&D process, reflected in higher expenditures. Protecting minority contains several issues: disclosure information to shareholders, director liability, use shareholder capital, governance, Board structure and control.

Hypothesis 2. Digital strengthens agency relations. Growth of the quality of Corporate governance is positively related with the digital parameters.Based on Digital Competitiveness estimation of 62 countries and Doing Business ranking in 2020, was built model, reflected strong relation. Technology and companies' performance are actively highly correlated. So, ease to start business, register property, protecting minority investors have strong interrelation with Future Readiness. It reflects significant factors likeAdaptive attitudes: E-Participation, Internet retailing, Tablet possession, Attitudes toward globalization. Russia remains low position except internet usage. The other factors – Business agility: Opportunities and threats, World robots distribution, Agility of companies, Use of big data and analytics, Knowledge transfer, Entrepreneurial fear of failure; and IT integration:E-Government, Public-private partnerships, Software piracy.

Table 1. Pearson correlation between Technology, Performance and Governance.

Technology factors	**Performance**		**Corporate governance factors**	
	Starting Business	Getting credit	Registry Property	Protecting Minority
Adaptive attitudes	0,48	0,05	0,39	0,14
Business agility	0,29	-0,15	0,31	0,1

4 Conclusion

The quality of Corporate governance directly affects the performance and strategies. This impact is not similar as in the country comparison, and for different stages of the life cycle. The characteristics of Corporate governance, based on ownership structure are different for Russian companies and companies in developed countries. For Russian companies, combinations of management types, according to I.K.Adizes, are not dominant in the development. The preferences of a large owner have recently being the main driver of development. Now they delegate functions to management and ecological, social, governance factors become more significant. According to the results of regression models, have found, that property rights and protecting minority investors are improving due to the higher stage of life cycle and age.

Digital also has a considerable impact on performance. It reflects changes in internal relations, modifies Corporate governance principles, referred to agency interaction. Based on empirical estimations of 62 countries in 2020 year, were identified significant correlations between Technology factors, companies performance and Corporate governance characteristics. So, an increase in Business agility, for example in case of use of big data and analytics or knowledge transfer, is associated with higher ownership protection. A decrease in Adaptive attitudes, presented by Attitudes toward globalization, deteriorates company's performance.

5. Acknowledgements

The article has been prepared with financial support of the Russian Fund for Fundamental Research. Grant of the RFFR № 18-010-01028 «a» (2018–2020).

References

1. A.A. Berle, G.C. Means, *The Modern Corporation and Private Property* (1932)
2. M.C. Jensen, W.H. Meckling, J. of Fin. Economic, **3**, 305 (1976)
3. D.W. Diamond, J. of Polit. Economy, **97**, 828 (1989)
4. M. Harris, A. Raviv, J.of Finance, **46**, 297 (1991)
5. G.L. Lippitt, W.A. Schmidt, Harv.Bus. Review, **45**, 102 (1967)
6. I.K. Adizes, *Corporate Lifecycles How and Why Corporations Grow and Die and What To Do About It* (1989)
7. I.K. Adizes, Organiz. Dynamics, **9**, 3 (1979)
8. L.E Greiner, Harv. Bus.Review, **4**, 37 (1972)
9. E.V. Krasilnikova, Ec. analyses: theory and practice, **475**, 704 (2018)
10. T.G. Dolgopyatova, *Institutional problems of the Russian economy* (2015)
11. F. Haque, C. G. Ntim, Bus. Strat. and the Environment, **27**, 415 (2017)

12. J. Lu, J. Wang,J.of Inter. Fin. Markets, Instit.& Money, **4** (2020)
13. V. Naciti, J. of Cl.Production, **237**, 117727(2019)
14. O.A. Esqueda, T. O'Connor, Res. in Inter. Bus. and Finance**, 51**, 101077(2020)
15. Official website of Accenture company, https://www.accenture.com
16. Official website of B the change, https://bthechange.com

Dynamics of the transition of corporate governance models of the Republic of Tatarstan from "oligarchic" to innovative ones through strategic regional management

Nazilya Zarifovna Sayfudinova[1,2,*] , *Daniya Kalimullovna* Shigapova[3], *Aleksej Vasilevich* Bolshov[4], *Aelina Askhatovna* Barieva[3], and *Irina Anatolievna* Akhmetova[4]

[1]The Center for Advanced Economic Research of the Academy of Sciences of the Republic of Tatarstan
[2]Kazan State Power Engineering University, Kazan
[3]Kazan State University of Architecture and Engineering, Kazan
[4]Kazan (Volga Region) Federal University, Kazan

Abstract. The article examines the dynamics of the transformation of corporate governance in the Republic of Tatarstan. The author's approach is presented, according to which the strategic regional development affects the corporate governance system of the organization. The analysis of enterprises from various industries of the Republic of Tatarstan has been carried out and the leading ones have been selected. Further, according to the indicators of these organizations, calculations were made using the Altman five-factor model for 5 years. The chosen model and time period allow to reliably trace the dynamics of the company's economic potential and the results of its development for the reporting periods. The obtained results of the analysis and calculations showed that the corporate governance of enterprises changed the trend and began to move to an innovative type. Regional development strategies have a positive impact on corporate governance of enterprises. Moreover, an important point is that by means of regional development strategies, organizations themselves are interested in improving the corporate governance system. Thus, the impact on enterprises and their changes do not occur "top-down" (authoritarian, edifying), but "bottom-up" (the organization itself seeks to meet modern requirements in order to be economically efficient).

1 Introduction

In recent decades, every year more and more interest has been shown in corporate governance of the enterprise. However, unfortunately, the problems with its formation and development did not diminish. As before, the efficiency and investment attractiveness of many companies remains low. Meanwhile, this situation is a consequence of the

* Corresponding author: nazilya_sf@mail.ru

underdevelopment of corporate governance in organizations or its absence at all. And as a result - the inconsistency of the interests of all participants in the enterprise [3].

In our opinion, many enterprises are faced with a problem in the form of an "oligarchic" corporate governance system, which was especially common about 2015. Under this model, governance is concentrated in the hands of a small group of citizens. In such a situation, power and capital serve personal and group interests, i.e. majority shareholders exploit the business through, for example, restricting access to the register of shareholders, buying up shares from minority shareholders, entering into alliances with "friendly" joint stock companies. This state of affairs leads to the formation of a contradictory corporate governance system [1].

In this regard, there is a need to adjust and improve: the efficiency of corporate governance at enterprises, control over its implementation, setting new tasks and goals in observing the interests of all parties (board, supervisory board, shareholders). As well as the transition from an outdated and obsolete model of company operation to an innovation-oriented one, which will increase investor confidence.

The foregoing determines the relevance of the study and its practical significance.

2 Materials and methods

Within the framework of this paper, we have investigated the change in corporate governance through the created programs and strategies both at the state and regional levels (on the example of the Republic of Tatarstan), which induce the transition to an innovative model of corporate governance.

To obtain the results, the five-factor Altman z-model was chosen, which allows us to determine the "disturbing" factors of the enterprise. Out of 22 financial coefficients, the author of the model singled out 5 main ones, on which the probability of bankruptcy most depends, according to which the weight coefficients are calculated. If they turn out to be in the "positive" limits or positive dynamics is observed, then, accordingly, this will show that the company is effective and follows an innovative path, since this is one of the defining indicators for investment.

The generally accepted Altman z-model is as follows:

$$Z=1,2*X1+1,4*X2+3,3*X3+0,6*X4+X5 \quad (1)$$

where [4]:

Z is the coefficient of the probability of bankruptcy;

X1 is the ratio of current assets to the total value of the company's assets;

X2 is the ratio of the company's retained earnings to its total asset value;

X3 is the ratio of operating profit to the total assets of the company;

X4 is the ratio of the market value of the company's capital to its total debt;

X5 is the ratio of the company's revenue to the total value of its assets.

If Z is less than 1.8, then the probability of bankruptcy is very high; from 1.8 to 2.7 - average;

from 2.8 to 2.9 is small;

above 3-bankruptcy is unlikely.

According to the five-factor model, bankruptcy can be predicted in one year with an accuracy of 90-95%, in two years - up to 70-80%, in three years - up to 48-50%.

Altman's index is a function of some indicators that characterize the economic potential of a company and its performance over the past period. The critical value of the Z index was calculated by Altman from the data of a static sample and amounted to 2.675. The calculated value of the creditworthiness index for a particular company is compared with this value. This allows you to draw a line between companies and make a judgment about the possible bankruptcy of some in the near future (2-3 years) (Z <2.675) and a fairly stable financial position of others (Z> 2.675).

Next, we will test this model for 2015-2019 based on indicators that are publicly available in the financial statements of selected enterprises.

Thus, in our case, we will analyze and compare the indicators of past periods, which will allow us to trace the dynamics over 5 years, that is, what was and what we came to.

3 Results and discussion

When analyzing the impact of strategic regional planning on the corporate governance system, we used the Altman model and selected the following companies - leaders in their industry:

1. PJSC "KAMAZ".
2. PJSC Tatneft.
3. PJSC Nizhnekamskneftekhim.
4. PJSC "Kazanorgsintez".
5. JSC "Network company".
6. JSC "Elekon".

We calculated the values of the variables that will be used to determine the Z coefficient based on financial indicators presented by organizations in the section "on disclosure of information" within the framework of corporate governance. The variables are presented for each facility in Tables 1-6.

Table 1. Values of variables for calculating the Z coefficient of Kamaz PJSC from 2015 to 2019. [5]

JSC "KAMAZ"						
Ser. No.	**Variable**	**2015**	**2016**	**2017**	**2018**	**2019**
1	X1	0.02	0.25	0.3	0.33	0.37
2	X2	0	0.01	0.02	0.02	0.04
3	X3	0.01	0.02	0.04	0.02	0.04
4	X4	0.33	0.28	0.27	0.2	0.2
5	X5	0.88	0.98	0.98	0.95	0.94

Table 2. Values of variables for calculating the Z coefficient of Tatneft PJSC from 2015 to 2019. [6]

PJSC "Tatneft"						
Ser. No.	**Variable**	**2015**	**2016**	**2017**	**2018**	**2019**
1	X1	0.07	0.08	0.08	0.09	0.09
2	X2	0.12	0.15	0.2	0.22	0.28
3	X3	0.2	0.4	0.3	0.4	0.6
4	X4	0.4	0.3	0.41	0.48	0.5
5	X5	0.8	0.85	0.84	0.91	0.98

Table 3. Values of variables for calculating the Z coefficient of PJSC "Nizhnekamskneftekhim" from 2015 to 2019 [7]

PJSC Nizhnekamskneftekhim"						
Ser. No.	**Variable**	**2015**	**2016**	**2017**	**2018**	**2019**
1	X1	0.05	0.07	0.06	0.08	0.09
2	X2	0.15	0.2	0.23	0.25	0.41
3	X3	0.32	0.46	0.4	0.44	0.51
4	X4	0.3	0.2	0.28	0.35	0.34
5	X5	0.71	0.78	0.85	0.91	0.9

Table 4. Values of variables for calculating the Z coefficient of Kazanorgsintez PJSC from 2015 to 2019 [8]

PJSC "Kazanorgsintez"						
Ser. No.	**Variable**	**2015**	**2016**	**2017**	**2018**	**2019**
1	X1	0.07	0.06	0.05	0.08	0.08
2	X2	0.2	0.2	0.4	0.51	0.5
3	X3	0.2	0.24	0.26	0.3	0.31
4	X4	0.35	0.38	0.44	0.42	0.4
5	X5	0.82	0.8	0.85	0.87	0.91

Table 5. Values of variables for calculating the Z coefficient of JSC "Network Company" from 2015 to 2019 [9]

JSC "Network company"						
Ser. No.	**Variable**	**2015**	**2016**	**2017**	**2018**	**2019**
1	X1	0.08	0.06	0.06	0.07	0.08
2	X2	0.18	0.22	0.27	0.34	0.42
3	X3	0.19	0.19	0.23	0.3	0.29
4	X4	0.4	0.32	0.35	0.42	0.5
5	X5	0.89	0.9	0.95	0.98	0.98

Table 6. The values of the variables for calculating the Z coefficient of JSC "Elecon" from 2015 to 2019. [10]

JSC "Elecon"						
Ser. No.	**Variable**	**2015**	**2016**	**2017**	**2018**	**2019**
1	X1	0.03	0.04	0.04	0.06	0.07
2	X2	0.1	0.18	0.19	0.2	0.25
3	X3	0.2	0.19	0.22	0.2	0.3
4	X4	0.31	0.3	0.34	0.3	0.39
5	X5	0.81	0.85	0.86	0.9	0.92

For each of these organizations, the Z coefficient was calculated (Z1, Z2, Z3, Z4, Z5, Z6 correspond to the ordinal number of enterprises represented in the list of enterprises), which are presented below.

2015:

Z1=1,2*X1+1,4*X2+3,3*X3+0,6*X4+X5=1,2*0,02+1,4*0+3,3*0,1+0,6*0,33+0,88=1, 14;

Z2=1,2*X1+1,4*X2+3,3*X3+0,6*X4+X5=1,2*0,07+1,4*0,12+3,3*0,2+0,6*0,4+0,8=1, 95;

Z3=1,2*X1+1,4*X2+3,3*X3+0,6*X4+X5=1,2*0,05+1,4*0,15+3,3*0,32+0,6*0,3+0,71 =2,21;

Z4=1,2*X1+1,4*X2+3,3*X3+0,6*X4+X5=1,2*0,07+1,4*0,2+3,3*0,2+0,6*0,35+0,82= 2,05;

Z5=1,2*X1+1,4*X2+3,3*X3+0,6*X4+X5=1,2*0,08+1,4*0,18+3,3*0,19+0,4*0,4+0,89 =2,09;

Z6=1,2*X1+1,4*X2+3,3*X3+0,6*X4+X5=1,2*0,03+1,4*0,1+3,3*0,2+0,6*0,31+0,81= 1,83;.

2016:

Z1=1,2*X1+1,4*X2+3,3*X3+0,6*X4+X5=1,2*0,25+1,4*0,01+3,3*0,02+0,6*0,28+0,9 8=1,96;

Z2=1,2*X1+1,4*X2+3,3*X3+0,6*X4+X5=1,2*0,08+1,4*0,15+3,3*0,4+0,6*0,3+0,82= 2,0;

Z3=1,2*X1+1,4*X2+3,3*X3+0,6*X4+X5=1,2*0,07+1,4*0,2+3,3*0,46+0,2*0,31+0,78 =2,76;

Z4=1,2*X1+1,4*X2+3,3*X3+0,6*X4+X5=1,2*0,06+1,4*0,2+3,3*0,24+0,6*0,38+0,8=2,17;

Z5=1,2*X1+1,4*X2+3,3*X3+0,6*X4+X5=1,2*0,06+1,4*0,22+3,3*0,19+0,6*0,32+0,9=2,1;

Z6=1,2*X1+1,4*X2+3,3*X3+0,6*X4+X5=1,2*0,04+1,4*0,18+3,3*0,19+0,6*0,3+0,85=1,96.

2017:

Z1=1,2*X1+1,4*X2+3,3*X3+0,6*X4+X5=1,2*0,3+1,4*0,02+3,3*0,04+0,6*0,27+0,98=2,29;

Z2=1,2*X1+1,4*X2+3,3*X3+0,6*X4+X5=1,2*0,08+1,4*0,21+3,3*0,3+0,6*0,41+0,84=2,46;

Z3=1,2*X1+1,4*X2+3,3*X3+0,6*X4+X5=1,2*0,6+1,4*0,23+3,3*0,04+0,6*0,44+0,85=2,76;

Z4=1,2*X1+1,4*X2+3,3*X3+0,6*X4+X5=1,2*0,05+1,4*0,4+3,3*0,26+0,6*0,44+0,85=2,45;

Z5=1,2*X1+1,4*X2+3,3*X3+0,6*X4+X5=1,2*0,06 5+1,4*0,27+3,3*0,23+0,6*0,35+0,95=2,37;

Z6=1,2*X1+1,4*X2+3,3*X3+0,6*X4+X5=1,2*0,04+1,4*0,19+3,3*0,22+0,6*0,34+0,86=2,11;

2018

Z1=1,2*X1+1,4*X2+3,3*X3+0,6*X4+X5=1,2*0,33+1,4*0,02+3,3*0,02+0,6*0,22+0,95=1,86;

Z2=1,2*X1+1,4*X2+3,3*X3+0,6*X4+X5=1,2*0,02+1,4*0,22+3,3*0,4+0,6*0,48+0,91=2,94;

Z3=1,2*X1+1,4*X2+3,3*X3+0,6*X4+X5=1,2*0,08+1,4*0,25+3,3*0,44+0,6*0,35+0,91=3,01;

Z4=1,2*X1+1,4*X2+3,3*X3+0,6*X4+X5=1,2*0,08+1,4*0,51+3,3*0,3+0,6*0,42+0,87=2,91;

Z5=1,2*X1+1,4*X2+3,3*X3+0,6*X4+X5=1,2*0,07+1,4*0,34+3,3*0,3+0,6*0,42+0,98=2,53;

Z6=1,2*X1+1,4*X2+3,3*X3+0,6*X4+X5=1,2*0,06+1,4*0,22+3,3*0,2+0,6*0,3+0,9=2,09.

2019

Z1=1,2*X1+1,4*X2+3,3*X3+0,6*X4+X5=1,2*0,37+1,4*0,04+3,3*0,04+0,6*0,2+0,94=1,32;

Z2=1,2*X1+1,4*X2+3,3*X3+0,6*X4+X5=1,2*0,09+1,4*0,28+3,3*0,6+0,6*0,5+0,98=3,76;

Z3=1,2*X1+1,4*X2+3,3*X3+0,6*X4+X5=1,2*0,09+1,4*0,04+3,3*0,51+0,6*0,34+0,9=3,43;

Z4=1,2*X1+1,4*X2+3,3*X3+0,6*X4+X5=1,2*0,08+1,4*0,5+3,3*0,31+0,6*0,4+0,91=2,93;

Z5=1,2*X1+1,4*X2+3,3*X3+0,6*X4+X5=1,2*0,08+1,4*0,42+3,3*0,29+0,6*0,5+0,98=2,92;

Z6=1,2*X1+1,4*X2+3,3*X3+0,6*X4+X5=1,2*0,07+1,4*0,25+3,3*0,3+0,6*0,39+0,92=2,57.

Having calculated the probability of bankruptcy for each enterprise in dynamics over 5 years, the summary information presented in Table 7 was generated.

Table 7. Summary table for calculating the probability of occurrence of enterprises from 2015 to 2019

Ser. No.	Name of the organization	2015	2016	2017	2018	2019
1	JSC "KAMAZ"	1.14	1.96	2.29	1.86	1.32
2	PJSC "Tatneft"	1.95	2.00	2.46	2.93	3.76
3	PJSC "Nizhnekamskneftekhim".	2.21	2.76	2.76	3.01	3.43
4	PJSC Kazanorg-Sintez.	2.05	2.17	2.45	2.91	2.93
5	JSC "Network company".	2.09	2.1	2.37	2.53	2.92
6	JSC "Elecon"	1.83	1.96	2.11	2.09	2.57

As can be seen from Table 7, the value of the probability of bankruptcy for most enterprises by 2019 exceeded the minimum allowable value - 2.675. The greatest success in corporate governance was achieved by PJSC Tatneft and PJSC Nizhnekamskneftekhim, whose Z coefficients exceeded 3. PJSC "Kazanorgsintez" and OJSC "Network Company" for 5 years got out of the critical situation and were able to become resistant to bankruptcy enterprises. During the period under review, JSC Elecon has shown a positive dynamics of transformation of the corporate governance system, as evidenced by the value of the coefficient, which in 2019 approached the threshold value of the financial stability of the enterprise. The only organization from the considered ones showed not the best results - PJSC "Kamaz". We are inclined to assume that the current position of the company in the automotive market has a great influence. Recently, the machines produced are in less demand, there is no need in progress, which only confirms our hypothesis that the innovative way of the company's development leads to a more successful development of the corporate governance system and the investment attractiveness of the organization.

Thus, we can conclude that corporate governance in the Republic of Tatarstan has significantly changed for the better over the past 5 years. The corporate value system is changing, as evidenced by the change in financial indicators and calculated values. Companies are beginning to look not to the present, but to the future - a strategic vision [11]. That is, not "pulling out" the maximum possible profit today, but working for the future, keeping up with the times.

4 Conclusion

In our opinion, on the basis of the work done, the following conclusions and recommendations are logical. Today, corporate governance plays a critical role in the development of an organization. The considered companies are trying to maintain leading positions in the industries of their activity, as evidenced by the analysis of indicators and their calculation. The probability of bankruptcy of enterprises decreased every year. Accordingly, it can be argued that the models of corporate governance are changing in the new economic conditions. And they change in a positive direction, but only for those who observe the “laws” of the market and are engaged in continuous development. This is largely facilitated by external influence in the form of new opportunities - this is the development of the region [2]. In the last decade, the Republic of Tatarstan has taken a competent course of strategic development, which contributes to the improvement of all areas of life. This directly applies to the enterprises that are motivated to meet modern requirements for companies. Therefore, we can conclude that strategic regional development contributes to the improvement and promotion of the corporate development system, which, using the example of the Republic of Tatarstan, showed how many organizations managed to improve their results from the "oligarchic" model over 5 years by switching to a new innovative path. And we recommend that enterprises wishing to achieve success turn to their corporate governance system and choose an innovative way of development.

References

1. A. Asghar, S. Sajjad, A. Shahzad, B.T. Matemilola, *Role of discretionary earning management in corporate governance-value and corporate governance-risk relationships*, 564 (2020)
2. M.R Safiullin, *Development of a development model for the Republic of Tatarstan* (2007)
3. M.R Safiullin M.R. Zainullina, *System-functional macroeconomic modeling of the development of the Russian Federation* (2012)
4. 4.A.S. Mitrofanova, D.A. Pantyukhova, S.N. Khobotova, *Application of Altman's model and methods of economic analysis to determine the financial risks of an enterprise on the example of PJSC "Lukoil"* (2017)
5. Site of Kamaz PJSC, section "on information disclosure", https://kamaz.ru/
6. Site of Tatneft PJSC, section "on information disclosure", https://www.tatneft.ru/
7. Site of Nizhnekamskneftekhim PJSC, section "on information disclosure", https://www.nknh.ru/
8. Site of Kazanorgsintez PJSC, section "on information disclosure", https://www.kazanorgsintez.ru/
9. Site of Grid Company JSC, section "on information disclosure", https://gridcom-rt.ru/
10. Site of Elecon JSC, section "on information disclosure", https://zavod-elecon.ru/
11. V.A. Malyshenko, K.A. Malyshenko, Economic and Social Changes: Facts, Trends, Forecast, **2**, 127 (2018)

Share repurchases, coronaeconomy and stakeholders' interests

Irina Tkachenko[1,*], and *Bela* Bataeva[2]

[1]Ural State University of Economics, st. March 8, 62/45, 620144 Yekaterinburg, Russia
[2]Financial University, Moscow, Russia

Abstract. The purpose of the paper is to consider the mechanism of share repurchases in the context of the coronaeconomy and the COVID-19 pandemic. The English- and Russian-language publications discussing share repurchases from the perspective of the stakeholder approach in corporate governance are reviewed. The practice of the Russian firms implementing share repurchase programs is analyzed, with the focus on the companies included in the Sustainable Development Ranking-100. It is concluded that share repurchases should be studied with regard to the impact they have on the interests of a wide range of company's financial and non-financial stakeholders, and to the mutual influence of buybacks and parameters of sustainable economic development.

1 Introduction

The amounts paid to shareholders as share repurchases, rather than dividend payments have increased dramatically over the last few decades. Until recently, corporate governance theory believed that corporations exist to earn profits for shareholders. With the spread of corporate social responsibility (CSR) and corporate sustainability concepts, the focus has shifted from shareholders' benefits to the benefits of different stakeholders and to long-term sustainability. Questions have been raised about whether companies are using cash resources wisely and whether future growth and job creation are still a priority when deciding whether to use excess cash.

Repurchasing shares, or share buybacks, is a procedure when the issuer buys back its own shares, which may lead to a decrease in the number of freely traded securities and an increase in share prices. Share repurchases were allowed in the United States in1960, but became popular when Rule 10b-18 was passed by Congress in 1982 and amended in late 2003. As a result, the S&P 500 dividend payout ratio slid from 55% in 1969 to around 25% in 2013 [1]. In the USA total share buybacks topped the one trillion mark for the first time in 2015 and had exceeded dividend payouts since 2005. In 2014 the value of repurchasing shares in the United States exceeded the value of new shares issued by five times [2].

In the United Kingdom, share buybacks were legalised in the 1980s. In the continental Europe, share repurchases became popular in the mid-1990s; in Asia - in the late 1990s.

* Corresponding author: Tkachenko@usue.ru

How popular are share buybacks in Russia? Share repurchases started to be used as the payment method by Russian firms to maintain shareholder value when the United States and EU countries introduced the sanctions related to the annexation of Crimea to Russia in 2014. In 2018 - 2019, buybacks were already considered the main trend. According to the National Rating Agency (NRA), transactions worth 39 billion rubles were conducted in the open market in Russian in the first half of 2018. The number of buybacks was expected to decrease in 2020, but the spread of the COVID-19 virus has made its own adjustments. In 2020, numerous Russian firms have been repurchasing shares, thereby affecting stock prices. Yet it takes time to see the long-term effect of this behaviour. The risks associated with buybacks may increase and the expected returns may not be gained.

The objective of the paper is to study the conceptual approaches to share repurchase behaviour in international and Russian literature and to see how this payout method meets corporate stakeholders' interests. The article also aims to identify benefits and shortcomings of share repurchases and to provide empirical evidence on the extent of using buybacks in Russian companies particularly under the threat of coronavirus. The authors are also making an attempt to analyze how the share purchase behaviour affects long-term sustainability.

2 Review of conceptual approaches

The majority of publications on share repurchases, both overseas and in Russia, dates back to the global economic and financial crisis in 2008-2009. The study of the *Google Scholar* bibliographic database dating from the early 1990s has shown that only one third of English-language publications on share repurchases considers corporate governance issues, and only 1/5 of the publications mention company stakeholders. The Russian-language publications on buybacks are few in number. Drawing on the purpose of our research, in the review we have focused on the publications that study various aspects of corporate governance, including the impact of buybacks on stakeholders' interests rather than those researching a conventional approach to repurchasing shares as an aspect of corporate finance.

We start our review with the work of Evgeniou T. et al. (2018) [3] where the authors tested the arbitrage and the market timing hypotheses about the presence of abnormal long-term excess returns following buyback announcements. They are calculated on the basis of the five factors Fama and French (2015) [4] and the four factors of the Stambaugh and Yuan (2017) model [5]. The authors draw attention to the fact that buybacks may target a specific group of shareholders, top management, rather than collective shareholders and stakeholders.

The broader context of issues related to the economic impact of stock buybacks on a wide group of stakeholders was analyzed in [6]. The author examines the impact of the Tax Cuts and Jobs Act (TCJA), passed in the USA in December 2017, on investment. The TCJA brought many changes to the US federal tax code, including lowering the corporate tax rate from 35% to 21%. The article argues that tax reduction has resulted in higher cash flow for many firms and some companies have made their choice in favour of returning portions of that cash to their shareholders through share repurchases.

The analysis of relationships between the short-term financial gains for managers arising from share repurchases and the long-term effects on the company's competitiveness [7] proves that an increase in the stock-based manager remuneration leads to shifting investments towards stock buybacks. This, while pushing the share price up in the short term, negatively affects the dynamic of firm's profits in the long run.

Some researchers [8] believe that buybacks are used to manipulate stock prices in the market, which may lead to a disproportional distribution of national income. Most buybacks occur when stock prices are high, in order to offset dilution of earnings per share (EPS) that results when the company offers stock options to its employees. Buybacks make no contribution to innovative growth of the economy, but are at the core of what Lazonick calls the "downsize-and- distribute" economy [8, p. 7]. Government policy should regulate the stock market for the sake of the middle class and the majority of the population, i.e. in the interests of broad stakeholders. In his further paper [9], Lazonick et al. state that massive repurchases are used by senior corporate executives to manipulate their companies' stock prices to their own benefit, since most of their remuneration comes from stock options and stock rewards. Buybacks enrich these opportunistic share sellers – investment bankers and hedge-fund managers as well as senior corporate executives – at the expense of employees as well as continuing shareholders. As a mode of distributing corporate funds to shareholders, buybacks surpassed dividend payouts in 1997. Since then, buybacks, which are much more volatile than dividends, have dominated distributions to shareholders when the stock market was booming. The authors argue that stock buybacks undermine the quest for fair and stable economic growth and should be banned.

Commenting on the connection between stock buybacks and dividends, Gruber & Kamin [10] tested the hypothesis that dividend and buybacks may be a consequence of pessimism about future demand and economic growth. Corporations defer capital spending, and simply return money to their shareholders to pursue more attractive investment opportunities.

Under the conditions of social and economic turmoil and a rapidly changing business landscape caused by the COVID-19 pandemic, the importance of the corporate governance paradigm based on the stakeholder approach has become relevant as never before. In 2020, many studies are addressing a changing attitude to stakeholders along with a continuous analysis of share repurchases popularity both in foreign countries and Russia. The search query for the keywords *buybacks, stakeholders, COVID-19* in *Google Scholar* has shown 159 English-language articles. By the beginning of November 2020, 81 articles have been published that are analyzing the impact of share buybacks on company's stakeholders. Definitely, the relationship between buybacks and the stakeholder model needs to be studied in the long run, but the problem statement itself and the application of the model of sustainable development and conscious capitalism are attracting recent research interest [11, 12].

Answering the questions how industrial companies with global manufacturing and supply chain can survive the impact of the COVID crisis, Sheth [13] argues that company management should support and protect all of its stakeholders and broaden the purpose of business. The leadership of the company should postpone the implementation of their share repurchases plans. The investor is just one of the stakeholders and not the only one, This will require senior executives to shift from a win-loss mindset to a win-win mindset that could be beneficial for everyone. .

Providing state support to companies during the pandemic, Western governments require that in receipt of this support firms would maintain all of their current employees, regardless of their contract type and the pay of government bonds, dividends, share repurchases and executive remuneration should be suspended until termination. In addition, the gains from future business recovery and equity gains should be distributed fairly among stakeholders. The latter will ensure changes in corporate governance and behaviour that are consistent with overall long-term environmental and ethical goals [14].

Discussing the incentives and consequences of buybacks, experts mainly highlight positive trends while neglecting cons of this process for a wide range of company's stakeholders, and not just for a narrow group of shareholders and top management: if the money indulgence in this elite group cannot be curbed, examples of managerial opportunism on a daily basis will be multiple [15].

The idea that governments have to accept a more active role in the economy is promoted in the report of the Labour Party [16]. Over the past decade there have been widespread calls for a more responsible capitalism. The mistakes of the 2008 financial crisis should not be repeated, when government support has facilitated even higher after-crisis gains. The corporate governance framework needs to be re-configured to support responsible business focused on long-term, sustainable success, shared by all their stakeholders.

The overview of publications allows us to conclude that most of the literature rarely aims at analyzing the impact of share repurchases on company's financial stakeholders. Still rare are works where the impact of stock buybacks is looked upon from the point of view of non-key, non-financial stakeholders and what could be done to reduce the ability of corporate insiders to use stock buybacks for personal gain [17].

The existing literature in the Russian language is even scarcer. Neither *Google Scholar* nor the *E-library* database has found any study, dated this year, of the relationship between such concepts as share repurchases, stakeholders' interests, sustainable development, and COVID-19. There are several analytical articles by stock market and BCS Express experts that identify the latest trends in buybacks practice in the Russian business, but deep analysis of benefits and shortcomings of share repurchases for all groups of stakeholders, especially in the context of the pandemic is almost non-existent. Consequently, the transition from verbal descriptions of the impact of the buybacks on corporate governance and all stakeholders to a real analytical endeavour and modelling of stakeholders' benefits seems to be an urgent task [18, 19].

The same trend – the focus on financial outcomes of stock buybacks – dominates the publications of Russian researchers. They argue that buybacks can mislead investors and incentivize senior executives only [20]. At the same time share repurchases can be used to increase company's economic value added [21] and can motivate top management to satisfy shareholders' interests [22].

3 Methodology

The conceptual and methodological basis of our research is to search and analyze studies performed in the field of share repurchases and their impact on the interests of all stakeholders. The following methods have been used: content analysis, comparative analysis, and synthesis. When analyzing analytical and expert data, an assumption was made that, on the one hand, market turbulence caused by the pandemic and the threat of a global recession make companies repurchase shares in order to maintain their price. On the other hand, buybacks are an ambiguous tool. This mechanism not only has a positive impact on equity, but also poses risks for certain groups of stakeholders, significantly increasing the wealth and earnings of some, and making others poorer. The authors studied the data of expert and rating agencies, raising the question not only about the welfare of stakeholders, but also about the presence (or absence) of the impact of buybacks on the company sustainable development.

4 Results and Discussion

Market decline and stock buybacks

The 2020 crisis has resulted in market decline, which has encouraged Russian companies to actively use buybacks to maintain the value of its shares. In March 2020, the Russian index fell by 25%. This prompted several Russian companies to announce the start of new buyback programs or the extension of the currently run programs [23].

On March 23-27, 2020, *Rosneft* repurchased 4 936 260 shares and 1 889 977 global depository receipts (GDR) on the open market, which totalled $25.7 million (about 2 billion rubles) and constituted 0.06% of the authorized capital. *Rosneft* buyback was accompanied by a quotation fall by an average 402 million rubles per day. The proportion of share buybacks to the trading volume accounted for 5.1% [24]. On March 18, 2020, *Detsky Mir* purchased 1 212 210 shares worth approximately 90 million rubles (0.16% of the equity capital). TCS Group acquired 650 thousand GDR from March 12 to 15, 2020.

Other companies, such as *Lukoil*, *Sistema*, *Yandex*, *Transneft*, *Novatek*, *Safmar*, that had approved share repurchase programs in 2019, took advantage of the situation and bought back larger blocks of shares than it had been planned. For example, *Sistema* that had approved a 3 billion rubles buyback program in September 2019, extended share repurchases till December 31, 2020. As of February 28, 2020, a subsidiary of *Sistema* repurchased 120.6 million shares worth 1.6 billion rubles. This is equivalent to 40.2% of the declared maximum number of shares and 54.8% of the amount allocated for the program [25]. *Novatek* had announced a buyback program in 2018. In 2020 the company increased the size of its buyback program; from the beginning of the year, *Novatek* has repurchased 2.5 times more shares than in 2019 and 1.6 times more than at the end of 2018. *Lukoil* approved a $ 3 billion share repurchase program in October 2019 and it will last till 2022.

Share repurchases activity of Russian companies from 2018 to 2020 is presented in Table 1. It also shows companies' positions in the sustainable development ranking

Table 1. Ranking of Russian firms with competed and non-completed repurchase programs

Company	Program timescale and rate of delivery	Size of repurchase	Ratio of buyback to total equity capital (%)	Sustainable Development Ranking-100
Panel A: Non-completed repurchase programs				
Rosneft	Rate of delivery: 18.48% ($ 369.5 ml) Start: August 06, 2018 Finish: December 31, 2020	$ 2 ml	3.2	2
Lukoil	Rate of delivery: 0% Start: October 01, 2019 Finish: December 30, 2022	$ 3 ml	5.1	5
Novatek	Rate of delivery: 61.22% ($367.3 ml) Start: June 07, 2012 Finish: June 07, 2021	$ 600 ml	–	8
MTS	Rate of delivery: 86.44% (RUR12.97 ml) Start: March 31, 2020 Finish: December 31, 2020	RUR 15 ml	–	11
Transneft	Start: June 2019 Finish: 2022	RUR 20 ml	–	40
TMK Group	Start: April 2020 Finish: not known		–	55
Sistema	Rate of delivery: 54.84% (RUR 1.65 ml) Start: September 17, 2019 Finish: December 31, 2020	RUR 3 ml	3.1	67
Globaltrans	Rate of delivery: 0.87% (76.88 th shares) Start: May 12, 2020 Finish: May12, 2021	8.84 ml shares	5	–
LSR Group	Rate of delivery: 0% Start: March 26, 2020 Finish: not known	RUR 5 ml	–	–
Safmar Financial Investments	Start: November 2018. Finish: November 2021	RUR 6 ml	–	–

Table 1. Continued

Detsky Mir	Rate of delivery: 0% Start: February 08, 2020 Finish Fenruary 07, 2023	–	–	–
RusAqua (Russian Aquaculture)	Rate of delivery: 49% (RUR 294 ml) Start: December 11, 2019 Finish: June 30, 2021	RUR 600 ml	–	–
Yandex	Rate of delivery: 52.1% ($156.3 млн); Start: November 18, 2019 Finish: not known	$ 300 ml	–	–
Etalon Group	Start: January 27, 2020 Finish: December 30, 2024	$ 400 ml	10	–
Globaltruck	Rate of delivery: 45.41% (1.33 ml shares) Start; November 11, 2019 Finish: May 31, 2021	2.92 ml shares	5	–
Obuv Rossii	Rate of delivery: 1.5% (RUR 13.51 ml) Start: October 26, 2018 Finish: December 31, 2020	RUR 900 ml	–	–
Panel B: Completed repurchase programs				
TCS Group	Rate of delivery: 100% (650 th shares) Start: Match 12, 2020 Finish: March 31, 2020	650 th shares	0.37	–
Magnit	Rate of delivery: 100% (164.71 th shares) Start: April 01, 2019 Finish: April 05, 2019	164.71 th shares	0.16	58
	Rate of delivery: 100% (RUR 22.2 ml) Start: September 29, 2018 Finish: March 01, 2019	RUR 22.2 ml	–	58
Lenta	Rate of delivery: 8.46% (RUR 980.8 ml) Start: October 29, 2018 Finish: April 02, 2019	RUR 11.6 ml	0.93	43
Lukoil	Rate of delivery: 100% ($3 ml) Start: August 30, 2018 Finish: August 20, 2019	$ 3 ml	4.96	5
Megafon	Rate of delivery: 100% Start: August, 2018 Finish: May, 2019	$1.1 ml	–	61

Note: Dashes indicate data not available
Source: BCS Express https://bcs-express.ru/buybacks; Expert online https://expert.ru/ratings/renking-ustojchivogo-razvitiya---100/

By comparing the information about the companies' buyback programs with Sustainable Development Ranking-100, it becomes evident that half of the 20 companies that are currently running share buyback programs are included in Sustainable Development Ranking-100. This suggests that share repurchases are subject to quite 'strong' companies.

The companies actively use buybacks to maintain their share prices, as evidenced by the contribution of share repurchases to stock performance. Table 2 illustrates this trend by providing data about four Russian gas and oil producers: *Rosneft, Lukoil, Gaspromneft* and *Tatneft*.

Table 2. Buybacks and stock performance of Rosneft, Lukoil, Gaspromneft and Tatneft (March-October, 2020)

Month	Ratio of buybacks to Rosneft stock trading volume of (%)	Stock performance (%)			
		Rosneft	Lukoil	Gaspromnef	Tatneft

Table 2. Continued

March	5.1	14.7	12.9	14.9	9.7
April	4.0	6.8	2.4	10.0	-0.5
May	1.6	12.1	8.6	-0.7	-3.3
June	1.0	-3.8	1.3	-0.1	4.6
July	3.4	-1.4	-4.2	-3.5	-0.3
August	2.1	4.6	-1.9	-0.9	-1.2
September	12.4	2.6	-10.4	-7.3	-15.4
October	7.8	1.5	0.1	-2.2	1.0

Source: [24]

As can be seen from Table 2, there is a noticeable difference between the performance of *Rosneft* shares and the growth of share portfolio of the company's competitors. For example, for four months (May, August, September and Octover) *Rosneft* shares were outperforming quotations of *Lukoil, Gazpromneft,* and *Tatneft*. For three months *Rosneft* was the second. These results are related to the volume of repurchases in the average daily turnover. Some Russian companies (e.g., *Lukoil*) in addition to share repurchase also used interim dividends as components of manager remuneration schemes [26]. In order to simplify the buyback procedure, the Central Bank of Russia introduced amendments to corporate legislation [27].

In contrast to foreign practice, the Russian literature lacks discussions about the relationship between share repurchases and manager remuneration, and bailouts. When the American aviation giants *Delta* and *United* asked for government rescue, senior executives had to suspend stock buybacks until federal loans that the companies were planning to receive as a form of bailout matured. *Boeing* was expected to receive government bailout support topped the scale $60 billion and also had to suspended buybacks and dividend payments. In 2020 the US Federal Reserve System (FRS) has set limits on dividend payments and share repurchases for 33 US banks with assets valued $100 billion. The European Banking Federation also called for the cancellation of dividend payouts and share buybacks for the entire period of 2020. In Europe, dividend and buyback programs are considered 'socially unacceptable' due to the pandemic impact. At the beginning of this year, the Central Bank of Russia also instructed banks to postpone buybacks and dividend payments till autumn.

Benefits and risks of share repurchases for shareholders and stakeholders

What are the benefits and risks that share repurchases can bring to shareholders and different stakeholders? Generally speaking, there are four main motives underlying the decision to buy back shares: (1) to maintain the value of shares at competitive level; (2) to reduce the number of shares in circulation (redemption of shares); (3) top management remuneration schemes; and (4) tax optimization. However, if in future shares are sold in a large investor's favour, this may pose risks in terms of redistribution of shareholder control. Another important issue is the objectivity of the repurchasing price when top management have share options. The benefits and shortcomings of buybacks are presented in Table 3.

Table 3. Benefits and shortcomings of share buybacks

Benefits	Shortcomings
Buybacks represent a positive net present value opportunity for a firm's shareholders if shares are being repurchased during periods when they are undervalued. When the market situation improves, shares will be sold at higher prices. Buyback are used to increase earnings per share. Corporations use buybacks as an alternative to dividend payments to increase shareholder wealth. Buyback are used as component of top management remuneration and employee incentive schemes. Buybacks are a more tax-efficient way to return capital to shareholders. Share repurchases are taxed at a lower rate and have a tax advantage over dividend payments. Buybacks are used for corporate tax optimization.	Buybacks decrease investment opportunities since they are funded by cash. Buybacks create an artificial demand for a firm's shares. Although the firm's market value might grow, it does not lead to improved efficiency. This situation is misleading for investors, since stock prices may decline in future. Repurchasing shares is found to transfer wealth from shareholders to top management. This is regulated by corporate compliance and share repurchases are required to be publicly announced. Funding buybacks is considered unethical, since it enriches senior managers and shareholders if used as an incentive, while these financial resources could target better working conditions. Holding company shares can be repurchased by subsidiaries when shares are not redeemed. If the holding company and subsidiary are located in different legal entities, tax authorities may consider buybacks as a scheme for avoiding taxes and charge additional taxes. Buybacks are believed to dilute shareholder value.

Source: https://cyberleninka.ru/article/n/corporate-stock-buyback-pro-et-contra; [1]; [28]

However, due to a relatively weak status of Russian stakeholders, public opinion and expert community do not give much consideration to negative impacts of share repurchases on stakeholders' interests. In this regard, future research may be aimed at assessing the impact of share buybacks on risks and benefits for financial and non-financial stakeholders; studying their impact on corporate governance, and determining the impact of buybacks on long-term sustainability.

4 Conclusions

Companies are increasingly using share repurchases to maintain the value of their shares. In all the above discussed cases buybacks were used to increase the firm's capitalization and to boost executive incentives. In times of continuing market volatility, companies have achieved an increase in share prices. There are different types of share repurchases [2], but the general observation is that firms engaged in share repurchase programs experience a better market position [24]. Nevertheless, share repurchases are considered as an ambiguous tool for changing the value of shares. Currently the Russian shareholders' interests prevail. However, the COVID-19 is posing new serious risks of further share value reduction. The question how the risks associated with buybacks will be dealt with is still awaiting the answer.

The overall conclusion can be made that Russian researchers relate share repurchases to only shareholders' interests rather than all stakeholders' benefits. This can be explained by the lack of influential non-financial stakeholders who have strong ability to make important financial decisions.

References

1. J. Kennon, Cash Dividends, Share Repurchases, Which is Better for Your Portfolio? The Balance (2020)
2. N.Wessona, B.W. Bruwerb, W.D. Hammana, South African Journal of Business Management, **46**(3) (2015)
3. T. Evgeniou, E.J. de Fortuny, N. Nassuphis, T. Vermaelen, Journal of Corporate Finance, **49** (2018)
4. F.E. Fama, K.R. French, Journal of Financial Economics,**116(1)** (2015)
5. F.R. Stambaugh, Y. Yuan, The Review of Financial Studies, **30(4)**, 1270 (2017)
6. E. York, The Economics of Stock Buybacks (2018)
7. H. Dawid, Ph. Harting, S. van der Hoog, Manager Remuneration, Share Buybacks and Firm Performance (2018)
8. W. Lazonick, Buybacks: from basics to politics (2015).
9. W. Lazonick, M.E. Sakinç, M. Hopkins, Harvard Business Review (2020)
10. J.W. Gruber, S. Kamin, *Corporate Buybacks and Capital Investment: An International Perspective* (2017)
11. J. Mackey, R. Sisodia, Conscious Capitalism. Liberating the Heroic Spirit of Business (2015)
12. I.N. Tkachenko, Economy of Region, **14,** 3 (2018)
13. J. Sheth, Industrial Marketing Management, **88** (2020)
14. E. Lonergan, M. Blyth, Beyond bailouts (2020)
15. W. Zhang, *Stock buybacks: some old norm should remain new*, Law and COVID-19 (2020)
16. C. Rees, D. Offenbach, *Towards Democratic and Sustainable Business: Possibilities for Corporate Governance Reform* (2020)
17. L. Palladino, International Review of Applied Economics, **34(2)**, 152 (2020)
18. B.S. Bataeva, I.N. Tkachenko, II International Scientific Conference GCPMED (2019)
19. I. Tkachenko, I. Pervukhina, Stakeholder Value Assessment: Attaining Company-Stakeholder Relationship Synergy, Corporate Governance in Central Europe and Russia
20. A.S. Firsanov, D.M. Petrova, A.A.Tsaytsev, Buybacks: international and Russian experience, Conference Proceeding (2019)
21. A.V. Kravtsov, V.M. Molokanov, Achievements and prospects for the development of youth science (2019)
22. V.A. Grebennikova, E.I. Polyakov, Sustainable Development Economy, **1(41)**, 190 (2020)
23. V. Gaidaev, *Issuers and top managers have come to the sale*, Kommersant (2020)
24. D. Puchkaryov, What buyback is the strongest in the Russian market, BCS Express (2020)
25. AFK Sistema extends the buyback program, BCS Express (2020)
26. Lukoil has decided on interim dividends, BCS Express (2020)

27. E. Vavina. "Rosneft" will facilitate the redemption of shares due to falling oil prices. Vedomosti (2020)
28. G. Guthrie, *The Dark Side of Share Buybacks*, SSRN (2020)

Influence of the presence of children taken care of in work motivation

Mariana Usheva, and *Philip* Ivanov*

South-West university "NeofitRilski", Faculty of economics, Department "Management and marketing", Ivan Mihailov 60, 2700 Blagoevgrad, Bulgaria

Abstract. Motivation is one of the most important aspects in human resources management. Motivated employees are often the difference between the good and great companies. The ever-complicating environment is demanding more and more from companies. In the current economical era, namely the human capital is the one thing, providing the requirements of the environment so that companies can afford to stay on the market. Therefore, everything linked to the proper human resources management should be examined in detail, which is the main purpose of the current study.
Here we aim to examine the link between the number of children for which the respondents are required to take care of and the most valuable motivators in workplace motivation. We firmly believe that the preferred motivators are directly linked to the number of children, with which we describe as the stage of the life cycle for each respondent.
First of all, in our article we would like to examine motivation as a part of the human resource management. Afterwards, we will examine the results from our survey, held among 595 respondents from various sectors of the Bulgarian economy. The survey was held in online format, guaranteeing the anonymity of the respondents. The larger part of the response group is part of a governmentally owned railway company.

1 Introduction

Examining motivation is one of the hardest tasks in human recourse management. Understanding what drives people to a certain behavior is maybe the most complex aspect of managing people, having in mind that motivation is directly linked to performance at work [9]. Motivation is one of the most important aspects in the company. Often, highly and properly motivated employees are the difference between the good and great companies.

This tendency is fully confirmed by the high interested among scholars. Numerous of authors have examined many aspects of motivation. From traits in the character of employees, to aspects from the environment. Our main goal in the current paper is to examine the link between the number of children, which have to be taken care of by our respondents and the most and least preferred motivators.

*Corresponding author: filip.ivanov.1@abv.bg

In the first part of our paper we will examine the problem of motivation. We will give some examples of other proven linkages between traits in the character and environment and motivation to work. Then we will examine the results from our empirical research, and finally make conclusions based on the results.

2 Materials and Methods

In the this part we will examine the complexity of workplace motivation in many aspects. Firstly, we will try to explain the term motivation and motivating. Then we will examine some past research on the linkage between motivation and other aspects of everyday life and the workplace.

The word "motivation" has a Latin route from the word "movere", which means – to move. The problematic of motivation has been examined since the dawn of our civilization, beginning with the Greek-Roman Empire, where many of the philosophers examined the reason of certain action in people's behavior. The framework of motivation in everyday life is ever since in the focus of many philosophers and later of psychologists.

The concept of motivation refers to internal factors that impel action and to external factors that can act as inducements to action. The tree aspects of action that motivation can affect are direction (choice), intensity (effort), and duration (persistence) [6]. Therefore, workplace motivation is the main source of directed to the company success actions. Once the person is properly motivated, his efforts towards the required direction would be greater, the persistence will be more sustainable.

In the first place, motivation is no other than a phycological process, which initiates, directs and maintains goal-directed behavior. It is a process with which the individual steps into action [5]. It describes the most basic model for motivation. Firstly, the person acknowledges a certain desire, then he directs his behavior to the given action required in order to acquire that desire. The goal maintaining, which was described also as duration of the effort, is completely no other than a subjective process. The person is valuing the given incentive and estimates the effort he should put into acquiring it, then he steps into action with the according effort and persistence.

Consequently, motivation is a psychological process, which is routed from the interaction of the individual and the environment [9]. In the aspect of human resource management, the manager is the one who is responsible for the good, or respectively, bad motivational environment. Although it is very hard to put in practice many of the postulates of today's theorists examining the motivation in the workplace, managers are called to do whatever they could in providing suitable environment. The foundation for the good application is a good knowledge of the theories of motivation, many of the aspects of motivation as a psychological process in the human brain and last, but not least, what motivation is linked to as concerns other aspects of human behavior.

Of course, motivating of employees must be economically efficient for the organization. Companies can't give incentives for every effort they have to put, in order to maintain their work obligations. Widely used in the economy of the company is the Pareto's law. It states that 80 % of the success for the firm comes from 20 % of the employees. Although this law is widely used in many economical processes, motivating can partly reject it. Mostly because self-efficient behavior suggests certain effort for a given outcome for the employee, which basically means that the members of an organization typically do not bear all the costs and benefits of the actions they take and the decisions they make [11].

The last suggests a detailed examination of motivation as a whole and in the organization; it is of utmost importance to know what motivation relates to. Some examples are – national culture [12], age and adult development [13],equity on the workplace [1], goal setting [9], job characteristics [10], self-determination in behavior[4], personality traits (neuroticism, extraversion, openness to experience, agreeableness, conscientiousness) [3], feedback [2] and many other research-based traits of human behavior and aspects of the environment.

Although from the above-mentioned researches is shown that authors had examined the adult development in the aspect of human motivation, it was from another aspect, different from the aim in this article. The wide variety of research-based conclusions linked to the correlation of motivation and other aspects of human behavior and the environment, boldly emphasizes the importance of human motivation to theorists and practitioners in human resource management. Therefore, we believe that the current article will provide new knowledge in the theory of workplace motivation. Consequently, here comes the time to formulate our main hypothesis:

H1: The number of children to whom people have to take care of, is directly linked to the most valuable motivators in work.

We will examine the results of the empirical research in the bellow section of the article.

3 Results and discussion

The research was held in the form of an online survey in the period between June 2020 and August 2020. The bigger part of the response group is part of a governmentally owned railway company in Bulgaria (n= 448), the other part of the response group is from a wide variety of economic sectors in Bulgaria (n=138), in total respondence n.586. We believe that the number of respondents is enough to let us make conclusions on the researched aspect of human motivation.

In order to examine the link between the number of children taken care from the respondents and the most valuable motivators in work, we divided the response group into 3 groups, respectively: no children taking care of; 1 child; 2 or more children.

In order to examine the workplace motivators, we gathered 38 examples from different theoretical authors and practice examples. The motivators were formulated into corresponding questions, using a 5-point Likert scale in order to examine the straight for each motivator subjectively for each respondent. Then we divided the motivators into 6 modules as follows: *1. Work Environment; 2. Remuneration; 3. Safety; 4. Personal development; 5. Affiliation and social aspects; 6. Respect and self-respect. Interest and challenge, autonomy and leadership.*

We evaluated the most and least important motivators, by calculating the middle averages for each motivator and for the modules overall. On this foundation we examined the most and least preferred motivators, compared the results and made conclusions.

We will examine the results for each group respectively. Afterwards, we will compare the results from each group and examine the eventual differences and the results.The deviation for each group is as follows:

1. No children to take care of – 240
2. 1 child to take care of – 191
3. 2 or more children to take care of – 155

It makes the No children group the largest and 2 or more children the smallest group in our research.

On Table 1 are shown the results from the group without children to take care of.

Table 1. No children to take care of, mean results

Module 1	*Module 2*	Module 3	Module 4	**Module 5**	Module 6
3,798833	*3,571357*	3,715749	3,65728	**3,919267**	3,671917

Source: Own research

As it is shown in the table, the least preferred module of motivators for the group of respondents without any children to take care of is Module 2 (Remuneration). Respectively, the most preferred module is Module 5 (Affiliation and social aspects).

Once it comes to examine motivators separately, the most preferred motivator for the current group is – "Comfortable working hours and brakes" part of Module 1 with a mean average of 4,137365. Correspondingly, the least preferred motivator for the examined group is – "Bonuses for diligence and results in work" part from Module 2 with a mean average of 3,175601.

On Table 2 we look through the results for the group with 1 child to take care of.

Table 2. 1 child to take care of

Module 1	*Module 2*	Module 3	Module 4	**Module 5**	Module 6
3,632161	*3,250047*	3,550136	3,460922	**3,896942**	3,623743

Source: Own research

In this group the tendency is basically the same, again Remuneration is the least preferred and Affiliation and social aspects is the most preferred module.

If we examine the motivators separately, here the most preferred is "Colleagues which are professionals", part of Module 5 and with an average of 4,085348. Respectively, the least preferred motivator is "Managerial responsibilities", part of Module 6, with an average of 2,932634.

And lastly, the results for the group with 2 or more children to take care of is shown in Table 3.

Table 3. 2 or more children to take care of

Module 1	*Module 2*	Module 3	Module 4	**Module 5**	Module 6
3,527776	*3,429019*	3,579781	3,613155	**3,842009**	3,634485

Source: Own research

Again, the tendency is almost the same. Module 5 is the most preferred among respondence from the group, and Module 2 the least preferred module of motivators.

Divided, the most preferred motivators for this group is – "Comfortable working hours and brakes" part of Module 1 with a mean average of 4,096083. Respectively, the least preferred motivator is – "Various forms of financial incentives" part of Module 2 with a mean average of 3,157228.

We will introduce the results from all 3 groups in Table 4, respectively, with most and least preferred modules.

Table 4 – Comparison between modules among groups

No children	**1 child**	**2 or more children**
Module 5 - 3,919267	Module 5 - 3,896942	Module 5 - 3,842009
Module 2 - 3,571357	Module 2 - 3,250047	Module 2 - 3,429019

Source: Own research

As we can see, as underlined already, the most preferred module is "Affiliation and social aspects", withthe highest average in the group without children to take care of. Respectively, the least preferred among the 3 groups is "Remuneration".

By our opinion the most preferred among the 3 groups of respondents is namely, Module 2, because of the social nature of the human being. From the dawn of our evolutional development, people were tended to live in groups, take care of each other, observe social norms etc. Therefore, it is absolutely logical that we observe this preference in all 3 groups of respondents. People will always tend to make social non-formal groups on the work place. Although it is from upmost value to give such possibilities in the work environment, managers have to be careful, when using this motivator in practice.

The least preferred module for our response group is the one linked to financial and material incentives. Which partly proves that on the workplace people don't look for only material stimulation. Almost in every resent research in motivation it is proven that money and material incentives, are not the most preferred by the workers in almost every country in the world. Unfortunately, the theoretical research and practice are in big contradiction with each other. Even though, it is proven that remuneration is not the only and most appropriate insensitive it is the most applied in practice. Even more, some authors have proven that in certain conditions, pay is a negative motivator, and actually demotivates and lowers the results [14].

For more detailed examination of the differences among the respondents in the 3 groups, we will examine also the divided preferences among the motivators. The systematization is shown on Table 5.

Table 5 . Comparison between motivators among groups

No children	1 child	2 or more children
Comfortable working hours and brakes - 4.137365	Colleagues which are professionals - 4.085348	Comfortable working hours and brakes - 4.096083
Bonuses for diligence and results in work - 3.175601	Managerial responsibilities - 2.932634	Various forms of financial incentives - 3.157228

Source: Own research

As we can see inTable 5, the most preferred among 2 groups (No children and 2 or more children) is the motivators "Comparable working hours and brakes". This may be a consequence of the necessity for more time with the family among the group with 2 or more children and the necessity for more free time among the group without children. Among the group with 1 child, the most preferred motivators are the professional colleagues, which is the first difference among the 3 groups. Therefore, this would be the first possible effect of the child presence in one's life.

We can see the tendency from examining the module preference also here. In 2 groups, the least preferred motivator is linked to remuneration. In the group with 1 child the least preferred motivator would be "Managerial responsibilities".

As we can see on Table 4 and Table 5, the tendency among the 3 examined groups is basically the same. Nevertheless, there is a small difference among the average attestation of the modules examined. There is also a small difference in the separate examination of the motivators.

This result cannot prove fully our hypothesis. Therefore, we cannot prove the direct and strong correlation of the presence of children in one's life and the most or least preferred motivators, having influence on worker's motivation. Nevertheless, there was a small difference among examined results. We observed difference in the average attestation for the modules in each group. Therefore, we are tended to say that there might me a small and weak correlation among the examined sign of human life and motivation to work. We examined a difference in the separately preferred motivators among the 3 groups, which partly confirmed the linkage between child presence and most and least preferred motivators.

4 Conclusions

In the current paper we examined the motivation at the work place. We underlined the importance of understanding motivation, once it comes to practically applying it. Managers have to examine motivation in their organizations, namely when there would be a proven correlation with some environment or individual trends.

Based on our results, we can cannot fully prove our hypothesis. Nonetheless, there were small differences in results among groups, which can partly prove a weak correlation between the examined variables.

We couldn't fully prove the correlation between the presence of children of one's life and work motivation, though. We are tended to say that managers have to be careful how they motivate people based on their life cycle as a whole and the number of their children. Although we haven't proven any direct link, other authors have proved such direct links.

Nonetheless, motivation will always remain one of the most important aspects in human management. Therefore, we need to have more detailed view on the theoretical work on many authors in the world and then we can formulate and apply motivational strategies. The necessity of theoretically based application of work motivation is mostly proved by the mismatch in using remuneration in motivating workers. In the current study we proved that remuneration is the least preferred among the modules. Herzberg has proved that pay is a hygienic factor in most kinds of jobs [7]. Rayan and Deci proved that pay can cause a negative effect on motivation and performance [15] [8]. Even though the mentioned authors have empirical prove for the mentioned tendency, practitioners are using more and more the pay for performance scheme in motivating employees. Such tendencies can be observed in many other aspects of human motivation, which we mentioned in the current paper.

Finally, although we could prove the direct link between the two examinate variables, we can conclude that motivation is one of the most complexed psychological trends in human resource management. Therefore, once it comes to motivation and examining motivation in a certain organization, the company needs to provide different forms of incentives among different kinds of workers, based on many trends, which we observed above.

References

1. S. Adams, New York: Academic press, **2**, 267 (1965)
2. F. Belschak, D. Den Hartog, App. Psych.: An inter. Rev., **2**, 274 (2009)
3. T. Bipp, Int.Jour.l of Sel. and Assess.t, **1** (2010)
4. E. Deci, R. Ryan, Psych.l inq., **4**, 227 (2000)
5. L. Deckers, *Motivation: Biological, psychological, and environmental (*2010)
6. Edwin A. Locke, Gary P. Latham, The Aca. of Manag. Rev., **3**, 388 (2004)
7. F. Hercberg, Harv. Bus. Rev., 53 (1968)
8. C. Ilardi, K. Leone, T. Kasser, M. Rayan, Jour. of app. Soc. Psych.(1993)
9. G. Latham, C. Pinder, Ann.Psych.Rev., 485 (2005)
10. G. Oldham, J. Hackman, Admini.Sci. quart., **1**, 66 (1981)
11. G. Robert, John Roberts, Princeton University Press, 56 (2012)
12. R. Robert, Irina Zinovieva, E. Dienes, App. Psych.: an intern. Rev., **4**, 658 (2000)

13. R. Kanfer, P. Ackerman, The Ac. of Manag. Rev., **3**, 440 (2004)
14. R. Ryan, E. Deci, Journ.of Person.,**74**, December, 1557 (2006)
15. M. Zuckerman, J. Porac, D. Lathin, R. Smith, L. Deci, Person. and Soc.Psych. bull, **4** (1978)

Assessing the Stakeholders' Contribution to Business Value

Elena Patrusheva, *Elena* Sapir, and *Igor* Karachev*

P.G. Demidov Yaroslavl State University, Sovetskaya Str., 14, 150003 Yaroslavl, Russia

Abstract. Stakeholder theory is the most attractive fundamental concept to ensure the necessary improvement of corporate governance in the modern world, integrated into the idea of value maximization as the main strategic goal of the enterprise. The proposed concept of total value considers this category as the sum of the organizational wealth and the values of the stakeholders, which together constitute the total contribution of the economic system to public welfare. Recognizing not only the financial implications of stakeholder relationships, the authors propose practical tools for value management, based on the agreed interaction of stakeholders – the modeling of resource flows to optimize the resulting total value and the algorithm of step-by-step management of interaction with stakeholders. The modeling of value creation reflects both the organizational benefits of interaction and explicitly identifies the value of the stakeholders' benefits, and also discloses a list of key management factors: the economic value of the resources used by parties, the fairness of the exchange, the relationship between interacting companies, the opportunity costs, etc. Critical aspects of value creation that go beyond profitability and economic returns are highlighted. Corporate values can only be understood by reference to the multiplicity of stakeholders' value perspectives.It is recommended that the process of managing interaction with stakeholders should be guided by an extensive interpretation of the target (value) setting of the business, taking into account the totality of material and intangible resource flows between the parties.

1 Introduction

As a strategic goal, most modern companies have adopted a focus on business value growth. The achievement of this goal within the framework of the Value Based Management (VBM) concept has traditionally been carried out mainly through the owners' welfare growth.

The crisis currently unfolding in the world economy shows that the previous model of corporate governance, which is oriented towards the interests of individual groups of owners and is highly related to unstable capital markets, needs significant adjustments necessary to ensure more stable economic growth in the long term.

For decades, value creation was understood as maximizing equity value, with a focus on outcomes rather than processes or intentions. Stakeholder Behavioral Theory focuses on the

*Corresponding author: karachev2011@yandex.ru

issue of «overlapping value» and, in particular, why stakeholders come together voluntarily to cooperate within the company [1-2].The purpose of identifying the overlapping value of stakeholders implies that the value derived by stakeholders from relationships or networks (e.g., supply chains) will affect whether they participate in transactions. When members of a network find overlapping points of value, they are interested in coming together and cooperating.

R.E. Freeman [2] argued that the success of an organization is determined by the degree of satisfaction of the stakeholders, which include shareholders, top managers and employees, consumers and suppliers, public authorities, as well as other interacting groups. Stakeholder theory considers any organization to be a network of interrelated elements, each of which contributes (voluntarily or compulsorily) to the results of the organization's activities and expects to benefit (or to avoid non-compensable harm) [3]. J.E. Post, L.E. Preston and S. Sachs emphasized the expediency of changing traditional views on the nature of the organization in the context of the rapid development of business networks, information services, and the knowledge economy [4].

Subsequently, in this theory, it was the increase in the value of the company that established itself as a criterion for strategic development [5]. The stakeholder theory formulated by M.C Jensen, which produced a measure of efficiency, became called enlightened stakeholder theory [6].

Value management requires the specification of the processes of interaction with the stakeholders and, in particular, the measurement of the financial contribution to its establishment by the various stakeholder groups, as well as the social impact of stakeholder interaction, to manage value maximization by understanding it as an integral value, the individual components of which are created in contact with one or another group of stakeholders.

The general hypothesis of the study is that effective interaction of an organization with its stakeholders leads to the excess of benefits over costs and the creation of positive net cash flows, which are the basis of the firm's sustained growth in value.

2 Methods

The methodology for estimating the value of an enterprise's relationships with individual stakeholders is based on model F. Figge, which determines the value as follows [7]:

$$StV = \sum_{n=1}^{N} \frac{R_n - C_n}{(1-d)^n} \tag{1}$$

Where **StV** –value created by the stakeholder; $\mathbf{R_n}$ –revenue from interaction with the stakeholder; $\mathbf{C_n}$ – stakeholder maintenance costs; **d** – discount rate; **N** – duration of the forecast period.

However, F. Figge [7] is of the view that the only party with which an enterprise has reciprocal cash flows is shareholders and creditors, while the direct cash flows of other stakeholders are only one-sided (for example, staff, government and suppliers are characterized by cash outflows from the enterprise in the form of wages, taxes, payment for raw materials, respectively).This position is somewhat limited. A number of modern researchers criticize the idea of creating only financial value for «arbitrary reduction of the potential range of factors» [8], arguing that «striving for financial success is necessary in such a way as to also bring social benefits» [9].

We believe that the relationship between the company and each group of its stakeholders is bilateral in nature and is expressed in cash flows, both received and spent.

On the one hand, it is clear that if an enterprise's value consists of the benefits it derives from its stakeholders, stakeholders are, in aggregate, the resource that makes the greatest contribution to the organization's wealth. On the other hand, stakeholders decide to engage with the enterprise only on the basis of the value of the benefits it generates, i.e., considering it as their resource. Adopting a resource-based approach to the organization's relationship vision and its stakeholders through cash flows requires a model description that removes the limitations of F. Figge's views.

This is the case with the concept of total value (CTV) of an enterprise, since it is not only intended to describe the individual elements of value arising from the relationship between the enterprise and its stakeholders, but also to show the natural relationships between them, as well as integrating parts into a coherent whole.

The overlapping value for stakeholders in networks reflects the converging interests of all participants in creating value, which includes economic considerations, as well as equity, trust and a sense of that the firm and its operations will allow all stakeholders to become more successful over time. Value created to one stakeholder depends on the behavior of other stakeholders in the firm.

Another important aspect of this idea, not found in the literature studied, is the solution to the multi-tasking problem of governance noted by M. Jensen, which implies the impossibility of maximizing in several directions. On the face of it, the presence, in addition to the organizational wealth, of the value of stakeholders whose interests must be taken into account poses a challenge to managers in identifying and setting priorities, as stakeholders may demand simultaneous, equivalent, different and continuous attention [10-11].

The conceptual approach, based on the identification of overlapping benefits in networks with stakeholders, assumes that different parties have a positive correlation between the economic value and the benefits of continuous cooperation. Where this is not the case, opportunism threatens the relationships among network members, especially where initiatives to ensure the sustainability of the network are costly and time-consuming.

One of the few (albeit substantial) weaknesses of our idea is that its understanding of the processes of creation and distribution of values is not yet sufficiently widespread in practice, which in turn leads to a lack of methodological approaches necessary for the realization of the stated ideas in management activities. Therefore, the next necessary step is to give the CTVan economic content and to identify, on that basis, the fundamental principles on which the enterprise value management in the context of the stakeholder theory should be built.

3 Results

The resource approach to value means that transaction participants continuously measure the benefits and costs of doing the transactions. They seek to maximize their income in order to cover their resources and efforts.For the exchange to be a non-zero-sum game, parties must increase transaction volumes before the available resources are exhausted.

In the case of diminishing marginal utility of resource flows, the parties will increase the exchange until the marginal income of one of them equals the marginal cost. It can therefore be concluded that maximizing the value of a company will involve maximizing the value of its stakeholders simultaneously, which together results in maximizing the contribution of the economic system to public welfare:

$$\begin{gathered} f_0(x_0) \to \max \Leftrightarrow f_i(x_i) \to \max, \\ D(x_0, x_i): \{0 \le x_0 \le X_0,\ 0 \le x_i \le X_i,\ i \in 1:M\} \end{gathered} \tag{2}$$

where $\mathbf{f_0(x_0)}$ and $\mathbf{f_i(x_i)}$ – target functions of the enterprise and stakeholders; $\mathbf{X_0}$and$\mathbf{X_i}$ – size of the enterprise's and stakeholder's stock of resources; **M** – number of the enterprise's stakeholders.

The application aspect of the CTV involves the identification of mechanisms for the creation and estimation of certain types of value.

We believe that the value of the company ($\mathbf{V_i}$) created through contacts with one or another stakeholder can be expressed as follows:

$$V_i = \sum_{n=1}^{N} \frac{R_{in} - C_{in}}{(1+d_0)^n} = \sum_{n=1}^{N} \frac{\sum_{k=1}^{K} \vec{X}_{ik} r_{0k} - \sum_{l=1}^{L} \vec{X}_{il} p_{0l}}{(1+d_0)^n} \tag{3}$$

where $\mathbf{R_{in}}$ – value of the resources received; $\mathbf{C_{in}}$ – value of the resources transferred; $\mathbf{d_0}$ – discount rate; **N** – duration of the forecast period; $\vec{X}_{ik}$ – natural volume of the resource k received by the enterprise; $\mathbf{r_{0k}}$ – value (profitability) of using resource k by the enterprise; $\vec{X}_{il}$ – natural volume of the resource l transferred by the enterprise; $\mathbf{p_{0l}}$ – value (price) of the resource l for the enterprise; **K, L** – number of names received and transferred resources respectively.

The same logic can be applied to the stakeholder value ($\mathbf{StV_i}$) created by the company:

$$StV_i = \sum_{n=1}^{N} \frac{StR_{in} - StC_{in}}{(1+d_i)^n} = \sum_{n=1}^{N} \frac{\sum_{l=1}^{L} \vec{X}_{il} r_{il} - \sum_{k=1}^{K} \vec{X}_{ik} p_{ik}}{(1+d_i)^n},\ i \in 1;M \tag{4}$$

where $\mathbf{StR_{in}}$ – value of the resources received; $\mathbf{StC_{in}}$ – value of resources transferred; **di** – discount rate; $\vec{X}_{il}$ – natural volume of the resource l received from the enterprise; $\mathbf{r_{il}}$ – value (profitability) of using the resource l; $\vec{X}_{ik}$ – natural volume of the resource k, transferred to the enterprise; $\mathbf{p_{ik}}$ – value (price) of the resource for the stakeholder i; **M** – number of stakeholders of the enterprise.

Thus, the mechanism of the formation and assessment of the company's values (**V**) and its stakeholders (**StV**), the maximization of which is the target functions of the parties, can be illustrated as follows:

$$V = \sum_{i=1}^{M} V_i = \sum_{i=1}^{M} \left(\sum_{n=1}^{N} \frac{R_{in} - C_{in}}{(1+d_0)^n} \right) \tag{5}$$

$$StV = \sum_{i=1}^{M} StV_i = \sum_{i=1}^{M} \left(\sum_{n=1}^{N} \frac{StR_{in} - StC_{in}}{(1+d_i)^n} \right) \tag{6}$$

Companies must be provided with the resources to create value for each stakeholder. However, the maximization of stakeholder utility functions that express stakeholder preferences for certain types of value is subject to different conditions and limitations.

It therefore seems advisable to focus on the possibility of achieving some optimal balance of interests between the parties involved, equivalent to the maximum possible stakeholders' value, provided that no harm is done to each business participant:

$$f_0(x_0) \to \max \Leftrightarrow \sum_{i=1}^{M} f_i(x_i) \to \max, \tag{7}$$
$$D(x_0, x_i): \{0 \le x_0 \le X_0;\ 0 \le x_i \le X_i;\ f_i(x_i) \ge 0\}$$

If a balance of stakeholder interests is not found, this, given resource constraints in the medium term, will lead to attempts to either restore the balance or sever the relationship. Therefore, the management of the firm should maintain a resource balance in order to build mutually beneficial cooperation with interested parties.

In addition, managers need to pay attention not only to the absolute amounts of benefits generated for stakeholders, but also to compare them with other options available to them, and vice versa:

$$f_0(x_0) \to \max \Leftrightarrow \sum_{i=1}^{M} f_i(x_i) \to \max, \tag{8}$$
$$D(x_0, x_i): \{0 \le x_0 \le X_0;\ 0 \le x_i \le X_i;\ \Delta(V_i) \ge 0;\ \Delta(StV_i) \ge 0\}$$

Where $\Delta(V_i)$ – company's rent; $\Delta\ (StV_i)$ – rent of stakeholder i.

Considering the fundamental impossibility of finding perfect maximizing techniques, it is advisable to consider the continuous increase of its value as the main strategic goal of the company, achieved through the gradual improvement of relationships with stakeholders and satisfaction of their interests.

Management of a firm in the financial interests of owners or shareholders is also important for creating value for all other stakeholders. It should be noted, however, that the concept of business value maximization does not imply that non-financial values, such as social and environmental aspects, are less important for creating value. Instead, these values are included in the financial value, taking into account their impact on the profitability of the firm.

Thus, the relationships described above form a model series, which fairly correctly reflects the relationship between the company and the stakeholders in terms of the creation of their values and therefore can be used further to develop practical tools for the approach to value management.

The particularity of the proposed CTV makes it possible to give a specific technological content to the approach to the company's value management: to formulate principles to ensure the achievement of management objectives, to determine the most appropriate actions of managers, to select and apply appropriate methods and tools.

For example, the key principles on which management should be based in the concept of CTV are:

– A broad interpretation of value. Values in business are so diverse that efforts to reduce this plurality of values to one common metric or measurement standard are unrealistic goals that ultimately reduce the total value created by firms.

– The need to take into account objective and subjective factors relevant to different economic agents.

– Improving the management process information field by researching stakeholder values.

– Development of a disclosure system that meets the interests of economic agents and ensures the required level of trust between them.

– Continuous monitoring of the overlapping values of stakeholders in order to find common interests and benefits as a basis for continuous cooperation.This would avoid the main networking problem of the perception that costs and benefits are not distributed equitably across the network.

These principles can be put into practice by applying the developed company value management algorithm based on the coordinated interaction of stakeholders, showing the sequence of the following interrelated steps:

1. Identification of stakeholders. As many external stakeholders as possible need to be identified and described.

2. Analysis and identification of the most important external relations for the company. Ranking of stakeholders according to their importance for the development of the company's promising economic activities.

3. Determination of possible uses of the organization's resources in terms of performance, stakeholder satisfaction, trust building, knowledge creation, etc.

4. Analysis of the volume and structure of the organization's resources from the point of view of developing interaction with the priority stakeholders and increasing the total value of the company.

5. Mobilization of the company's internal reserves to increase the resource volume and improve the resource structure.

Thus, it can be concluded that the completion of all stages of the enterprise value management process provides opportunities for the necessary analysis and monitoring of essential indicators related to the external environment of the enterprise. It becomes possible to make specific management decisions aimed at maximizing the use of all stakeholder resources and, as a result, increasing the company's value.

4 Conclusions

1. The traditional model of corporate governance organization aimed at maximizing the income of owners does not reflect the critical transformations that have occurred in business in recent years: industrial revolution 4.0, transition to high information technology, digitalization and deep cooperation based on global value chains.

2. An adequate theoretical concept capable of providing the necessary improvement of corporate governance is the concept of the total value of stakeholders, redefining the value of the company in a broad context – as a resource, created by all stakeholders and business participants.

3. The proposed concept of total value, as opposed to the classical theory of stakeholders, considers the value of a company to be the total wealth of the organization and the sum of the value of all stakeholders, which together contribute to the overall public welfare of the established business ecosystem.

4. Thus, the concept of total value highlights the objective relationships that exist within the stakeholder network and that support its efficient functioning, thereby eliminating the contradiction between the increase in the company's value and the satisfaction of the interests of its stakeholders, since it unites them in the common task of maximizing the social welfare.

5. The process of corporate governance, in the light of the concept of total value, is based on a broad interpretation of the business objective and takes into account the total resource of the company's value growth, namely: all value flows, generated directly or indirectly by stakeholders and their productive interactions.

6. The practical lesson to be drawn from this concept is the objective need for continuous monitoring and analysis of stakeholders in order to provide information support of the management process, focus on the future, long-term engagement and trust of the parties, as well as the willingness to provide the most favorable conditions for all stakeholders in realizing their interests and achieving the stated business goals.

References

1. R. H. Coase, The nature of the firm. Economica,**4**, 386 (1937)
2. R.E. Freeman, *Strategic Management: A Stakeholder Approach* (1984)
3. T. Donaldson, L.E. Preston, Academy of Management Review, **20(1)**, 65 (1995)

4. J.E. Post, L.E. Preston, S. Sashs, *Redefining the Corporation: Stakeholder Management and Organizational Wealth* (2002)
5. R.E. Freeman, Business Ethics Quarterly, **4**, 409 (1994)
6. M.C. Jensen, Journal of Applied Corporate Finance, **14**(3), 8 (2001)
7. F. Figge, S. Schaltegger, *What Is Stakeholder Value? Developing a Catchphrase into Benchmarking Tool* (2000)
8. R. K. Mitchell, G. R. Weaver, B. R. Agle, A. D. Bailey, J. Carlson, Academy of Management Review, **41**(2), 252 (2016)
9. M.R. Kramer, M.W. Pfitzer, Harvard Business Review, **94**(10), 80 (2016)
10. F. Bridoux, J. Stoelhorst, Strategic Management Journal, **35**, 107 (2014)
11. C. Tantalo, R. L. Priem, StrategicManagement Journal, **37**, 314 (2016)

Modeling the stakeholder value and stakeholder risks

Alexander Zlygostev*, and *Elena* Makarova

Ural State University of Economics, 8 March str., 62, 620144 Ekaterinburg, Russia

Abstract. The aim of the paper is to expand understanding of stakeholder value and consider possibilities of its assessing for minimizing stakeholder risks. The study makes contributions to achieving a balance between interests of stakeholders and implementation of sustainable business development principles. The paper introduces a practical toolkit for modeling stakeholder value and stakeholder risks. New methods, developed by the authors to model and analyze stakeholder value and risks, determine the novelty of the study. The paper provides the results of new methodology validation performed with the use of mathematical and econometric methods applied to open source data. Models of contributions and benefits of stakeholders are examined, levels of the "gap" between contributions and benefits by groups of stakeholders are outlined, the map of "stakeholder risks" is presented. The analysis of the results allows to make conclusions and provide recommendations on establishing balanced relationships with stakeholders and minimizing stakeholder risks in order to harmonize interests for sustainable business development.

1 Introduction

The relevance of the study is determined by the demand for a new paradigm of corporate governance, central to which is the concept of stakeholder value. Being the focus of controversy, modern capitalism and issues related to its development, need to be reviewed. According to most economists, including World Economic Forum Founder and Executive Chairman Klaus Schwab, shareholder capitalism needs reforming and must be replaced by "stakeholder capitalism" ("conscious capitalism"). Coronavirus pandemic, environmental and social issues contribute to exacerbating of existing contradictions, and, as a result, enhance theoretical and practical relevance of the research topic.

Despite being central in many studies, the concept of "stakeholder value" remains a subject of discussion, as most theoretical formulations lack recommendations for practical implementation of stakeholder approach. This paper attempts to contribute to the development of practical tools for assessing and modeling stakeholder value.

There is a large volume of published foreign theoretical studies and literature reviews on the concept of "stakeholder value" [1, 2]: R. K. Mitchell and his co-authors highlight the need of developing an interdisciplinary theory, comprising the theory of entrepreneurship,

* Corresponding author: letrus.alex@gmail.com

accounting, risks, and corporate governance to create new accountability which takes into account stakeholder value based on sharing of risks by stakeholders [3]. A recent study by R. Tapaninaho, who provided an overview of "stakeholder value" narratives, introduced classification of modern approaches to the study of stakeholder value [4]. Previous study related to the current research area [5], which discusses and analyses literature on the concept of "stakeholder value", was based on the classification mentioned above.

A number of studies, conducted by I. N. Tkachenko and other authors, address the concept of stakeholder value from practical implementation perspective [6, 7].

A. B. Ankundinov considers the possibilities of assessing stakeholder value creation and distribution [8]. The study reports that research in this area should be conducted within the framework of assessment of the company's interaction with individual groups of stakeholders; however, A. B. Ankundinov predicts difficulties in applying econometric analysis in this context.

I. V. Ivashkovskaya argues that economic indicators of stakeholder value creation and the company's growth sustainability index allow to assess corporate strategies based on value creation for stakeholders [9].

The interdisciplinarity of the present study is based on theoretical and methodological foundations of modern management, economic theory, main concepts of corporate finance, enterprise economy, economic and mathematical modeling.

The paper first attempts to develop innovation and unique toolkit for managing organization's relationship with stakeholders on the bases of minimizing stakeholder risks through achieving a balance of interests (contribution and benefits of stakeholders).

The purpose of the study is to provide suggestions to develop organizational and methodological support for the effectiveness of corporate governance based on modeling the contribution and benefits of stakeholders and minimizing stakeholder risks.

The paper provides the results of new methodology validation performed with the use of mathematical and econometric methods applied to business object's open source data.

The findings add substantially to understanding the concept of stakeholders in corporate governance. Implementation the authors' methodological approach and toolkit allows to increase efficiency and quality of assessing stakeholder value on the basis of stakeholder contribution/benefit model, minimizing stakeholder risks and developing recommendations for balanced management interests of stakeholders.

2 Materials and methods

To diagnose stakeholder risks, "gap" analysis of stakeholders' contribution and benefits was performed. To carry out the task efficiently, models of contributions and benefits of stakeholders were designed. Modeling stakeholder value, conducted by the author, had been presented previously [10]. Mathematical and econometric methods were applied to design and construct stakeholder value models. Revenue served as a proxy for stakeholder value.

Data for modeling was obtained from open web-based sources, company sites, and informational and analytical reports (2008–2019). The constructed models allow to represent imbalances in the structure of the stakeholder values' creation and distribution, identify the "tension" and risks in stakeholder relationship. The results of the study have numerous important implications which can be recommended to object companies for minimizing stakeholder risks and achieving a balance of stakeholders' interests in order to ensure long-term sustainable development.

To achieve the purpose of the study the following research methods were used:

1. Identification and segmentation of stakeholder groups, as well as justification of their segmentation (to simplify, we address "classic" division of stakeholders' groups, such as "employees", "owners", "customers" and the others).

2. Collecting and analyzing information presented in object company reports (open joint stock company "Zhirovoy Combinat" ("Fat Processing Plant").

3. Correlating collected data with categories of "benefits" and "contributions" of stakeholders.

4. Modeling and calculating missing information (the missing information implies "contributions" of employees and suppliers, which were calculated on the basis of econometric model. The factor multiplied by regression coefficient gives a contribution to stakeholder value, see below for more details).

5. Substitution of the obtained data into the model, calculation of "gaps", visualization of stakeholder risks

6. Providing guidance and recommendations on adjusting the distribution of contributions and benefits to achieve a balance of interests and sustainability of stakeholders' collaboration, optimization of stakeholder risks, improving a stakeholder relations policy, searching for ways to harmonize interests and relationships in the network of stakeholders.

3 Results and discussion

Table 1 shows a model, illustrating the dependence of revenue (proxy indicator of stakeholder value) on the indicators of employees and suppliers' benefits (remuneration and payment to suppliers). The model was designed on the basis of econometric modeling of employees and suppliers' contribution. We assumed that contribution of these categories of stakeholders is proportional to their benefits. To identify the effect, which ratio of all these factors has on revenue, a cross factor (m_to_labor: m_to_suppliers) was introduced.

Table 1. Econometric model of stakeholder contribution

Formula	Revenue = m_to_suppliers + m_to_labor + (m_to_labor:m_to_suppliers)				
Coefficients:	Estimate	Std. Error	t value	Pr(>|t|)	
(Intercept)	1,420,000.00	623,800.00	2.28	0.05	.
m_to_labor	2.53	1.12	2.25	0.05	.
m_to_suppliers	0.58	0.18	3.33	0.01	*
m_to_labor:m_to_suppliers	0.0000002026	0.0000000663	3.05	0.02	*
Residual standard error: 570900 on 8 degrees of freedom					
Adjusted R-squared: 0.9969; F-statistic: 1162 on 3 and 8 DF, p-value < 0.000					
Confidence interval	2.50%		97.50%		
(Intercept)	-18,283.270		2,858,671.000		
m_to_labor	-0.059		5.125		
m_to_suppliers	0.180		0.989		
m_to_labor:m_to_suppliers	0.0000000496		0.0000003555		

The model assessment was based on the following: F-statistic <0.05 (number of observations are sufficient), there is a close linkage between the indicators. The indicators, which were chosen by intuition as they seemed interconnected, account for the high value of the adjusted coefficient of determination - 0.997. Moreover, this coefficient is over grading, since the analysis for multicollinearity showed correlation between the factors. This led to instability of parameter estimation. Anyway, for the purpose of the analysis under the conditions of insufficient information the model was adopted. The Durbin-Watson test showed the absence of residuals autocorrelation.

Analysis of the collected data and application of the model allowed to calculate values of stakeholders' contribution and benefits and gaps between them (Table 2).

Table 2. Contributions, benefits, stakeholder gaps

Stakeholder contribution, RUB			
Stakeholder group	**Defined by**	**2019**	**2018**
Owners	Present amortized cost of authorized capital + owners' contributions to capital in the current period	204,850	809,580
Employees	2.53 * Remuneration (econometric model)	5,210,414	2,212,672
Suppliers	0.58 * Payments to suppliers (econometric model)	20,130,544	6,311,163
Government	no data	0	0
Investors	(short-term and long-term debt) Risk-free rate (e.g. weighted average interest rate of CBR)	163,399	201,366
Stakeholder benefits, RUB			
Owners	Dividends + increase in market value of shares	0	1,545,100
Employees	Salary	2,057,013	873,538
Suppliers	Payments to suppliers	34,434,731	10,795,694
Government	Tax	90,680	114,311
Investors	Interest paid	101,882	95,785
Gaps (Contribution - Benefits), RUB			
Owners		204,850	-735,520
Workers		3,153,401	1,339,134
Suppliers		-14,304,187	-4,484,531
Government		-90,680	-114,311
Investors		61,517	105,581

The graph showing "stakeholder risks" is based on the data presented in Table 2. The "gap" is determined as the difference between contribution and benefits of stakeholders for each group and is plotted along the vertical axis of the graph (Fig. 1). Positive gap values indicate that the group of stakeholders contributes more to value creation (revenue) in comparison to what they receive in the form of benefits (wages), having the status of "donors" (employees, in our case).

The analysis of the graph proves the dominant role of suppliers and workers in creating value (abstracting from stakeholders as clients). The business of the object company is most dependent on stakeholders as suppliers. Moreover, the "suppliers" receive significantly more benefits than they invest in creating value, while the benefits of "employees", according to the model, on the contrary, are less than their contribution: this is clearly seen from their position in the positive zone of the vertical axis. Further research should concentrate on analyzing the possibility of increasing benefits of the "employees" group by temporarily halting the growth of the benefits of the group of "suppliers". This will contribute to achieving a balance of interests and minimizing stakeholder risks.

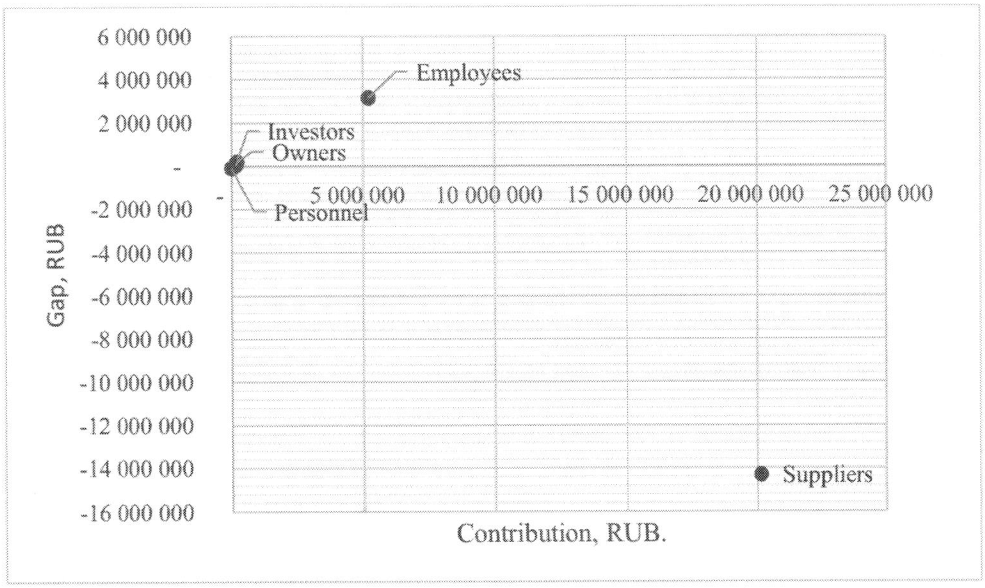

Fig. 1. Stakeholder risks map (Contribution and benefit gap map), 2019.

4 Conclusion

The purpose of the present study was to consider and identify stakeholder interests and risks on the basis of the designed monetary "objective" model of stakeholders' contributions and benefits and a "gap" map (stakeholder risk map). Analysis of contribution to the revenue allowed to identify categories of stakeholders.

Modeling contributions and benefits of stakeholders is significant for assessing possible stakeholder risks and damage (by assessment of the contribution of a stakeholders' group). Risk probability depends on a number of factors, including the gap between contributions and benefits of stakeholders. Applying sociological methods (polls, questionnaires, focus groups) to subjective stakeholders' assessments can help gain better understanding of the process of assessing risks' probability and specifics of risks in comparison to modeling the gap between monetary contribution and benefits.

The paper introduces a new method of assessing contributions and benefits of stakeholders in order to identify stakeholders' groups and imbalances in created and distributed value. The analysis toolkit, presented in the paper, will assist in minimizing stakeholder risks (gap level) and will contribute to harmonious and long-term development of the company's stakeholder network by achieving a balance of stakeholder interests.

Acknowledgements

The reported study was funded by RFBR and Sverdlovsk oblast, project number 20-410-660032 p_a.

References

1. F. Figge, S. Schaltegger, *What is stakeholder value? Developing a catchphrase into a benchmarking tool* (2000)
2. E. Garriga, A. Werner, AOM Proc., **1**, 17201 (2018)

3. K.R. Mitchell, H.J. Van Buren III, M. Greenwood, R.E. Freeman, JOMS, **52**, 851 (2015)
4. R. Tapaninaho, J. Kujala, *Social responsibility and sustainability* (2019)
5. A.A. Zlygostev, Theoretical Economics, **8** (2020)
6. I.N. Tkachenko, *New trends in the development of corporate governance and business*, 93 (2017)
7. I.N. Tkachenko, *Institutions and values in the system of corporate relations, Effective development models* (2000)
8. A.B. Ankudinov, KFU Bul., **3**, 70 (2011)
9. I.V. Ivashkovskaya, *Integrated management system of the company's value* (2010)
10. I.N. Tkachenko, I. V. Pervukhina, A. A. Zlygostev, The Manager, **2**, 2 (2020)

Information content of corporate reports of US medical companies

Julia Kuznetsova*

Branch of Kuzbass State Technical University named after T.F. Gorbachev in the city of Novokuznetsk, Ordzhonikidze Str., 8a, 654000 Novokuznetsk, Russia

Abstract. The document shows that corporate reporting is not only a source of information about the current state, results and trends of the company's development, but also a thing that transfers its image to the external and internal environment. It is determined that the transfer of a charitable image to external recipients of such information about the companies' activity operating in the social services is especially important. Based on the analysis of 10-K form reports in the 10 largest medical companies in the United States there were revealed information blocks, focusing on which the consumer will pay attention to a particular company, and not a competitor's company.

1 Introduction

Corporate reporting is a channel that connects the company with all parties interested in its activities, both in the external and internal environment, as well as one of the main directions for improving the corporate governance system. The corporate reporting system differs in Russia and abroad, both in the composition of documents and their information content. The main requirement for reporting in the United States is the information utility for the user to make certain decisions. Reporting regulations affect a limited number of organizations: this includes joint-stock companies and enterprises in regulated industries. For others, the form and composition of reporting is determined by management independently. However when submitting financial reporting it is important to obtain an appropriate auditor's opinion [1]. A distinctive feature of corporate reports of American companies is not their form, but their qualitative and quantitative information content [2]. According to the Association of Chartered Certified Accountants (ACCA), competent corporate reporting can influence stock prices, help businesses gain access to financial resources, enter into contracts with customers and suppliers, attract employees [3], and its quality indicators are: 1) up-to-dateness and relevance of the material; 2) completeness of the information provided; 3) reliability; 4) data comparability; 5) ability to verify data; 6) timeliness; 7) understandability. The following criteria are separated to evaluate narrative reports: future orientation, strategic focus, dependence on the object (any information that directly relates to the management object), correlation with previous years reports, research of all important aspects of the activity, and laconicism.

*Corresponding author: acanaria2005@yandex.ru

One of the reporting forms is form 10-K, which is provided by American companies to the Securities and Exchange Commission. Preparation of this document is necessary for all US companies. This report consists of 4 parts and 15 sections: 1) business characteristics (risk factors are highlighted separately); 2) location of offices; 3) court proceedings; 4) disclosure of security information; 5) issues related to the stock market; 6) selected financial data; 7) analysis of financial condition and results of operations; 8) financial statements and additional data; 9) information about changes and disagreements with accountants on accounting issues; 10) Directorate and corporate governance; 11) management remuneration; 12) security of property of certain owners and management; 13) transactions with related persons and independence of the Director; 14) basic accounting fees and services; 15) reports and graphs on financial activities.

It appears that in order to build a positive reputation in the market of goods and services, the company should pay special attention to the quality of information parameters provided to consumers. This is even more important for those companies whose activities directly affect the quality of life of the population, i.e., those that operate in the social services. According to many foreign and domestic researchers social corporate reporting has a semiotic property to a greater extent [4, 5], but the same property can be attributed to global general corporate reports. J. H. Spangenberg tries to trace the feedback, suggesting that corporate reporting should contain a criterion similar in content to the human development index, formed however for the internal environment of the company (the corporate human development index) [6].

In this regard the purpose of this document was to identify and analyze the sections of the report on form 10-K of the ten largest medical companies in the United States that can significantly influence the decision of consumers to contact this, and not another company.

2 Research methods

To achieve this goal, we studied the information content of the report on form 10-K of the ten largest medical companies in the United States for the last year of the report, which is available in the public domain. Evaluation of the corporate reports content of US medical companies was carried out in 2 stages: expert evaluation of the significance of sections of form 10-K in terms of attracting and retaining potential and existing customers; evaluation of information content of significant sections.

The significance of the form 10-K sections was evaluated by a group of 17 Russian experts in the field of medical equipment supply and medical services. Significance is evaluated by assigning a point from 0 to 1, where 0 is no significance, and 1 is the highest level of significance. During the assessment there was a condition that no more than three sections can receive an identical rating. The final outcome evaluation is defined as the arithmetic mean value of 17 experts' assessments.

The information content of the relevant sections was evaluated by male and female citizens at the ages from 30 to 55 who are active users of medical products and services. The number of experts was 63 people. The survey involved respondents who speak English. It was conducted in August-October 2020 via email and orally in two stages. At the first stage, respondents answered the question: what are the most important characteristics of the information provided in the report of American medical companies on form 10-K? At the second stage, the same respondents were asked to fill out a table for evaluating the information content of corporate reports of designated medical companies. The characteristics that were most frequently mentioned by respondents in the first stage of the survey were evaluated based on assigning one of the following criteria to each characteristic: B – high level of quality, C – medium level, H – low. As a result, the occurrence of each characteristic for all medical companies for each respondent was calculated.

3 Evaluation of the reports information content of US medical companies

As a result of the first stage implementation of the corporate reporting survey of US medical companies on form 10-K, the following results of expert evaluation were obtained (table 1).

Table 1. Expert evaluation of the significance of form 10-K sections

Section №	Expert evaluations																	Outcome evaluation
	1	2	3	4	5	6	7	8	9	10	11	12	13	14	15	16	17	
1	1	0,5	0,7	0,9	1	1	1	0,6	0,8	0,9	1	0,9	0,6	0,6	0,7	0,7	0,8	0,81
2	0,9	1	0,5	0,6	0,6	0,7	1	0,9	0,8	0,9	0,8	0,7	1	0,9	1	0,9	1	0,84
3	0,8	0,9	0,9	0,8	0,8	0,8	0,7	0,8	0,9	0,7	0,8	0,6	0,8	0,6	0,8	1	0,9	0,80
4	1	1	1	1	0,7	0,8	0,9	0,6	0,7	0,8	0,9	0,8	0,9	0,7	0,9	1	1	0,86
5	0,5	0,6	0,8	0,7	0,6	0,5	0,5	0,4	0,6	0,5	0,6	0,5	0,5	0,5	0,7	0,7	0,5	0,57
6	0,6	0,6	0,7	0,9	0,8	0,6	0,6	0,7	0,7	0,5	0,9	0,8	0,7	0,6	0,7	0,7	0,8	0,70
7	0,5	0,7	0,7	0,7	0,9	0,5	0,5	0,6	0,7	0,6	0,7	0,7	0,6	0,7	0,6	0,6	0,7	0,65
8	0,4	0,6	0,6	0,5	0,7	0,6	0,6	0,7	0,6	0,8	0,9	0,8	0,6	0,7	0,6	0,5	0,7	0,64
9	0	0,1	0	0,2	0	0	0	0	0	0	0,2	0,1	0,1	0	0	0	0	0,04
10	0,9	0,7	0,6	0,9	0,5	0,9	0,9	0,5	0,5	0,6	0,7	0,7	0,7	0,6	0,6	0,9	0,8	0,71
11	0,5	0,4	0,3	0,3	0,2	0,3	0,4	0,4	0,3	0,2	0	0	0,2	0,2	0,3	0	0	0,24
12	0	0	0	0	0	0,1	0,1	0	0	0	0	0	0	0	0	0,1	0	0,02
13	0,1	0	0	0	0,1	0,1	0,1	0	0	0	0	0	0	0	0,1	0	0	0,03
14	0	0	0	0	0	0	0	0	0,1	0,1	0	0	0	0	0	0	0	0,01
15	0,1	0,2	0,3	0,2	0,3	0,1	0,2	0,1	0,2	0,1	0,3	0,1	0,2	0,1	0,1	0,2	0,1	0,17

In general the experts' evaluations are consistent for each of the sections described. According to experts the sections included in the first part of the report are of the greatest importance. The greatest value from the point of view of forming the commitment of consumers to the company's products are:

1) section 4 "Disclosure of security information" (0,86);
2) section 2 "Location of offices" (0,84);
3) section 1 "Business characteristics" (0,81).

It is determined that information about safety characteristics is important from the point of view that the consumer can be confident in the high quality of products, the process of obtaining them, as well as the absence of negative consequences of their use in the long term. The high significance of section 2 is due to the consumer's need to get information about the availability of the company's products and possible difficulties in obtaining it from the point of view of geographical and territorial features. As for the first section, based on the information provided here, the consumer forms a complete picture of the product in his mind and makes a conclusion about whether to use it or the products of another company.

The sections on accounting fees and services, operations related to the company's activities, the independence of the Director, and the security of management's property are considered the least valuable for the consumer.

At the second stage of the survey information from these sections was analyzed in the reports of the 10 largest medical companies in the United States (table 2).

Table 2. Expert evaluation of the significance of form 10- K* sections

Characteristics of the information provided	Quest Diagnostics [7]	Laboratory Corp. of America Holdings [8]	Express Scripts Holding Co. [9]	Cerner Corp. [10]	DaVita, Inc. [11]	HCA Holdings [12]	Premier, Inc. [13]	Tenet Healthcare Corp. [14]	Universal Health Services, Inc. [15]	Community Health Systems, Inc. [16]
Completeness of information provided	39H 50C 58B **	40H 63B 60C	45H 38C 51B	30C 45C 54B	35H 30C 45B	30H 45C 60B	45C 40C 59C	39H 30C 48B	46C 44C 59C	34B 40C 47B
Accessibility of the presentation to the mainstream population	46C 55B 60B	40H 50C 61C	39H 45C 50B	35C 50C 50C	39H 30C 50C	34H 49C 58C	40C 40C 41C	41B 39C 30H	37B 39C 54B	42C 40C 43B
Absence of the possibility of dual understanding of information	47C 50C 50B	35C 30C 48C	30C 31H 40C	31C 30C 42C	36C 30C 41C	45C 39C 32H	39C 40C 39C	42C 49C 31C	39C 48C 37C	43C 50C 34B

* for each company, the results of evaluating the characteristics that scored the highest number of points among respondents are given (for example, 39H means that 39 people described the completeness of providing information by Quest Diagnostics as low)
** the first value is the evaluation of section 4, the second - section 2, the third – section 1

Of the three analyzed characteristics the completeness of providing information about companies' activities is the highest. Most often, security information is provided by a separate link that consumers either don't open or have technical difficulties navigating. In their assessments experts stand in full solidarity with the third characteristic for all companies: during the analysis there was no ambiguity in the interpretation of the information received. This aspect is important because the analysis of the reports was carried out by native Russian language speakers. In addition, medical information has certain specifics, which, however, is also easily perceived by experts. We cannot say that the characteristics of the information presented in the report of US medical companies on form 10-K have high evaluations, but a large number of low evaluations are not recorded. In general there are wide opportunities for current and potential consumers to form a proper image of the company to choose one or another as a priority supplier of products.

4 Conclusion

Corporate reporting is an important element in the activities of modern companies. Not only the effectiveness but also the overall success of an activity depends on its qualitative performance . It appears that another function of corporate reporting can be attracting and retaining of potential and current consumers of products. This is especially important for those that operate in the social services. The results of the survey allowed us to determine that the most valuable information for the consumer is information about the safety of the company's activities, existing offices and divisions, including in other regions, as well as characteristics of the company's activities and products provided. The information content analysis of the most important sections for the consumer showed that not all medical companies in the United States have high quality information content. It seems that the presented conclusions can be of practical value for the activities of organizations in terms of organizing activities to attract consumers.

References

1. N. Prodanova, N. Plaskova, L. Popova, I. Maslova, I. Dmitrieva, J. Kharakoz, V. Sitnikova, Journal of Advanced Research in Dynamical and Control Systems, 8, **411** (2019)
2. D. Crowther, Urban Studies, 10, **1837** (2000)
3. *Tenets of good corporate reporting* (2018)
4. M. J. Milne, Accounting & Business Research, 1, **83** (2003)
5. V. Sukhonos, I. Makarenko, Problems and Perspectives in Management, 1, **166** (2017)
6. J. H. Spangenberg, Journal of Cleaner Production, 134, **414** (2016)
7. Quest Diagnostics. Action from Insight. 2019 Annual Report on Form 10-K., https://www.annualreports.com/
8. Laboratory corporation of America holdings. United States securities and exchange commission. Form 10-K, https://ir.labcorp.com/
9. Express Scripts holding company. United States securities and exchange commission. Form 10-K, https://web.archive.org/
10. Cerner Corp. 2019. United States securities and exchange commission. Form 10-K, http://pdf.secdatabase.com/
11. DaVita Inc. 2019. Form 10-K Annual Report. U.S. Securities and Exchange Commission, https://www.sec.gov/
12. HCA Holdings, Inc. 2018. Form 10-K Annual Report. U.S. Securities and Exchange Commission, https://www.sec.gov/
13. Premier, Inc. 2015. Form 10-K Annual Report. U.S. Securities and Exchange Commission, https://www.sec.gov/
14. Tenet Healthcare Corporation. 2017. Form 10-K Annual Report. U.S. Securities and Exchange Commission, https://www.sec.gov/
15. Universal health services, Inc. FORM 10-K. Retrieved 13 May 2020. U.S. Securities and Exchange Commission, https://www.sec.gov/
16. Community Health Systems. FORM 10-K, https://last10k.com/

Searching for mechanism of correlation of interests of the transport holding's managers and employees

Natalia Chevtaeva[1], *Olga* Shestopalova[2], *and Tatyana* Okuneva[2,*]

[1]Ural Institute of Management, branch of the Russian Presidential Academy of National Economy and Public Administration, 8 March str., 66, 620144 Ekaterinburg, Russia
[2]Ural State University of Railway Transport, Kolmogorova str., 66, 620034 Ekaterinburg, Russia

Abstract. Digital transformation is the current development paradigm of any modern organisation. Digitalisation, automation and robotisation actively affect the activities of both large and small enterprises and have an impact on all spheres of life. Studying the correlation of corporate interests of employees and managers of Russian Railways, JSC transport holding allowed us to identify areas of their correlation, as well as areas where we defined their inconsistency. At the same time, this area of interaction between employees and employers requires active research since it does not always fall into the research field. Based on empirical research, analysis of corporate reports on the company's business activities, as well as content analysis of interviews with managers, the authors show that alignment of interest can be achieved by means of systematic development and meeting the needs of staff in training and development; availability of a clear program of the company's development and provision of a wide range of social benefits and guarantees contribute to the achievement of stable position on the market of logistic services, stable work teams, high level of staff involvement in the activities of the company, higher job satisfaction, greater confidence of staff in developing and implementing changes, as well as improving technical, technological and operating processes. On the other hand, the study shows that in the opinion of managers and potential employees (based on the analysis of the generation Z's system of values) the corporate management model of Russian Railways, JSC may create barriers to the development of the company and fulfilment of employees' labour potential. This problem can be avoided by improving corporate management systems and increasing the independence of the staff.

1 Introduction

Dynamic changes that are associated with modern trends of digitalisation, automation and robotisation and determine the transformation of all spheres of life, cannot but affect current relations between employees and employers. Such transformations are especially

*Corresponding author: tat-okuneva@list.ru

reflected in the activities of large organisations where the interests of various stakeholders such as employees, company managers, business owners, partners, and clients, may come into conflict[1]. In general, organisations, including large ones, where these trends are particularly acute due to the need to maintain competitiveness, are interested in meeting contemporary social and economic requirements, implementing technical and technological changes, and developing personnel. At the same time, employees strive to ensure a steady job and a comfortable workplace. The environment contributes to the employee-employer relationship in the form of such global trends as work-life balance, self-fulfilment in the profession, and opportunities to influence the administrative reality. As for the companies, they announce such trends as an increase of involvement, staff planning, social protection of employees, and well-being as part of work [2]. In this situation of conflict between the requirements of the external environment, the needs of employees and the interests of the company, the search for mechanisms for their coordination, through which each of the parties will be able to achieve its goals in constantly changing and unpredictable circumstances, is of particular importance.

2 Literature Review

The common trend of digitalisation, robotisation and automation of work processes, which was declared and supported by world leaders in connection with the announced achievement of Industry 4.0 by advanced economies, is studied in relation to such sectors as industry, finance, and services and has been described by many researchers. [3,4]

They pay a lot of attention to the transport industry [5] and its modernisation in the light of these trends.

At the same time, as part of the study of corporate governance issues, researchers most often consider the alignment of corporate interests of companies and groups of stakeholders, especially partners and clients. [6, 7, 8, 9] The problems of interaction between employees and managers are also studied, but most often they are considered from the position of forming favourable relations for work, changes on the labour market, transformation of employment conditions for employees [10, 11], and creating a social and psychological climate and a comfortable environment. [12]

In our opinion, the issues of correlation of interests of employers and employees have not been sufficiently studied, which allowed us to determine the purpose of the study: to identify areas of correlation and inconsistency of interests of managers and employees in achieving the strategic objectives of the transport holding, as well as to search for mechanisms for their coordination.

This goal allowed us to define the following research problems:

1. To identify the area of correlation of interests of the transport holding company and the main groups of stakeholders, specifically the correlation of interests of employers and employees
2. To identify the area of the inconsistency of corporate interests and suggest possible mechanisms for their coordination taking into account the current trends of transport, industrial, social and economic development.

3 Materials and Methods

The interconnection of the relationship between employees, company, and environment has a systemic and fundamental nature and determines the structure of interaction between these interests during the transformation of technology, as well as the position of the individual as an employee and a member of the social environment. The company's activities in the field of meeting its operational and tactical interests lie in the zone of implementation of adaptive potential while its strategic interests lie in the zone of the company's innovative potential.

The indicators for implementation of adaptive potential are the existence of sound policies for the development and introduction of innovations; monitoring operative, technical, social and economic indicators; wage level, staff stability and turnover; availability of social guarantees; and level of job satisfaction of the company's employees.

Skill improvement, training and staff development can be considered as a way to guarantee jobs under the changing working conditions and adapt the qualification level of personnel in accordance with the requirements of the production, as well as to create opportunities for employees to grow professionally, identify workplace issues and offer solutions for them thereby changing the technical, technological, social and economic conditions and revealing innovative potential.

In order to implement the innovative potential, the company should ensure conditions under which employees can receive timely information about the strategy and tactics of the company's activities, should create opportunities for development and implementation of innovative projects in its activities and support innovators. The presence of such platforms, as well as the desire and active participation of initiative workers in innovative activities, are the indicators of the implementation of the innovative potential.

The object of our research is employees and managers of Russian Railways, JSC in the conditions of technical and technological changes. The purpose of the study is to determine the priorities of the company's development as a leading employer and leader of the transport industry in the Ural region, to consider whether the employees' system of values and needs correspond to the main approaches and principles of the corporate management.

The research methodology involved analysis of the company's corporate social responsibility reports, other publications, analytical materials and reviews of its activities in the corporate publications, as well as interviewing employees of Russian Railways to identify their system of values, attitudes to work, and professional preferences.

Information sampling for the study was done by selecting articles that include interviews with managers and leading specialists of the holding published in The HR-Partner. Vremya. Vyzovy. Resheniya (HR-Partner. Time. Challenges. Solutions) and The Pult Upravleniya (Control Board) magazines, The Gudok (Whistle) and The Uralskaya Magistral (Ural Highway) newspapers from 2014 to 2020 (included 126 publications) and data from the reports of Russian Railways on its activities in the field of sustainable development for the period from 2016 to 2019 [13]

According to the company's corporate reports, we ensured that representatives of various categories of employees (managers, white-collar and blue-collar workers) were present in the sample (Table 1). Since there is no official information regarding the region, we were guided by the ratios that are typical for the company as a whole.

Table 1. Characteristics of the sampled interviewed employees of the transport holding

	Managers, %	White-collar workers, %	Blue-collar workers, %
2016			
Total for the company (N=774.0 thousand people)	7.6	29.1	63.3
Representation in the sample (N=768 people)	7.9	33.4	58.7
2019			
Total for the company (N=743.1 thousand people)	7.7	29.2	63.1
Representation in the sample (N=856 people)	8.7	34.6	56.7

We ensured that employees from different types of cities where the company has its structural divisions were presented in the sample (Table 2).

Table 2. Territorial distribution of respondents by cities of the company's presence

Types of cities in the Sverdlovsk region	Cities of the Sverdlovsk region	Territorial distribution of respondents (people / % of the total number of respondents)	
		2016	2019
Million-plus city	Ekaterinburg	278/ 36.2	318/ 37.1
Major city	Nizhny Tagil	78/10.2	84 / 9.8
	Tyumen	71/9.3	86 / 10.1
	Chelyabinsk	67/8.7	71 / 8.3
Large city	Kamensk-Uralsky	61/ 7.9	81 / 9.5
	Kurgan	58/7.5	58 / 6.8
Medium-sized city	Serov	63/ 8.2	57 / 6.7
	Shadrinsk	52/5.2	54 / 6.3
Small city	Alapaevsk	40/ 6.8	46 / 5.4
Total		768/100%	856 / 100%

4 Results and Discussion

In our opinion, in its activities, the company primarily implements its adaptive potential which allows it to maintain and keep its position and at the same time, take into account the trends of technical, technological, social and economic development. This is the area of relations between employees and the company where the corporate interest correlate.

The company developed and approved the Strategy of Railway Transport Development in the Russian Federation until 2030, Long-term Development Program of RZD, JSC until 2025, and the Strategy of Digital Transformation of RZD, JSC. In particular, the company has defined 11 areas of development that describe technological and technical innovations in the transportation process, customer digital services, etc. These strategies are supplemented by the result-based monitoring system and offer an opportunity to improve the innovation activities. Such projects as New Link, Leader of Changes, and Meeting of Youth of Russian Railways offer competitions the purpose of which is to develop engineering and technical potential and involve young people in solving corporate problems and innovative development. [13, p. 153-155] Such projects help to overcome some of the limitations of the Single Corporate Competencies Model adopted in May 2019, which sets clear but limited requirements for the competencies of employees at various hierarchical levels. *'First, we show what an employee should strive for so that the company could achieve its goals – this is the first step. Then we say how to achieve these goals – this is the second step. And the third step is the key qualities that an employee needs today.' (D. Shakhanov, 2019)*

Such systematic activity results in the company's leading position as a cargo carrier (market share of 87,2%), as well as a socially significant provider of passenger transportation services (they account for 23.4%), although in this segment the primacy remains with air transport (56.6%). The following data show that management directs its attention to mobility and flexibility with a focus on retaining staff: the company has a consistently low level of staff turnover (7.5% in 2016 and 6.7% in 2019), a large per cent of young employees (39.8% under the age of 35 and 39.4% aged 36-45), while more than half of employees have 5-10 years of experience or more. The salary of employees is competitive and indexed annually; the holding is in the 23rd position among 88 main economic activities, and at the end of 2019, the salary of employees of the company was by 22% higher than the average salary in the country. [13]

Russian Railways, JSC implements a number of projects committed to the social welfare of its employees and their families and provision of medical services for employees and passengers, supports sports and culture and implements programs to support working and non-working pensioners, etc. According to the Russian Public Opinion Research Center, Russians named the company among the three most desirable places to work for. As it was accurately noted in one of the interviews, '*If we compare the infrastructure today and, let say, ten years ago, these are two different railways. We managed to achieve high performance and technological results thanks to the stability of our team.' (A. Lesun, 2019).*

A survey of the holding's employees in the Ural region showed a high level of employee satisfaction. 28.4% of employees of the holding estimated their income at the level of "High" and "Very high", the overwhelming number of respondents (76.4 in 2016 and 78.2 in 2019) noted that they liked their current job, and if they had such an opportunity, less than a quarter of the surveyed employees would leave their jobs (22,5 in 2016 and 19,3 in 2019).

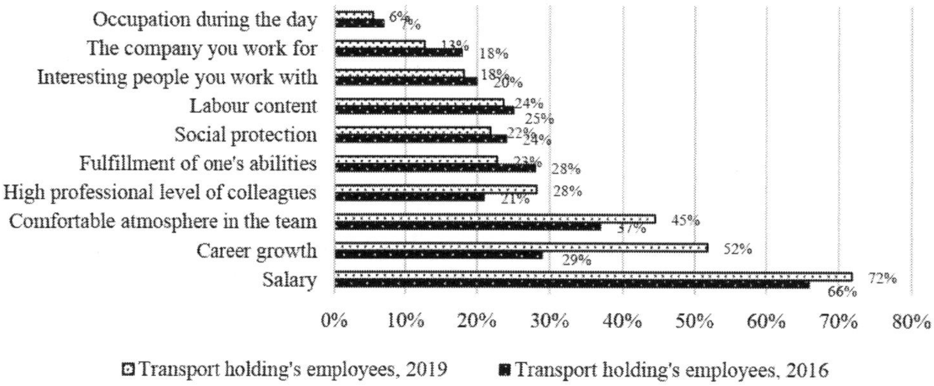

Fig. 1. Distribution of responses to the question "What is most important for you in your work?", % of the number of respondents

In 2019, compared to 2016, the importance of such factors as "Fulfilment of one's abilities", "Social protection", "Labour content" and a number of other factors decreased (Figure 1). We attribute this to the fact that the company adapts not only to the needs and interests of employees, and they do not feel the significance of these factors due to their high level of representation in the company. At the same time, the importance of such factors as "Career growth", "Comfortable atmosphere in the team" and "High professional level of colleagues" increases, which reflects the desire of employees to develop, work among "advanced" colleagues in a comfortable environment. As a result, there is a large number of those who are proud to work for this company. (Figure 2)

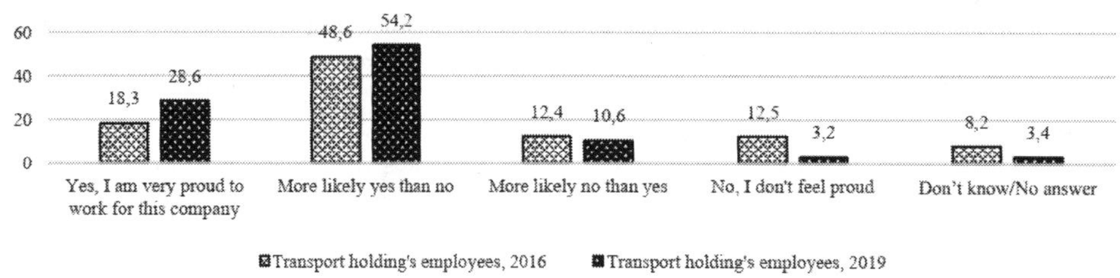

Fig. 2. Distribution of responses to the question "Are you proud to work for this company?", % of the number of respondents

In 2019, more than 86 thousand people were trained in the successfully operating corporate university of RZD; more than 150 educational programs of different categories were implemented; the company successfully implements and upgrades existing educational programmes, and introduces courses for employees to learn digital technology. More than 51.5 thousand workers were educated for leading professions, including 32.7 thousand employees who were trained for a profession for the first time, 6.1 thousand employees who were trained under the professional development programs, and 12.7 thousand employees who were trained for a second profession; more than 175.8 thousand people raised their qualifications by taking courses in training centres, technical schools and colleges. [13].

Special attention should be paid to the Quantum Communications project aimed at the development of the high-tech areas. The roadmap of the project was submitted to the Government of the Russian Federation in May 2020 [14]. By the degree of ambition of the tasks to be solved and on condition of their successful completion, the holding will ensure a leading position in the sphere of *'implementation of quantum communications for the digital transformation of the industrial landscape' (G. Goltsman, 2019),* which means the ability to determine the direction of technical and technological development at the global level. [15]

However, this requires the company to shape a new worldview in order to achieve its goals and mitigate undesirable social consequences. The rhetoric of managers clearly shows a desire to promote changes from the top: *'offer new solutions, initiate new projects and motivate employees to look at work processes in a new way.' (D. Shakhanov, 2020)*. At the same time, managers admit that there is a certain amount of conservatism which *'creates additional barriers for a person who wants to change, is ready to take the initiative, take responsibility, or is focused on making the most rapid decisions that the market requires.' (D. Shakhanov, 2019).*

A rigid system of innovation implementation, which restricts the initiative of those employees who do not fully fit into the format of the proposed events, where they can express their views on improving work and interaction, is the main area of inconsistency of the corporate interests. In particular, this is due to the fact that the functions of the change management lie with the top management and are not even partially performed at the rest levels of management.

In our opinion, it is more strategically correct to build relationships with employees not from the position of hierarchical distribution of roles, but to provide for more flexible management of linear enterprises by delegating more powers to middle-level and lower-level managers, as well as offer an individual approach to career development. Such a step towards managerial freedom opens up the company's prospects for more comprehensive use of the potential of employees and creates conditions for meeting a greater range of employee needs and increasing their satisfaction with the jobs and the company in general. It is important to take into account the features of professional and career paths, as well as the place of professional education in them. [16, 17]. The choice of profession is based on a high level of remuneration and a guarantee of a stable workplace and does not ensure full involvement of employees in the operational processes, as well as reduces the opportunities for the fulfilment of their labour potential.

It is necessary to take into account the system of values of the young generation - representatives of Generation Z - who are becoming and will be the main labour source in the future. Their priorities include autonomy, independence, and openness to change while respect for tradition ranks the bottom place in their hierarchy of values. Taking into account that the management approach of Russian Railways, JSC is focused on hierarchy, it may cause a significant mismatch of interests in the future.

5 Conclusion

In summary, we would like to note that the company maintains a confident position in implementing its interests in the field of operational work with personnel and solving tactical and strategic tasks by holding a leading position in the provisions of logistics services. The company is a leading employer in Russia and systematically implements successful measures that contribute not only to the stability of the staff but also to a high level of employee satisfaction with their work and the company. It is safe to say that Russian Railways, JSC is a socially responsible company which has achieved a high degree of correlation of corporate interests between the main subjects of organisational interaction.

The issues that are well worth another look include the effectiveness of the implemented innovative measures, incentives for participation of employees and, in particular, the degree of formalism in the development and presentation of projects on the part of employees. It is also important to consider the risk of productivity loss as the powers of executive and operative administration employees are extended on condition of their increased responsibility for their activities and performance results of the department, as well as the active motivation of employees to implement innovations.

References

1. A.S. Ogorodov, S.Yu. Saranchuk, N.G. Chevtayeva, RUDN Journal of Sociology, **1**, 83 (2017)
2. International trends in personnel management — 2020, https://www2.deloitte.com/
3. A.I. Pozmogov, Management Issues, **39** (2019)
4. I.V Naumov, The Manager, **5** (2020)
5. V.A. Sadovnichii, G.V. Osipov, A.A. Akaev, A.S. Malkov, S.G. Shulgin., Mathematical Simulation and Forecast, **14** (2018)
6. J. Fernández-Guadaño, J.H. Sarria-Pedroza , Open Access Journal, **6** (2018)
7. A.Hein, M.Jankovic, W.Feng, R.Farel, J.Yune, B.Yannou. Journal of Cleaner Production, **148** (2017)
8. I.N. Tkachenko, I.V. Pervukhina, A.A. Zlygostev, The Manager, **2** (2020)
9. A.S. Vatoropin, N.G. Chevtaeva, S.A. Vatoropin, Management Issues, **4** (2017)
10. N.G. Chevtayeva, A.S. Vatoropin, *The image of the future 2019*, Materials of the scientific conference of the XIII Kovalev readings (2019)
11. I.A. Koch, V.A. Orlov, Management issues, **1** (2020)
12. A. Fedorova, 10th International Days of Statistics and Economics (2016)
13. Report on activities in the field of sustainable development (2019), https://company.rzd.ru/
14. A new round of development of quantum communications at Russian Railways: the program was sent to the government, https://digitalrzd.ru/
15. Quantum communications: what is it and why are they Russian Railways need it?, https://www.rzd-partner.ru/
16. O.N. Shestopalova, T.V. Okuneva, Herald of the Ural State University of Railway, **30** (2016)
17. T.V. Okuneva, O.N. Shestopalova, Herald of the Ural State University of Railway, **43** (2019)

Digital technologies and their impact on the innovative development of the region

Tamara Egorovna Rodina*

Bryansk state engineering-technological University, Lenin Str., 26a, 241000 Bryansk, Russia

Abstract. The article discusses the processes of digitalization of the economy at the regional level. The subject of the research is the work of the region to find opportunities for implementing digital technologies in various fields of activity. Research methods-analysis and synthesis, tabular, historical past. The significance of the introduction of digital technologies as a factor of self-development of the region is proved. The article substantiates the need to create conditions for improving the quality of education through the development of online courses in the framework of the project "Modern digital educational environment". The practical significance is to identify the problems and prospects of testing digital technologies in the development of the region's economy.

1 Introduction

The development of the digital economy is a priority and one of the most significant trends in the modern world. In Russia, the term "digital economy" received an official state definition in 2017, which is reflected in the country's information society development Strategy. First of all, the digital economy is considered from the point of view of forming new markets based on the use of ICT, which will strengthen the growth of the Russian economy on the basis of high-tech industries and industries, available opportunities.

At the present stage of development of the Russian Federation, issues of economic and financial security of regions are becoming a priority [1]. The Bryansk region has sufficient strategic opportunities for development in the context of digitalization. Industry, agriculture, transport and telecommunications are well developed in the region. Products of regional enterprises are in high demand both in the region and abroad.

The Bryansk region is actively working to implement the main platforms of the Federal project "Digital economy of the Russian Federation".

The main customer of such events is the Ministry of digital development, communications and mass communications of the Russian Federation. On the territory of the Bryansk region, the implementation of the idea is determined to be carried out by PJSC Rostelecom.

The government of the Bryansk region, in accordance with the powers assigned by the Ministry, provides comprehensive support to the regional Telecom operator.

The regional project is of great social significance for residents of localities, since the

*Corresponding author: ribnikoff@yandex.ru

https://doi.org/10.1051/shsconf/20208903001

availability of communications will ensure the development of the network for individuals and legal entities in the near future.

The result of this work is the connection to the network of the school and FAPA in the village of Novoe Kaplino, Zhiryatinsky district, together with the Bryansk branch of PJSC Rostelecom.

High information technologies allow changing business processes in other sectors of the economy [2, 3]. According to official data, the labor market is experiencing a shortage of employees with it skills. Companies ' need for it specialists is growing rapidly.

In this regard, the introduction of digital educational platforms is an important and urgent task. Online education can become both basic and additional, this is due to the fact that it involves a convenient, flexible schedule and the ability to take courses from anywhere in the world at any time of the day.

Whether all these trends are advantages or disadvantages for online education is up to each student to decide for themselves, but it is clear that science and online education are significantly ahead of classical education in terms of technology.

High information technologies allow changing business processes in other sectors of the economy [4, 5]. The labor market is experiencing a shortage of employees with it skills. Companies ' need for it specialists is growing rapidly.

Statistics show that the share of relevant ads on HeadHunter increased by 6% between 2016 and 2018 alone. However, demand outstrips supply. According to the estimates of analysts, to meet the needs of the market in it personnel, it is necessary to increase the recruitment and graduation of students of the relevant specialties by at least 2.5 times. This trend is also confirmed by the goals of national projects. In accordance with the task, by 2024, universities should offer the market 120 thousand rubles. graduates in the field of it, the proportion of the population with digital skills should be 40%.

In this regard, the introduction of digital educational platforms is an important and urgent task. This area of work includes not only the process of optimizing the educational process, but also a lot of work on the development of digital platforms [6]. Today, teenagers are more adapted to the introduction of new modern technologies, at the same time, their parents and teachers have to master electronic services. A qualitative example is the exercises in the electronic «Yandex.Textbook». Tasks are set by 37 thousand teachers and solved by 600 thousand students. Indeed, digital tools can make learning more effective, allowing you to discover the abilities of each individual child.

In addition, the experience of acquiring knowledge in the online system shows children that the computer is not only cartoons and games, but also an opportunity to learn.

It companies also help solve the personnel problem to a certain extent. They open laboratories, basic departments and entire faculties in higher education institutions, which helps to reduce the gap between theory and practice. School educational projects of specific companies are also being implemented, in particular, for example, "Yandex.Lyceum", where children are taught to program in Python.

Of course, a lot of work is being done in various areas. However, even with the assistance of representatives of educational institutions, the needs of the digital economy will not be fully met soon. The educational system is quite conservative, and this is a normal phenomenon.

In comparison with other countries of the world, in our country, employers pay much less attention to the higher education of applicants, for example, than in America. Young professionals in employment rarely limit their specialization in search of a job. According to VTsIOM, about 48% of Russian specialists work outside their specialty.

2 Materials and Methods

According to the materials of the Federal project "Personnel for the digital economy", which is an integral part of the national program "Digital economy", the total cost of implementing the relevant measures in 2019-2021 will amount to 50.7 billion rubles.a significant part – over 46 billion rubles – will be provided from the Federal budget. Most of the activities included in this Federal project will be fully funded from the Federal budget [7].

In most cases, the curators of these events will be the Ministry of economic development, the Ministry of education or the Ministry of Education.

According to calculations, the most expensive event will be the creation of a network of international scientific and methodological centers aimed at spreading the best international practices in training personnel in the field of the digital economy.

7.4 billion rubles will be allocated from the Federal budget for the implementation of this direction. The performers will be the Global universities Association and the Association of leading Russian universities.

The Federal budget will allocate 6.2 billion rubles to create a network of centers based on higher education institutions aimed at developing "Digital University" models using digital technologies.

The costs of the Federal project "Personnel for the digital economy" of the national program "Digital economy" for the implementation of individual tasks for 2019-2021 are shown in table 1.

Table 1. Expenditures of the Federal project "Personnel for the digital economy" of the national program "Digital economy" for the implementation of certain tasks for 2019-2021, million rubles.

Areas of activity	Total cost	Sources		Cost		
		federal budget	extra-budgetary funds	2019	2020	2021
Creating conditions for the support and development of talented youth	22528	22528	-	3755	6615	12158
Creating conditions in the education system for training competent specialists	7802	3402	4400	1022	2805	3975
Introduction of a system of additional professional education for citizens	5944	5929	15	1615	2199	2130
Creating conditions for the implementation of leading educational projects	5932	5932	-	1500	1500	2932
Providing assistance to citizens, including older people, in mastering the competencies of the digital economy	3034	3019	15	478	454	2102
Implementation of the requirements of key competencies	3000	2960	40	1195	340	1 465
Development and testing of personal profiles of citizens ' competencies	874	874	-	501	361	12
Development and testing of the concept of competencies for effective interaction between business, education and society	644	644	-	325	244	75
Providing motivation for highly qualified foreign citizens	458	458	-	244	214	0
Ensuring the functioning of the competence center for solving Federal project tasks	154	154	-	50	49	55

Continuation of Table 1. Expenditures of the Federal project "Personnel for the digital economy" of the national program "Digital economy" for the implementation of certain tasks for 2019-2021, million rubles.

Enabling the use of competence profiles and personal development trajectories	142	142	-	112	30	0
Development and testing of a mechanism for independent assessment of citizens ' competencies in the education system and labor market	105	105	-	45	60	0
Implementation of digital economy tasks in the qualification requirements of employees	85	85	-	19	32	34

Considerable attention is paid to the implementation of the national program "Digital economy" at the regional level.

A digital economy cluster has been created in the Bryansk region to implement the state policy and operation of the region's information technology companies. The agreement on its creation was signed between the Bryansk regional business incubator and the enterprises of the new structure.

The cluster is aimed at implementing complex projects to find partners and clients among the members of the Association. It provides well-coordinated work aimed at developing the region's digital economy and small and medium-sized businesses.

3 Results and Discussion

Everyone interested in implementing complex and interesting technologies is interested in developing a digital economy cluster in the Bryansk region [8]. They are representatives of both small and medium-sized businesses, as well as large companies and universities in the region. The cluster currently has ten companies in the region: industrial structures, high-tech enterprises and organizations, and educational institutions.

For students at cluster enterprises, it is necessary to organize an industrial practice. Educational activities of the cluster are designed more for students.

Creating conditions for the sustainable development of the digital economy cluster in the Bryansk region requires effective interaction between participating companies. It is necessary to promote effective interaction of participants in the field of web development. The main representatives of the cluster in the field of web development-IP Ananyin A.V., IP Malygin N. O. when implementing the strategy should interact with each other to solve common problems of market promotion and reaching larger customers.

A virtual brand is an Association of web developers of cluster companies under a common name that allows them to be quoted in industry ratings and receive larger orders.

Commercial projects - the launch of joint Internet projects (online stores, services), in which the profit of web developers appears as a result of the commercialization of Internet projects, and not as a payment for development.

Cluster companies can rely on the use of two existing trends to advance to the Russian market:

- import substitution - the policy of replacing foreign software systems, electronic components and equipment used in the management of technological processes in significant sectors of the economy;

- investment in agriculture-the trend of investment growth against the background of import restrictions, which increases the size of the market for automation of agricultural facilities and creates opportunities for cluster participants.

It is planned to implement several ways of promotion: comprehensive proposals for the formation of commercial offers to large customers by combining the competencies of several web developers in the cluster.

The second direction is Internet marketing: development and support of advertising companies on the Internet. Yandex.Direct and Google Adwords. Configuring CRM systems

Creating a Cluster means increasing tax revenues to the budget of the Bryansk region, expanding the number of jobs at small and medium-sized businesses.

Office technologies LLC has the strongest competencies and market positions in the digital industry. Thousands of people have become users of the company over 10 years of operation on the market. Automation systems developed by Office technologies LLC allow you to automate any organization:

- large multi-branch organizations;
- small and medium-sized enterprises;
- unified e-health system in the region.

The products of Office technologies LLC are successfully used for automation of more than 100 corporate clients - public and private medical organizations, public and private pharmaceutical networks and pharmacies, educational institutions:

- specialists of Office technologies LLC and Rusbitech-Astra LLC conducted joint test tests for compatibility and correct operation of computer software products;
- Office technologies LLC, in partnership with Rostelecom, is implementing a project to automate the University clinic at Oryol state University, etc.

The implementation of management systems increases the efficiency of personnel at any level-from ordinary employees to managers.

In order to intensively implement digital technologies and platform solutions in the agricultural sector of Russia, the Ministry of agriculture is implementing the departmental project "Digital agriculture", designed for 2019-2024.

During the course of the year, the Ministry of agriculture of Russia plans to launch the smart contract concept to formalize contracts with grant recipients during the implementation of the plan for 2019-2024. It is planned to introduce domestic integrated digital agricultural solutions everywhere. Among them, the most popular are "Smart greenhouse", "Smart farm", "Smart field", "Smart herd", "Smart agrooffice". Previously, more than 35 regions of the Russian Federation actively participate in the Ministry of agriculture's Digital agriculture project.

Thus, the availability of such software products is popular among the society and effectively affects the performance of functions by public authorities [9].

Innovative technologies for the development of the agricultural sector suggest that information and communication technologies will be used [10].

Two platforms can become key and complementary for digital ecosystems in the field of agricultural business: "Technologies of the food and processing industry" and "Eurasian agricultural technology platform". In addition, other complementary technologies can be applied - "Space and geoinformation technologies-products of global competitiveness" and others.

Today, some Russian agricultural complexes successfully use the latest technological processes of farming in animal husbandry and crop production. In the Bryansk region, for example, the latest technologies in vegetable growing are being successfully introduced [11]. They were discussed at the international agricultural exhibition "day of the Bryansk field", at the Slavic international economic forum.

The Moscow region applies the reconstruction of farms with the introduction of the latest technologies for keeping, feeding and milking animals, as well as the introduction of energy-saving agriculture. Thanks to high technologies, it became possible to freely maintain housing for cattle. Individual farms (for example, enterprises of the Krasnodar territory, the Republic of Tatarstan, Belgorod, Rostov, Kursk and other regions) have significantly succeeded in the so-called arable farming, having built livestock complexes of the highest level.

In 2018, the law on organic agriculture was adopted, which provides farmers with excellent opportunities not only to work in the domestic market, but also to win over foreign consumers.

It is advisable to form the budgets of cluster companies taking into account their orientation (and differentiation) for the purposes of operating activities and project financing. The main principle of promoting the development of clusters on the part of the authorities is trust in the decisions made by the cluster participants both in terms of managing its development and in justifying certain projects.

The operational activities of cluster companies in conditions of a lack of trust between their participants should be guaranteed by the allocation of budget funds.

Creating a platform for fruitful interaction between cluster members and forming trusting relationships between them is a basic condition for intra-cluster cooperation, which determines priorities and the sequence of actions.

The main results of the cluster development strategy implementation are shown in table 16 and relate to possible growth scenarios for the main activities of the cluster member companies.

Three scenario scenarios for cluster development with the following average annual growth rates are proposed: conservative - less than 10 %; moderate - up to 20 %; target-up to 40 % (Table 2).

Table 2. Main effects of implementing the cluster strategy.

Indicator	**Actual value**	**The value of the indicator for the years 2019-2022**		
		Conservative scenario	**Moderate scenario**	**Target scenario**
The number of personnel involved in the field of web development, people	10	11	12	14
Number of personnel employed in the field of automated process control, people	66	72	79	92
Number of employees engaged in business programming and information technology infrastructure, people	84	92	100	117
The total number of staff employed in the companies of the Cluster, people	160	175	191	223

The basis for calculations is the number of employees involved in the main areas of cluster activity. The output per employee is calculated at the rate of 1 million rubles / person per year, which includes salary with deductions, the company's costs for organizing the workplace, purchasing equipment, overhead costs, taxes, and enterprise profit.

The implementation of the functional strategy for the development of the digital economy cluster in the Bryansk region is aimed at the following activities:

1. Development and coordination of the current development plan (directions).
2. Implementation of communication functions.
3. Defining the format of management and participation in joint projects.
4. Formation of budgets of cluster companies, taking into account the differentiation for the purposes of operating activities and project financing.

4 Conclusions

Digitalization of business processes will enable Russian businesses to move to a new stage of their development and provide them with long-term competitive advantages [12]. In addition, activities aimed at large-scale practical implementation of digital technologies and acceleration of modernization, development of domestic practical experience in priority sectors of the economy have been intensified [13].

1. Analysis of the activities of regional IT companies has shown great potential in the information and communication sector of the Bryansk region. Both large companies and small and medium-sized businesses that develop software operate in the region, but the Bryansk region is characterized by insufficient development of information networks. The development of the information and communication technologies sector is hindered by low demand for products and services from the state, organizations and the population. In particular, the demand for providing state and municipal services in electronic form is highly estimated by regional experts, and this is the second most important opportunity after increasing the availability of mobile communications and the Internet in the Bryansk region.

2. A significant part of the identified problems can be solved on the basis of state support, implementation of effective IT projects, expansion of the information field about the demand for specific products on the national market, and reduction of administrative barriers. In the business sector, it is also important to build up knowledge and experience.

3. It is necessary to create conditions that will ensure sustainable economic growth, create high-performance jobs and accumulate opportunities for the development and implementation of digital transformation projects. To this end, a program for creating and developing a digital economy cluster has been developed.

4. The digital economy cluster of the Bryansk region is in the initial stage of development. Information interaction between the participants has been established, but its coverage is insufficient, and constructive cooperation with the regional authorities has been established. There is practically no necessary link for the cluster-the scientific base. Interaction with educational institutions is developing quite actively. The existing cluster chains are still few and the principles of intra-cluster cooperation need to be streamlined, and cluster management needs to be developed.

The project will allow the cluster to the digital economy of the Bryansk region to obtain synergetic effect from the cooperation of the involved enterprises, reflected in the economic growth of cluster members and increase the efficiency of their functioning, obtaining economic benefits to all participants, improve the social climate in the region, as well as receiving the Bryansk region budget effect due to the growth of the tax base.

Acknowledgements

The article has been prepared as part of the research work of the Bryansk state University of engineering and technology, contract No. 171 of 25.11.2019.

References

1. M. A. Sabelnikova, G. I. Abdrakhmanova, L. M. Gokhberg, O. Yu. Dudorova et al., *Information society: main characteristics of the subjects of the Russian Federation* (2018)
2. G. Cainelli, V. Marchi, R. Grandinetti, Economics of innovation and new technology, 48 (2020)

3. M. Faravelli, K. Kalayci, C. Pimienta, Experimental economics, 468 (2020)
4. K. Williams, Economia, 404 (2018)
5. O. Afonso, P.C. Neves, T. Pinto, Economic systems (2020)
6. G.I. Abdrakhmanova, K.O. Vishnevsky, L.M. Gokhberg et al.; Indicators of the digital economy (2019)
7. V.L. Abashkin, S.V. Artemov, E.A. Islankina, etc.; *Cluster policy: achieving global competitiveness* (2017)
8. Website of the Bryansk region administration, http://www.bryanskobl.ru/
9. S.Y. Glazyev, Great and the digital economy, http://www.nlr.ru/
10. A. Chursin, A. Tyulin, Springer International Publishing (2018)
11. E. Islankina, T. Thurner, Entrepreneurship & Regional Development (2018)
12. On the strategy for the development of the information society in the Russian Federation for 2017-2030: decree of the President of the Russian Federation, http://www.consultant.ru/
13. A. Jacqueminet, L. Trabelsi, Advances in strategic management, 169 (2018)

Impact of digitalization on the evolution of internal corporate institutions in the northern resource-extraction region

Vadim Faruarovich Islamutdinov*

Yugra State University, st. Chekhov, 16, 628011 Khanty-Mansiysk, Russian Federation

Abstract. The article examines domestic and foreign sources that study the impact of digitalization on the evolution of intracorporate institutions. Then the state of internal corporate institutions in the northern resource-extracting region - KhMAO-Yugra is considered. The author's features of the classification of intracorporate institutions are proposed. Institutional traps of intracorporate institutions are identified and described, both all-Russian: the trap of the authoritarianism of the head and the trap of information asymmetry, and specific to Yugra KhMAD: the trap of bureaucracy, the trap of incentive payments and the trap of low unemployment. The influence of digitalization on the evolution of internal corporate institutions in the northern resource-extraction region is shown, including on overcoming old institutional traps and on the emergence of the new ones: the trap of ease of interaction and the trap of trust in the algorithms. The impact of the coronavirus pandemic on the pace of digitalization of internal corporate institutions in -Yugra KhMAD, which lags behind the global ones, was especially noted.

1 Introduction

The development of digital technologies has an increasing impact on all aspects of economic activity, including the evolution of intracorporate (intrafirm) institutions. On the one hand, these processes are proceeding within the framework of the general trend towards digitalization, on the other hand, intracorporate institutions are a more hidden field of study from the view of scientists, and a number of processes are almost asymptomatic. The evolution of internal corporate institutions accelerated especially strongly during the coronavirus pandemic and the massive transfer of employees to a remote format. This, in turn, revealed the contradictions and opportunities hidden up to this point in labor collectives, when it turned out that a significant part of the work can be quite acceptable to be performed remotely or replaced by information systems, and some of the employees, having gone to the "remote work", are no longer ready to return to the office for offline work.

In addition, the specificity of the northern resource-extraction region leads to additional features of the evolution of internal corporate institutions under the influence of

* Corresponding author: isvad74@gmail.com

digitalization. In particular, the remote nature of work in the context of the coronavirus pandemic forced many employees of large and small companies in Yugra KhMAD to think about the objective need to physically be in difficult natural and climatic conditions. Also, "remote" experience has increased the self-esteem of many employees and now allows them to apply for positions in large federal and international companies. At the same time, the companies themselves have greatly revised their needs for employees and their requirements for them, especially in terms of physical presence on the territory of Yugra As a result, we are already seeing a decrease in the number of vacancies and an increase in unemployment in Yugra.

In article [1], we analyzed the qualitative change in the institutional environment under the influence of the spread of new digital technologies, such as artificial intelligence, digital platforms, blockchain, and the Internet of Things. The positive effects, as well as the threats and risks of digitalization of the economy, have been identified. It was revealed that digital technologies significantly reduce transaction costs and thereby destroy existing institutional traps. Extrapolation of this trend into the future has shown that the "death of institutions" does not threaten us, but at the same time the effect of the Coase theorem should be fully manifested. However, existing institutions are under pressure from digital technologies, transforming or dying. At the same time, there are areas of an institutional vacuum, when new digital technologies remain insufficiently institutionalized for quite a long time, and their capabilities are not fully used. It is shown that the influence of digital technologies on the institutional environment can be direct and indirect. At the same time, the essence of the qualitative changes taking place in the institutional environment under the influence of digitalization can be summarized in the following options: digitalization (algorithmization) of the institutions themselves, increasing the efficiency of existing institutions, destruction of existing institutional traps and the emergence of new ones, the emergence of an institutional vacuum, destruction and decrease in the effectiveness of existing institutions (structures, agreements).

As an example of domestic complex work on the study of the impact of digitalization on internal corporate institutions, you can take the dissertation of Yu.I. Gribanov. [2], where he examined the formation of key competencies of personnel in the process of digital transformation and digitalization of the economy, and also studied the business model based on the digital platform and its role in digital transformation, and based on this substantiated the project mechanism for the digital transformation of enterprises with the participation of a service integrator.

Another Russian author [3] examined how digital economy technologies are changing the living conditions and economic behavior of people, focusing on business behavior and new opportunities in the business environment. He also analyzed organizational forms and new methods of doing business in the context of digital transformation and digital economy.

In his article [4], Parshin A.V. considered the possibility of developing elements of the institutional system of a firm in the context of accumulating corporate knowledge of an explicit and implicit form. He especially drew attention to the problems of implementing the in-house training institute, as well as options for overcoming contradictions in the formation of information and cognitive spaces of enterprises.

T.V. Kramin considered intra-firm institutions as an element of company value management. At the same time, he notes, "that innovations in the information sphere give companies the opportunity to reduce transaction costs of agency relations, as well as increase the transparency of relations with partners and investors. Consequently, the role of in-house institutions is growing even more "[5, p. 2].

Among the foreign studies, it is worth noting the article [6], in which the authors studied how digital communication tools are used for internal communication in transnational corporations (TNCs). In particular, the study illustrates the role of digital channels in internal communication, the benefits they bring, and the difficulties associated with using them. Based on this, the authors conclude that digital tools can facilitate internal communications in TNCs, although some problems may arise in connection with the planning of their use.

Another group of foreign researchers in their article [7] notes that the increase in digitalization has influenced various activities, including the business models of companies, allowing the creation of various new forms of cooperation between the companies and leading to the emergence of new products and services, as well as new forms of relationships of the companies with the customers and the employees. To assess the impact of digitalization, they collected quality empirical data from 12 key respondents across two different industries: media and automotive. Research has been conducted to examine the differences and similarities between how digitalization affects value creation, supply, and company takeover, and how firms cope with the challenges posed by increasing digitalization.

Urbach N. and Röglinger M. in the chapter of the monograph [8] talked about best practices and lessons learned from organizations that have succeeded in solving problems and seizing the opportunities of the digital economy. They showed how 21 organizations used their capabilities to create disruptive innovations, develop digital business models, and digitally transform themselves. These cases relate to different industries and countries and cover many aspects that digitalization can have, including in terms of the evolution of internal corporate institutions.

The authors of this work [9] analyzed the impact of technology development on organizational behavior. The most important digital technologies were identified and an analysis of their impact on employees and management was carried out.

The purpose of the study [10] was to identify and demonstrate the key factors in the development of strategic business management in light of the recent waves of digital transformation faced by companies around the world. The above study showed that digital transformation is a worldwide phenomenon; however, regional impact varies significantly depending on the regulatory framework and the level of economic complexity. In fact, the disruptive nature of this trend can undermine the foundations of industries, requiring firms to fundamentally change their approach to strategic management.

A review of the references showed that despite a fairly wide range of studies on the impact of informatization on intracorporate institutions, the issues of the formation and destruction of institutional traps, as well as the specifics of intracorporate institutions in the northern resource-extraction region, have not been sufficiently studied, which determines the relevance of this study.

2 Materials and methods

2.1 The state of internal corporate institutions in Yugra KhMAD

Intracorporate institutions are inherently a kind of relational routines, since they regulate the relationship between the employees of the firm and their main function is to reduce the transaction costs of interaction within the corporation by coordinating actions, smoothing out conflicts, engaging in joint projects, etc. Internal corporate institutions can be classified according to the following criteria:

1. By the degree of formalization: formal and informal
2. By functions performed: coordinating, stimulating, smoothing out the conflicts

3. By the degree of digitalization: offline and online.

Formal internal corporate institutions include:

- internal rules of conduct;
- regulations on compensation;
- provision on incentive payments;
- regulations on corporate culture;

Informal intracorporate institutions function on the basis of interest groups, which can technically be formalized in the form of groups in social networks and various messengers. Informal intra-firm institutions include as follows:

- informal leadership;
- rules for the organization of corporate events;
- the rules of the "behind-the-scenes" struggle for the resources and the attention of the leader.

The analysis of intracorporate institutions in Yugra KhMAD showed that, with rare exceptions, they are very poorly formalized. Many local companies do not have local corporate culture regulations. Basically, issues of internal corporate culture and rules are monitored only in those organizations that are subdivisions of larger companies: vertically integrated oil companies, federal networks, etc. Also, certain rudiments of intracorporate culture can be observed in regional institutions, but there they are mainly limited to the adoption of local regulations, and no one really monitors the compliance with the rules and regulations prescribed therein.

2.2 Description of institutional traps of intracorporate institutions of Yugra KhMAD

Internal corporate institutions in Yugra KhMAD are subject to approximately the same traps as in Russia as a whole, although there are certain specificities associated with the socio-economic specifics of the region.

All-Russian institutional traps of internal corporate institutions:

1. The leader's authoritarian trap

The essence of this trap lies in the fact that the tendency towards paternalism and passivity inherent in Russian employees leads to the fact that the manager has to constantly turn on the administrative resource in order to achieve any results from his subordinates, and for this he has to arrange periodic “hassles” for them. As a result, subordinates get used to the fact that initiative is punishable, that the leader will do it his way anyway, which means that there is no point in taking the initiative, but you just need to get the solution out of the leader and execute it.

2. Information asymmetry trap

This trap follows from the previous one and is manifested in the fact that the tougher the leader behaves towards his subordinates, the less information about their activities the subordinates try to disclose to the leader. As a result, the manager receives less and less feedback, loses understanding of the state of affairs and begins to face sudden blockages and rush jobs. And attempts to establish a network of informers lead to the fact that they become information brokers and filter information in their own interests, using it in the behind-the-scenes struggle.

Specific institutional traps that are typical for firms in Yugra KhMAD to a greater extent than for other regions are:

1. The bureaucracy trap

Bureaucratization is inherent in all large corporations, however, subsidiaries of vertically integrated oil companies (VICs), headquartered in Moscow or St. Petersburg, are very common in Yugra KhMAD. As a result, the solution of many rather small and operational issues is delayed due to the need for agreement "at the top". Moreover, this practice has spread to other sectors of the economy as federal structures take over local companies. In particular, this has already happened in the banking sector, in the pharmaceutical industry, in telecommunications, in retail, etc.

2. *Incentive payouts trap*

This trap is very specific, and it is connected with the fact that the labor resources of Yugra KhMAD were formed and continue to form from among the most mobile and active residents of other regions of Russia and the Near Abroad. And most of them came to Yugra KhMAD for a "pretty penny", although now, of course, the difference in the level of wages is not so big. As a result, corporate executives have to set high salaries and even higher incentive payments to attract high-quality employees. However, from some point on, the increase in wages begins to be taken for granted, and does not lead to a comparable increase in the productivity and efficiency of workers.

3. *The trap of low unemployment.*

Due to the fact that the population of Yugra KhMAD is highly mobile, at the slightest deterioration in its socio-economic situation, or at the slightest opportunity to find a job with a comparable salary on the "mainland", highly qualified specialists "break down" and leave very quickly. As a result, there is almost always a shortage of personnel and very often the "cool" positions are forced to hire employees who either do not have sufficient qualifications or do not have sufficient experience. That is, there is an unnecessarily fast career advancement.

3 Results and discussion

3.1 Modification of internal corporate institutions under the influence of digitalization

Digitalization has a significant impact on internal corporate institutions, since itself significantly reduces transaction costs, of course, after the costs of its implementation have been overcome.

The impact of digitalization on internal corporate institutions is manifested in the following:

- large formalization and algorithmicization, that is, many norms and rules need to be detailed more carefully in order for them to work within information systems, and these norms and rules should be set forth in the form of consistent algorithms.
- increase in the amount of information, incl. information noise, that is, the amount of both useful and useless information increases.
- reduction of informal communication, because on the one hand, there is less time for informal communication due to the increase in the amount of information; on the other hand, formal interactions penetrate into those channels that were previously used for informal communication and replace it.

Digitalization makes it possible to overcome some of the institutional traps of intracorporate institutions. For example, the trap of information asymmetry can be overcome by increasing the amount of information and partial anonymity of participants in the discussion of a particular problem in social networks and instant messengers. As a result, the signs of a problem are more likely to appear than in the absence of information technology.

The trap of bureaucratization is weakened by the transfer of approval processes to electronic mode, thanks to which it is immediately clear at what of the approval stages the document is "stuck" and which of the employees is slowing down this process.

The trap of incentive payments can be smoothed out due to the algorithmization of the payroll process, when the decision on the amount of incentive payments is made based on the achievement or non-achievement of certain objective indicators, i.e. through the introduction of a "smart contract" [3, p. 42].

The low unemployment trap can also be partially overcome by actively engaging the remote workers.

At the same time, informatization leads to the emergence of new institutional traps of intracorporate institutions:

1. *The ease of interaction trap*

This trap is due to the fact that thanks to information technology, it has become very easy and simple to organize interaction with any employee, both in your own company and in other organizations. However, since the need for interaction can find a person in the most unforeseen situations (driving, in a restaurant, at night, etc.), he can answer "automatically" without thinking too much about the essence of the problem.

2. *Trap of confidence in the algorithms*

Here the problem is somewhat different and is connected with the fact that with the maximum formalization and algorithmization of intracorporate institutions, people begin to trust them too much. For example, the problem of copy-paste in documents, when the text wanders from one document to another without making any changes. But more serious situations arise where algorithms replace a person when performing monotonous work: sending letters, calculating salaries, etc. Since no algorithm can take into account all possible options, this leads to incidents and unpleasant situations. Also, employees who are well versed in the mechanism of the algorithm can artificially "wind up" their indicators and receive unjustified additional incentive payments.

3.2 Impact of the coronavirus pandemic on the pace of digitalization of internal corporate institutions

The coronavirus pandemic has unambiguously accelerated the processes of digitalization of internal corporate institutions, which had already been going on before, but not at such a pace.

The most obvious manifestation of the acceleration of digitalization of intracorporate institutions was the widespread transfer of employees to work remotely. Firms en masse began to provide the employees with the equipment and software necessary for full-fledged remote work: corporate mail programs (if had not existed before), equipment (laptops, webcams) and software for video conferencing and remote electronic document management, electronic digital signatures, etc. Naturally, it also required retraining of workers, especially the older ones.

Less noticeable is the acceleration of the processes of replacing the workers with robots and algorithms. Even before the pandemic, Sberbank began replacing the lawyers and call center employees with bots, and now other companies have picked up this trend. However, the mass introduction of robots in physical form has not yet taken place in Yugra, apparently shift workers, even taking into account the additional costs of providing the sanitary standards and sick leave, cost oil and gas companies cheaper than quadrocopters, drones and robotic dogs.

A new trend in digitalization, which manifested itself precisely during the pandemic, was the growing interest in contactless methods of interaction and transfer of objects. For example, many companies began to install non-contact temperature measurement systems, and there was also interest in installing the postamats.

4 Conclusions

Thus, it should be noted that the impact of digitalization on the evolution of internal corporate institutions in the northern resource-extraction region is ambiguous. On the one hand, some internal corporate institutions have become significantly more efficient thanks to digitalization, which helps to overcome a number of old institutional traps. On the other hand, digitalization leads to the emergence of new problems and new institutional traps.

At the same time, in the northern resource-extraction region, the evolution of intracorporate institutions proceeds within the framework of all-Russian trends, but has its own specifics, which is also influenced by digitalization.

The coronavirus pandemic has had an impact in the sense that the pace of digitalization in general has accelerated, but in Yugra KhMAD they still lag behind the global trends.

Acknowledgments

Participation in the conference and publication was funded by the Science Foundation of the Yugorsk State University, grant No. 17-02-07/57 dated February 14, 2020.

References

1. V. F. Islamutdinov, Journal of Institutional Studies, **12(3)**, 142 (2020)
2. Yu. I. Gribanov, *Digital transformation of socio-economic systems based on the development of the institution of service integration*, Doctoral thesis, 355 (2019)
3. M.V. Safronchuk, Digital Economy, **2(3)**, 38 (2018)
4. A.V. Parshin, Social policy and sociology, **1**, 179 (2019)
5. T.V. Kramin, *Intrafirm institutions as an element of company value management*, https://www.researchgate.net/
6. H. Lipiäinen, K. Karjaluoto, M. Nevalainen, Corporate Communications: An International Journal, **19(3)**, 275 (2014)
7. M. Rachinger, R. Rauter, Ch. Müller, W. Vorraber, E. Schirgi, Journal of Manufacturing Technology Management (2018)
8. N. Urbach, M. Röglinger, Introduction to Digitalization Cases: How Organizations Rethink Their Business for the Digital Age, Chapter (2019)
9. B. Grab, M. Olaru, R.M. Gavril, The impact of digital transformation on strategic business management, https://www.researchgate.net/
10. U.S. Foerster-Metz, K. Marquardt, N. Golowko, A. Kompalla, Ch. Hell, Journal of EU Research in Business , 340873 (2018)

Influence of digitalization upon formation of corporate strategy and new business models of modern organizations

Raisya Akmaeva[1,*], *Ravil* Arykbaev[1], *Nelli* Epifanova[2], and *Edvard* Glinchevskiy[1]

[1]Astrakhan State University, Tatischev Str., 20a, 414056 Astrakhan, Russia
[2]Caspian Pipeline Consortium, Kuibyshev Str., 62, 414056 Astrakhan, Russia

Abstract. This paper asserts that the issue of change of power in a corporate management configuration is quite urgent; it emphasizes that corporate management should identify the right combination of social interests and corporate business interests within the context of social, economic, and political processes currently taking place in Russia. The pandemic and digitalization exert a profound influence upon those processes; the paper states the fact that the Russian practice of corporate management – how decisions are made and how their execution is controlled – is characterized with a low quality, owing to the practice of "manual management" that prevails at all the hierarchical levels. The main idea of this article is that Russian managers should gradually substitute the traditional hierarchy with a "natural" one, when one's authority depends not on their official position, but on their contribution to work, when special emphasis is placed on self-control, on establishment of a horizontal control system based on the Agile approach. Managers ought to realize that they should adopt new business models by integrating cutting-edge technologies (digitalization and other forms of work) in all the aspects of their corporate business activities.

1 Introduction

The world will undergo substantial changes in the post-coronavirus era, and the national economy will endure a heavier recession. Yet, the main issue of corporate management in Russia will remain as urgent as today: the model of efficient management of any corporation must be determined, first and foremost, with such top-priority factors as orientation towards social aspects related to a complete unleash of corporate human resources' intellectual potential, stimulation of processes of self-organization and creativity; flexibility of internal mechanisms of corporate management and its compliance with the adopted strategy; adaptability and sensibility to permanent changes in the external environment [1]. Hence, assessment of impact exerted by digitalization upon the formation of corporate strategies and new business models of present-day corporations is quite a topical issue.

2 Materials and Methods

The paper applies research methods of economic, historic, and statistical analysis; comparison and analogies; the empiric approach in the development of social and economic processes in Russia in general and in Astrakhan Region in particular. Findings of field researches conducted by Astrakhan Region's scholars have been used as empiric data.

3 Results and Discussion

The most essential issue of corporate management that is still urgent in present-day Russia concerns change of power in a corporate management configuration. Should it be referred to the system of management, to its workforce, or should it be addressed to business owners, who hold the control packet of shares? What configuration should the workforce and particular members of staff be related to? Obviously, the main role in protection of rights and interests of all the participants is played by managers, who ought to identify the right combination of social interests and corporate business interests within the context of social, economic, and political processes that are currently taking place in Russia. The pandemic and digitalization exert a profound influence upon those processes. In our view, the issue of correlation between corporate centers of power (who plays the main role – managers or owners?) is becoming pointless, since Russian business owners and managers are still the same people. A well-known disadvantage of being both an owner and a top manager is absence of any external control of a General Manager's activities and thereby frequent violations of the workforce's rights.

Earlier findings gained by the Russian Bureau of Economic Analysis (BEA) confirm that most frequently, extra-large packets of shares in the Russian industries were held by less efficient categories of owners [2, p. 112]. Those companies whose control packets of shares were held by their managers proved to be more efficient upon their privatization, e.g. Severstal, Cherkizovo, etc.

At the current stage of development, stakeholders and top managers are equally interested in maintaining their businesses' competitiveness by mastering modern digital technologies and management innovations. Besides owners-stakeholders and top managers, who have a real power at Russian enterprises, other stakeholders (the regional authorities, the public, consumers, suppliers, and investors) have some authority as well. One should consider with interests of such a powerful stakeholder as the state, which ought to act as a guarantor of compliance with rules of economic behavior, not as an active economic subject. Specific features of the Russian national system of corporate management, which reflect both its development stages and the national mentality, quite often result in the fact that actions performed by a number of representatives of Russian state interests are based on their personal (or affiliated with other people) interests and tasks, which can influence corporate development negatively.

An increasing number of business representatives begin to realize that their main common goal (mission) is not only gaining profits, since an organization is an open system and it can survive only provided it meets one's external needs, whereas making profits is a company's internal problem. A business goal (mission) should include not only profits, but also social responsibility, which may be defined as a company's commitments to protect and enhance the social wealth considering with its own interests at the same time [3].

A manager's target interest is efficient application of capitals both at present and in the future, as well as sustainable revenues independently on the capital origin (be it their own or loan capital). Disagreement between managers' and owners' interests may only be due to their different horizons: a typical manager has a short-term vision, while an owner tends to regard their business in a longer-term run. In addition, the problem of the field of activities is more acute for a manager (owing to their professional preferences, since a manager is more interested in a growing scale which their increased personal authorities provide), while an owner is more interested in a growing enterprise value.

The developing nature of management activities also requires appropriate changes owing to changing conditions of business activities in the present-day crisis economy; it concerns especially the Russian practice of corporate management – how decisions are made and how their execution is controlled. It is well-known that Boards of Directors of Russian corporations do not cope with their functions; their level of management is low owing to the practice of "manual management" widely spread at all the levels in Russia [1].

Russian managers are mostly autocrats, and the very system of management is of authoritarian nature. One could agree with I. Adizes that Russian managers have no time to get used to present-day realities, to say nothing of changing actively. Such factors as unjust egalitarianism, nepotism, and corruption hamper their adaptation. Post-communistic Russia inherited not only the authoritarian style of management, but also fear of chiefs and authorities, even though such qualities as good education, erudition, creativity, and enterprise are real features of Russian managers [4, p. 40].

The world's best general managers, according to the ranking published in Harvard Business Review – Russia, state unanimously that success is provided mostly by a team; it is impossible to manage in a strictly centralized way, so a flat structure is essential; the most important thing in a General Manager's activities is to motivate people and sustain their team spirit, with a large number of staff members having the right to make decisions. Actually, the best managers have confirmed the importance of processes and tendencies, which have already emerged as a response to current changes in the world economy, in practice: (1) the end of the authoritarian style of management, since one chief cannot assess all the risks in the chaos of non-stop changes (so well-agreed actions of a team are necessary); (2) the end of a hierarchy, as systems fully denying it are developing – those are ecosystems, holacracy, network organizations, flat structures, and the agile methodology; (3) the force of team spirit, which helps tackle complicated tasks by developing interaction between teams and units, disregarding internal organizational boundaries [5, p. 13-18].

These tendencies coincide with the 25 large-scale tasks of management in the 21st century listed in the concept "Management 2.0: New Version for New Age" [6]. The first and foremost task is to set high goals that should aim to tackle socially demanded and noble tasks (not to maximize profits, as it is now). A second task is to adopt the idea of corporate responsibility: management of the future ought to consider with interests of not only its top managers and stakeholders, but also its workforce and local communities, i.e. be sensitive to social needs in general. A third task implies that businesses of the future should easily adapt to any new conditions, provide a favorable environment for their employees and set high goals for them, develop innovations continuously, be socially responsible, and gain new knowledge in various fields. A fourth task is to reject the traditional hierarchy: one should substitute the traditional organizational pyramid with a "natural" hierarchy, which implies that one's status and authority depend not on their official position, but on their contribution to work.

A little more than 10 years have passed, and cutting-edge companies are already implementing these revolutionary ideas in practice; they adopt management innovations actively and introduce the most modern management techniques (agile, holacracy, etc.). Many Russian managers not only set socially demanded and noble goals for their companies; not only do they try to consider with their stakeholders' and top managers' interests, but also with their employees' and local communities' needs so as to gradually substitute the traditional hierarchy with a "natural" one, when one's authority comes from one's contribution to work (not from one's official position in that hierarchy), when everyone's self-control is of top priority, when a control system is applied horizontally, with one's colleagues' opinions (as the agile approach advises) of top importance [7].

All the managers agree that only digital transformation can give them a new chance to eliminate Russia's technological backwardness. Digitalization of economy is becoming of top priority for not only large Russian corporations, but also for small businesses, which requires a far wider application of new digital tools by Russian companies: cloud computing, 3D printing, the Internet of Things tools, virtual and augmented reality, Big Data, the blockchain technology, high-performance platforms to store and process data (such as S/4 HANA), etc.

Many industrial corporations are already optimizing their business models and production based on digitalization. Their range includes not only Sber, but also KamAZ, Petrozavodskmash, Kalashnikov, etc. In particular, KamAZ has already implemented automatic planning based on an ERP system; it has developed a production monitoring and operational management system; it interacts with its clients based on the SAP Hybris Cloud for Customer cloud platform; it applies almost 900 robots. As for Petrozavodskmash, thanks to its complex digitalization, it monitors manufacturing of particular details and components; one can easily see whether their quantity is sufficient to manufacture this or that product, how busy employees are at each stage of the production process; one can also obtain information related to planned and actual self-cost, which helps apply a more flexible approach to management of the entire enterprise [8]. One should note that a number of large corporations that have the necessary resources are already introducing expensive digital technologies, such as SAP/4 HANA (Severstal, the Alcohol Siberian Group, Cherkizovo, etc.). Nevertheless, other large companies assume a let-us-wait-and-see position; thus, they risk losing their competitiveness. As for small companies, they do not even dream about digitalization owing to their insufficient assets and resources, except for startups, which were lucky to find their niches at the market of digital innovations.

Besides the current pandemic, an incredibly high growth of technological modernization is becoming a new tendency. That was the response given by over 72 % of top managers of the world's 500 large companies, when the Fortune conducted its 2017 survey [8]. In our opinion, technological innovations are the key to new business models, new markets, and new growth opportunities for Russian corporations.

Erased boundaries between industries are becoming an essential trend. Here are examples of how digital transformations require reconsideration of business models and diversification of activities: the Russian Post has become a retail bank, while Rostelecom has established a partnership with SAP to provide services in financial transactions, transformation into a universal supplier of cloud-based business solutions together with other telecommunication companies [9].

Thanks to Big Data technologies and information related to its previous purchases, Rostselmash has introduced a SAP-based client-oriented approach and automated its dealers' workstations; as a result, the company has developed a unique strategy oriented towards a particular client, being aware of what that client needs. Production of unmanned KamAZ trucks equipped with machine vision has already been planned for 2025, with intellectual transport systems to control them. The degree of automation and robotization at Cherkizovo, which manufactures summer sausages, is also impressive.

Large corporations are adopting SAP/4 HANA, which is a more progressive platform of new generation, successfully to automate their business processes compared with ERP. Severstal presented its report "SAP/4 HANA May Also Be Agile" at the forum "SAP NOW online" in April 2020. Severstal provides a good example as a company that implements digital projects regularly, with SAP/4 HANA becoming its digital heart. Therefore, digital transformation results in changing business models owing to introduction of breakthrough technologies into all the aspects of business activities.

Large companies should consider experience gained by small innovative enterprises, which now adapt and customize easily to meet their consumers' new individualized needs, since customization is becoming a new trend of the objective reality [10]. Makers (i.e. people involved in small-scale inventions, or making) can create highly customized products as fast as standard ones, with their costs remaining the same; they can manufacture self-developed products in small quantities and with a high quality thanks to application of digital cloud technologies in their startups. A. Auzan states that it will be possible to produce a lot of either 10, or 1 million units at high-capacity facilities, with logistic costs making the only difference [10].

4 Conclusions

Therefore, digital innovations and new organizational forms of works such as agile, holacracy, and formation of open ecosystems for consumers, innovators, and partners of modern efficient organizations are becoming a key to new strategies and business models, new markets and new opportunities of economic growth.

The identified tendencies and challenges set the problem of competitiveness in a new way. Present-day ecosystems must change as fast as their external environment; digital tools must be introduced by each member of staff on a daily basis, as they make manufacturing of many products cheaper and easier.

In present-day efficient companies, which are involved in digital transformation and which implement new organizational forms such as Agile, annual planning and budgeting cycles are substituted with shorter ones. Practicing managers applying the flexible Agile technology are well-aware of the fact that an annual cycle restrains innovations and that a consumer needs frequent changes represented by cutting-edge ideas. Elon Musk states that approximately 20 enhancements are implemented into Model S every week [11, p. 59]. Thus, new development strategies must be updated more frequently under conditions of digital transformation; they ought to consider application of digital tools, especially in the field of the Industrial Internet of Things and augmented reality, which will influence all the spheres of economy by changing the processes of education, decision-making, and interaction [12, p. 30].

In its turn, an obligatory condition to implement digital innovations successfully is new organizational forms of corporate ecosystems which unleash the staff's potential in its striving to create values for clients and realize its opportunities to develop as active entrepreneurs, which set everyone free to make their dreams come true [13, p. 54]. For Russian companies, the current coronavirus pandemic is a good reason to consider their renewal and update their business models.

Therefore, Russian companies should digitalize their activities and create their own ecosystems; a number of companies (Agrosignal, the Internet of Things Association, Rostelecom, Farvater, etc.) are already doing that. For instance, the Russian agriculture is being digitalized as a part of the general system of economy digitalization; in the future, it is planning to create platform ecosystems. There are already quite many organizations that offer business-oriented digital solutions in various fields: CodLix, Agrosignal, etc. In other words, there is a real supply of such solutions; yet, demand for those digital solutions is still rather low. According to a special survey conducted by the Internet of Things Association, the main factor that hampers introduction of digital innovations is absence of specialists who have completed at least a primary training program to understand digital tools.

In our opinion, the most essential task for practicing managers is to search and select particular digital projects to implement in their organizations based on efficient application of those projects in other companies. The main source of that information is the Internet, as well as webinars and seminars delivered by special professional organizations such as the

Internet of Things Association established at the initiative of the Bauman Moscow State Technical University and the Internet Initiatives Development Fund, as well as by other similar associations established by the Higher School of Economics — National Research University, Peter the Great St. Petersburg Polytechnic University, etc. Managers do not have a full list of such professional organizations that might aid them; if there were such lists, it would save much time for searching the relevant information related to digital tools that might be implemented in practice. In our opinion, the legal authorities could make a decision to create self-regulated organizations in the sphere of digital transformation, which would have all the necessary information to share with managers. Joint efficient actions by corporate management and a special state information system to search and select efficient and relatively inexpensive digital projects to implement in Russian companies could make it possible to cope with this task. In this case, one cannot get along without legal authorities. This idea was expressed at the Marathon of Ideas arranged by the Agency for Strategic Initiatives. The appropriate units and departments of branch ministries and agencies are supposed to create that information system, which would include information about successful implementation of digital innovations at particular enterprises and in particular industries, similar to Bureaus of Scientific and Technical Information (BTI) in the Soviet era. Soviet enterprises used to receive BTI Bulletins about various innovations regularly. One could apply the Soviet experience and develop a system to collect the relevant information in various industries and regions. Support of the state represented by the Ministry of Economic Development and its regional units is necessary to promote this information for companies and corporations.

Nowadays Agrosignal acts as a guide for agricultural enterprises at its own initiative; it offers relatively inexpensive digital solutions to agricultural companies upon their requests. Besides solutions that it developed itself, it offers them solutions developed by other organizations and implemented in Russia successfully. The state represented by the Ministry of Economic Development and its regional units should contribute to active promotion of this information.

References

1. R. I. Akmaeva, *Development of Efficient Management at Modern Russia's Industrial Enterprises* (2007)
2. R. I. Kapeliushnikov, N. Demina, Russian Journal of Management, **3(3)**, 110 (2005)
3. A. E. Khachaturov, Management in Russia and Abroad, **2**, 10 (2004)
4. I. Adizes, HBR – Russia, 6 (2017)
5. R. I. Akmaeva, Bulletin of Dagestan State University, Ser. 3 Social Sciences, **32(4)**, 13 (2017)
6. *Management 2.0: New Version for New Age*, HBR – Russia, 91 (2009)
7. D. Rigby, J. Sutherland, H. Takeuchi, HBR – Russia, 39 (2016)
8. M. Belzer, *Digitalization of Industry: a Fashionable Trend or a Necessary Condition to Maintain Competitiveness?* https://promdevelop.ru/
9. S. Ragimova, Forbes Brand Voice, http://www.forbes.ru/
10. A. Auzan, HBR – Russia, 81 (2016)
11. D. Rigby, J. Sutherland, A. Noble, HBR – Russia, 59 (2018)
12. M. Porter, J. Heppelmann, HBR – Russia, 30 (2017)
13. G. Hamel, M. Zanini, HBR – Russia, 47 (2018)

The method of the best performer selection based on a competency-based approach

Anna Bakanova[1,*], and *Aleksey Nikolaevich* Shikov[2]

[1]ITMO University, 197101, KronverkskyPr.49,Saint Petersburg,Russia
[2]The Russian Presidential Academy of National Economy and Public Administration (North-West Institute of Management, branch of RANEPA), 199178, Sredny prospect VO 57/43, Saint Petersburg, Russia

Abstract. One of the important areas that determine the company's strategic development is human resource management. Information technology in this area is a crucial tool that implements many management functions, eliminating the human factor, facilitating the work of managers and improving performance indicators of work activities. The paper presents a method for finding the best performer based on a competency-based approach. Because of the ontologies use, it was possible to implement a complex multi-criteria search, which became the basis for organizing corporate training and monitoring competencies in the company.

1 Introduction

The development of information tools in the field of human resource management is an important and urgent issue in modern companies. Intellectual support of such processes as the assignment of an employee to a work task or project, training and career planning, human resource planning expands the company's ability to achieve strategic goals. Understanding the relationship of these processes and a wise approach to themanagement organization increases the productivity of employees and has a positive effect on the organization development as a whole. An increase in human resource management by 1% gives an increase in organizational innovation by 16%. [1]

The main bond connecting such administrative and executive functions of human resource management as control, organization, adaptation, personnel recruitment, development and training, planning of personnel and work process is competence. This term is interpreted in differetly, but the greatest semantic value is its relationship with the work task performed by the employee. The main feature lies in the symmetric relationship: "the employee performs the work task because he possesses the necessary competence for this" and "the employee has the necessary competence, which means he is able to complete thework task." From the first expression, the idea arises that an employee can acquire a new competence through the performance of a work task, training takes place on the job.

*Corresponding author: bakanova55anna@gmail.com

From the second expression, the idea arises that the employee first learns, gains competence, and then gets the opportunity to perform a work task. Both ideas, together with the task of finding the best performer, formed the basis of the method for finding the best performer which relies on a competency-based approach.

2 Materials and methods

2.1 Existing solutions review

Competency management process informatization in connection with various processes in the organization, including the workflow, is represented by many papers. In particular, in [2], an example of the graphs use in project management for resource allocation is given, but in this case it becomes impossible to compare all employees in terms of training. In works [3-6], the use of competencies is considered for recruiting a team and developing a career. However, these works do not consider the possibility of assigning an employee to specific tasks, and his development based on motivation to acquire competence for solving specific problems. The set of employee competencies is compared with the required set for the position or role, and only on this basis, the distribution of tasks occurs [3, 7, 8]. The issue of motivated learning and improving productivity is partly realized via gamification, when the acquisition of new knowledge and skills is supported by a goal that is understandable for the employee. For example, an increase through training in a level in a particular rating will have an impact and lead to positive changes in the working environment. While most often the purpose of human resources allocation is to minimize the time required to complete a work task or project [9-11].

Since the concept of competence is directly related to the training field, it was used to create systems dedicated to corporate training and the accumulation of knowledge [12-16]. This kind of systems makes it possible to identify gaps in knowledge and form a recommendation for employees training, however, the relationship between training and employee development in relation to the work process and competencies control in the company remains unrealized.

Most HR systems use an ontological approach [17-23]. Ontologies are the most functional tool for the classification and aggregation of any information. They help combine different directions of interaction with data and form a knowledge base for control systems. The ontologies use in the design of the method combines the problem of creating a competency base in a company that is used to find a suitable performer and organize corporate training.

The proposed method allows one to define the requiredlearning goals, eliminate manipulability when assigning an employee, reduce discussion time when assigning tasks, and is suitable for both a project approach and a product (Scrum) approach.

2.2 Method ideas in personnel management, problem statement

The main idea of the method is to support corporate training as a result of constant competencies monitoring in the process of selecting the best performer for a work task. There are three urgent management problems of the company's competencies, which are solved using the method:

- Creation of a critical situation in the absence of a key specialist,
- Formation of the company's competence base,
- Support for effective corporate training

When searching for a suitable contractor for solving a work task, various ratings are formed that clearly represent the situation in the company with respect to specific competencies and situations with the solution of specific work tasks. The method algorithm lies in finding the best performer and analyzing the information obtained by comparing the values of the criterion compatibility for the employee with the reference value for the criterion.

The search organized within the framework of the method is aimed at solving a task/problem (E) which relies on a competence-based approach. To solve various types of tasks / problems, two types of search are set: single-purpose (E_1) and multi-purpose (E_2). In a single-purpose search, the best performer is selected, taking into account the competencies that he or she possesses, the conditions that must be met for the implementation of activities and restrictions on the number of simultaneously performed tasks. In multipurpose search, in addition to selecting the best performer, the situation is analyzed with the possibility of performing a specific task by other employees and competencies monitoring in the company. Because of the formation of ratings, employees are distributed according to the levels of training regarding the solution of a specific task.plann

The method solves the following tasks / problems:

- Search for a specific employee for a position (E_1),
- Search for a specific employee to complete a work task (E_1),
- Search for a specific employee to be assigned to a project (E_1),
- Finding a group of employees to complete a work task (E_2),
- Search for a group of employees for assignment to the project (E_2),
- Team / department competency analysis (E_2),
- Company competency monitoring (E_2).

Employee training is recommended based on search results and is aimed primarily at controlling competencies in the company and supporting work activities. But,simultaneously, thanks to the visual representation of one's training in relation to work activity through the formed ratings, the mechanics of influence and status,reward, development andcompetition are implemented. These game mechanics create learning motivation and personal development preferences in employees. The information management structure of a corporate training organization is shown in Figure 1.

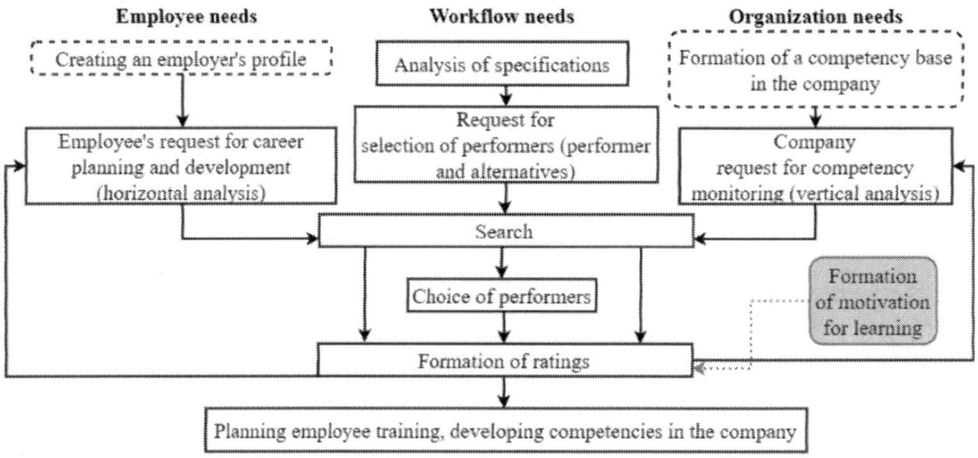

Fig. 1. The information management structure of corporate learning

The correct goals setting is important not only for the educational process, but also for work activities. When combining the required learning objectives, both the organization's requirements for the training of the employee and his personal and professional motives are taken into account.

2.3 Method development for choosing the best performer

The developed search model is based on a partially automated method for analyzing hierarchies, which, in line with the linear convolution method, allows solving the problem of multi-criteria choice and building a consistent hierarchy of estimates for all participants in the workflow.

For this, the reference value of compatibility is calculated for the selected set of competencies and conditions (criteria) corresponding to the task / problem. Then the reference value is compared with the compatibility result for each employee [24]. Compared are those employees who are pre-selected according to external conditions: from the same team, department, etc.

The set of criteria is represented by formula (1), where $K(E)$ is a set of competencies necessary to solve a search problem, $U(E)$ is a set of conditions necessary to solve a search problem, respectively, formulas (2) and (3):

$$C(E) = \{c_1, c_2, ..., c_m\} = K(E) \cup U(E) \quad (1)$$

$$K(E) = \{K_1(E), K_2(E), ..., K_l(E)\} \quad (2)$$

$$U(E) = \{U_1(E), U_2(E), ..., U_z(E)\} \quad (3)$$

Criteria are assigned with importance and weight values. The weight is assigned by means of a peer review method. When searching for the best performer for a work task or project, a set of criteria and their weights are generated automatically by analyzing the specifications. For other tasks, the formation of search criteria sets and their importance is carried out by an expert: an employee of the HR department, head of a department, etc.

The search model looks like the following formula (4):

$$y = (P_{exp}, D, y_1, y_2, y_3, Z) \quad (4)$$

$$y_1 = (P_{High}, P_{Dev}, P_{Low}) \quad (5)$$

$$y_2 = (SI_{High}, SI_{Dev}, SI_{Low}) \quad (6)$$

$$y_3 = (SK_{High}, SK_{Dev}, SK_{Low}) \quad (7)$$

wherey_1– the plurality of sets reflecting the ratings of employees on request is represented by the formula (5)
y_2– plurality of sets reflecting horizontal analysis, formula (6).
y_3– plurality of sets reflecting vertical analysis, formula (7).

The P_{exp} expert for a task / problem is the employee with the maximum compatibility score, which is calculated for each of the considered employees. Such an expert is the key employee for a specific work task or project. The choice of the best performer D occurs when solving tasksE_1, taking into account the maximum number of tasks simultaneously performed by an employee. Accordingly, this employee may be busy or absent, and there is a need to determine alternative replacement options. The plurality of sets y_1 represents alternative options for choosing a suitable executor (P_{High}) a set of potentially trained employees (P_{Dev}) and those for whom compatibility by criteria is minimal for performing a work task (P_{Low}). The plurality of sets $y_2$is responsible for horizontal analysis, which makes it possible to distribute the employee's competencies by compatibility levels. The plurality of sets y_3 is responsible for vertical analysis, which makes it possible to divide employees according to the level of proficiency in the criterion and identify the lack of competencies in the company. Searchrestrictionsarerepresentedbytheset$Z$.

The analyzed information on competencies, in addition to being used when choosing the best performer, is reflected in the employee's profile. Thus, everyone can see their weak competencies and performance level of work tasks. The results of the vertical and horizontal analyzes are used to create customized learning paths.

2.4 Method realization

In order to implement the method, a module of a corporate e-learning system was developed for accounting and working with employee competencies based on an ontological approach. The module interacts with third-party services and ontologies. Interaction with third-party services is carried out through the analytical subsystem, which is responsible for the semantic analysis of specifications and the assignment of sets of competencies and conditions to work tasks and projects. The latest specifications are provided by third-party services, which describe the necessary requirements for completing work tasks / projects. Thanks to this approach, there is a constant cycle of interaction between the workflow and employee training.

Competencies and conditions are assigned to a work task (specification analysis), according to the assigned set of criteria, the search for the best performer for the work task is carried out, the employee receives a new competency (training / work performance). This cycle is carried out through the use of ontologies.

Due to the data reusability, the use of ontologies is most effective for building intelligent search engines and creating corporate information services. Convenient storage, updating, classification and aggregation of knowledge have improved the field of e-learning knowledge management in the organization. The ontological model of competencies used to implement the method is organized in such a way as to create a hierarchical structure of competencies in the company. The competences of the previous level together with the conditions form new competencies [22, 24]. Using a condition as a component of a competency expands the functional use of competencies. There is an opportunity to create intra-organizational competencies used to organize the work process and search for performers, taking into account the specifics of the company's activities. Such a structure is also more understandable for the employees themselves, since the relationship between different competencies is structured through their component parts. At the same time, the classical representation of competence (KLM) is preserved as a set of knowledge, skills and abilities [17, 23].

Figure 2 shows an algorithm for assigning a competency to an employee after completing a work task.

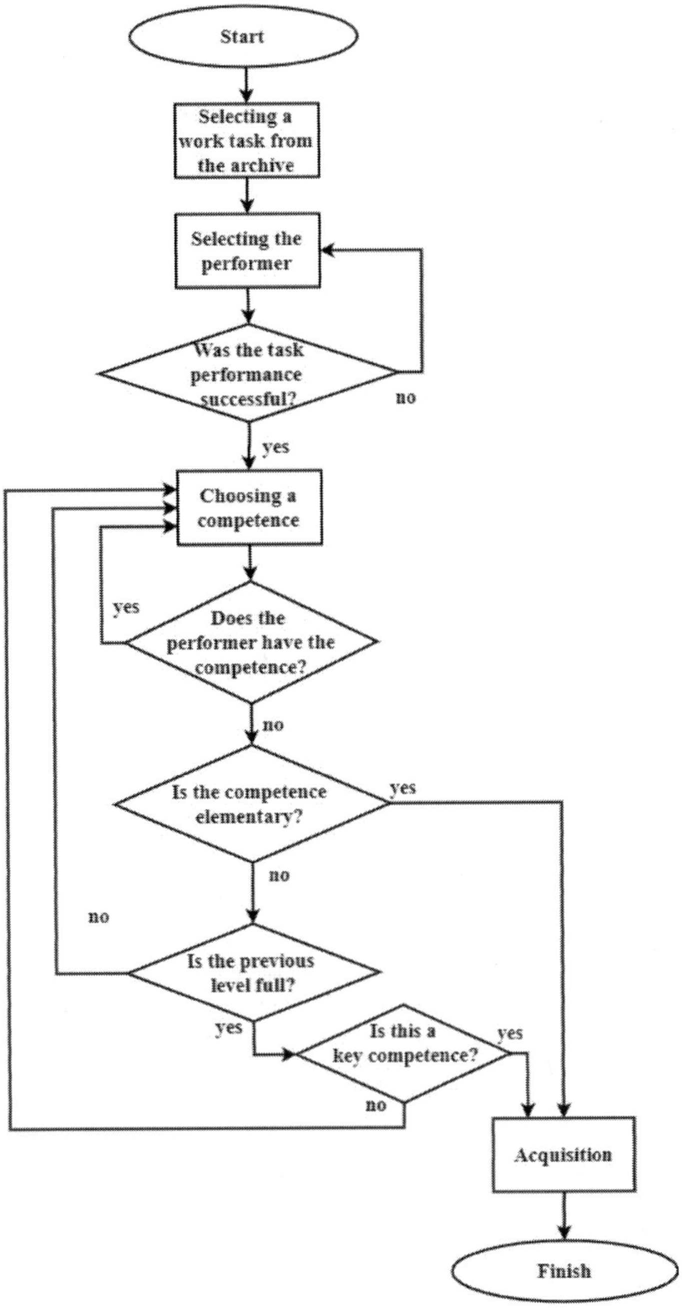

Fig. 2. Algorithm for assigning competencies to an employee based on the results of a work task

3 Results and discussion

The method of finding the best performer based on the competence-based approach was developed for SCOUT Group of Companies and showed good efficiency results. To compare automatic and expert solutions, a number of special evaluation criteria have been developed. The experiment was carried out in several stages and included the assignment of competencies and conditions to work tasks, the distribution of their weight and the selection of performers. The accuracy of selection and a positive assessment by experts of the automatic competencies selection was 85.6% and 91.6%, respectively. Given the hierarchical structure of competencies, a high rate of selection accuracy affects the correct selection of a performer. Since the correctness of the expert choice for a work task is from 80 to 90%, it indicates that at least 8 out of 10 competencies are involved in the appointment of the required performer. And these 8 competencies are the main ones for the assignment. The weight value for matching competencies falls within the confidence interval of 95%. The performer's automatic selection and the distribution of training levels were rated positive by 88% of employees. More than half of the 15 survey participants surveyed found the rating-based training recommendations helpful.

The method's practical significance is reflected in the increase in productivity from 13% to 27%, the number of successfully completed tasks in the customer support department has doubled. The employees got the opportunity to implement internal organizational tasks aimed at improving the company's performance.

It should be noted that an important condition for applying the method is the company's willingness to develop competencies and an accurate attitude to the knowledge base. In order for the method to work correctly, it is necessary to conduct an accurate initial assessment of the employee's competencies using proven information collection methods. It is worth excluding the possibility of an employee's self-preparation of a competency map, and using high-quality expert assessment methods. Since providing false information makes work activities, personal training and employee development ineffective.

The next factor influencing the positive use of the method is the structure and composition of the specification for work tasks and projects. It should contain the following main sections, avoiding duplication of information and preserving the maximum amount of information:

• Introduction.
•General characteristics.
• Description of requirements.
• Direct description of the work task / project.
•Additional requirements.

Conclusions

The development of information systems in the field of company's human resource management faces many challenges. Duplication of functions in various structures of the organization leads to redundancy of information, difficulty in combining numerous factors and analysis of aexcessive amount of information. The presented method showed that when combining competency monitoring with workflow and training, it is possible to increase productivity efficiency and affect motivation to learn. It made it possible to implement various management functions and support the adoption of managerial decisions. The use of competence as a key element enabled combination of the search for the best performer for a work task with the organization of corporate training. This approach to the control of competencies in the company contributed to development of new recommendations for training and career planning. Through the compilation of ratings, deficient competencies in the company are identified and personal training paths for employees are formed.

References

1. S.A.Hashemi, F. Dehghanian, Eng. Technol. Appl. Sci. Res.,**7**, 2200 (2017)

2. B. H. Burkov, A. Ju. Zalozhnev, D. A. Novikov, *Teorijagrafov v upravleniiorganizacionnymisistemami*, 124 (2001)
3. C. Kimble, J.B. de Vasconcelos, Á. Rocha, Inf. Syst.Front.,**18**, 1119 (2016)
4. C.Dorn, F. Skopik, D. Schall, S. Dustdar,Data Knowl. Eng.,**70**, 866 (2011)
5. J.Decius, N.Schaper, Procedia Manuf.,**9**, 376 (2017)
6. F. L. Krause, W. Hacker, U. Debitz, C. Kind, M. Strebel, CIRP Ann, Manuf. Technol., **55**, 135 (2006)
7. L. Fernández-Sanz, J. Gómez-Pérez, A.Castillo-Martínez, Comput. Stand. Interfaces.,**51**,30 (2017)
8. S. N.Neborskij, Rep. Bel. Univ. of Inf. and Radioelectronics,**2**, 73 (2009)
9. L.C.e Silva, A.P.C.S.Costa, Int. J. Proj. Manag.,**31**, 100 (2013)
10. R.V. Dopira, R.Ju. Kordjukov, A.A. Beglecov, S. V. Sergienko, Programmnyeproduktyisistemy,**33**, 22 (2014)
11. A.V.Kataev, T.M.Kataeva, E. L.Makarova, Izv. Sarat. Un.. New series. The Ec. Ser.. Manag.. Right, **16**, 294 (2016)
12. J. M. Silva, J. H. Rosa, J. Barbosa et al., Brazilian Comput. Soc.,**16**, 163 (2010)
13. T. Ley et al., J. Knowl. Manag.,**12**, 31 (2008)
14. S. Oliveira, J. Carlos Gluz, Adv. in Exp. Med. and Biol.,41 (2010)
15. A.Garro, L.Palopoli,L. N. in Comp. Sc. (incl. subs. L. N. in Art. Intell. and Bioinf.),283(2003)
16. J.L. De Coi et al.,Webist 2007 - 3rd Int. Conf. on Web Inf. Sys. and Techn, Proc.,9 (2007)
17. W. Zaouga, L. B. A. Rabai, W. R. Alalyani, Procedia Comput. Sci.,**151**, 417 (2019)
18. H. Ma, H. Sven, V. Panrawee, Enterp. Model. Inf. Syst. Archit.,**13**, 182 (2018)
19. M.Mochol, R.Oldakowski, R.Heese, Proc. of the GI2004 Conference (2004)
20. S.Strohmeier, F.Röhrs, AIS Trans.on H-Comp. Inter.(2017)
21. J.L.V. Barbosa et al., Comput. Ind.,**72**, 27 (2015)
22. S. Miranda et al., Data Knowl. Eng.,**107**, 51 (2017)
23. J.H. Rosa et al., Int. J. Artif. Intell. Educ.,**25**, 455 (2015)
24. A.Bakanova, A.N.Shikov, CEUR Workshop Proceedings.,**2590**, 1 (2020)

Optimization of resource use in the formation of a network interaction model

Andrey Plakhin*, *Maxim* Kocheryan, *Elena* Davydova, and *Natalia* Shuraleva[1]

Ural State University of Economics, 620144 Ekaterinburg, Russia

Abstract. The article is based on the research materials of the network interaction of universities in the context of physical culture and sports activity. The research is based on the methodology of network theory and the resource approach in the strategic management of organizations. The results of the study show that the formation of a network model can significantly increase the coverage of the population with projects of physical culture and sports activity using already existing resources. The results obtained indicate the effectiveness of strategies for networking organizations.

1 Introduction

The formation of a network strategy for managing a university is a priority task at the present stage of development of the education system of the Russian Federation. Its use is based on resource, relational and network theories, the provisions of which are adapted for the university as an object of strategic management in the works of A.A. Gres'ko, M.S. Rakhmanov, K.S. Solodukhin [1], K.S. Eleneva, Yu.Ya. Yeleneva [2], V.N. Lupanov [3]. Network strategic management in general involves the use of various integrative forms in order to form and effectively use the infrastructural, intellectual and resource potential of network participants [4].

These multifaceted connections are explained within the framework of the concept of "network", which is a set of individual subjects of physical culture and sports activity and multidirectional connections between them. In the work of M. Castells [5] it is noted that in the context of the theory of networks, three key terms are used: network formation or networking, networks and network structures. Networking is a general term for activity, as a result of which connections are created between people and organizations through regular meetings, conferences, new communication technologies. Networks are formed when connections between organizations or individuals are formalized, i.e. become regular and subordinate to the principle of satisfying mutual interests. Network structures arise when the activity of separately acting network elements is no longer sufficient. Keast R. et al. [6] noted that network structures arise when the actors participating in the network realize that they are only small fragments of the big picture. Network structures may require independent actions of individual members, but at the same time, network members, when they take on complex tasks that are beyond the capacity of independently operating

* Corresponding author: apla@usue.ru

organizations, are transformed into a new whole. Moreover, as noted in W. Mastenbrook's book [7], people depend on each other and have, at the same time, their own interests ... in a certain sense, we could even talk about a coalition of divergent interests.

At the same time, the subjects within the network organization of physical culture and sports activity in Russia have their own hierarchical characteristics [8]. So at the primary level, professional and amateur teams are distinguished at educational organizations and enterprises whose activities are related to physical culture and sports, amateur physical culture and sports teams, sports schools, sports clubs and sections.

The subjects of the primary level are the basis for the subjects of the second level of the hierarchy, whose activities are related to the issues of methodological, coordination and financial support of the primary level. The subjects of the second level of the hierarchy include sports associations, sports leagues and sports confederations.

On the basis of the subunits representing the primary public organizations of physical culture and sports in the Russian Federation, larger organizations are formed, designed to improve the efficiency, organizational and methodological or commercial leadership of the field of activity that unites these people. In the Russian Federation, such public organizations are: sports organizations, physical culture and sports societies, sports associations, sports federations, sports leagues, sports confederations, sports unions.

It is quite obvious that effective network strategies make it possible to resolve issues of coordination of activities, to overcome the resource constraints of individual universities through the formation of partnerships, unions, and associations. In turn, the theory of strategic management considers networks as a way of organizing the activities of relatively independent subjects, which in the framework of the subject of this study are institutions of higher education. Subjects remain independent while pooling resources and forming stable relationships with respect to objects of interaction. And for the physical culture and sports activities of the university, the objects are joint projects, including both educational activities and socially significant events with the participation of universities.

2 Methods

In the context of the research, the following conceptual model of the formation of a network strategy for the interaction of physical culture and sports activities of a university within the framework of the project approach is presented in Figure 1.

The implementation of the conceptual model essentially corresponds to the solution of a large-scale optimization task, where the target function is to maximize the number of participants included in the projects of physical culture and sports activity, and resource constraints provide for the analysis of the following:

- Sport facilities;
- Sport equipment;
- Financial resources;
- Staff.

According to the proposed model, it is necessary to analyze the totality of projects of physical culture and sports activities of the university, which are planned to be implemented in the coming period. The main factors of the analysis are determined by the needs for the resources involved for the implementation of these projects. At the same time, the possibilities of external counterparties - potential participants in network interaction are being studied, their characteristics and resource potential are determined, which can be used in the implementation of joint projects.

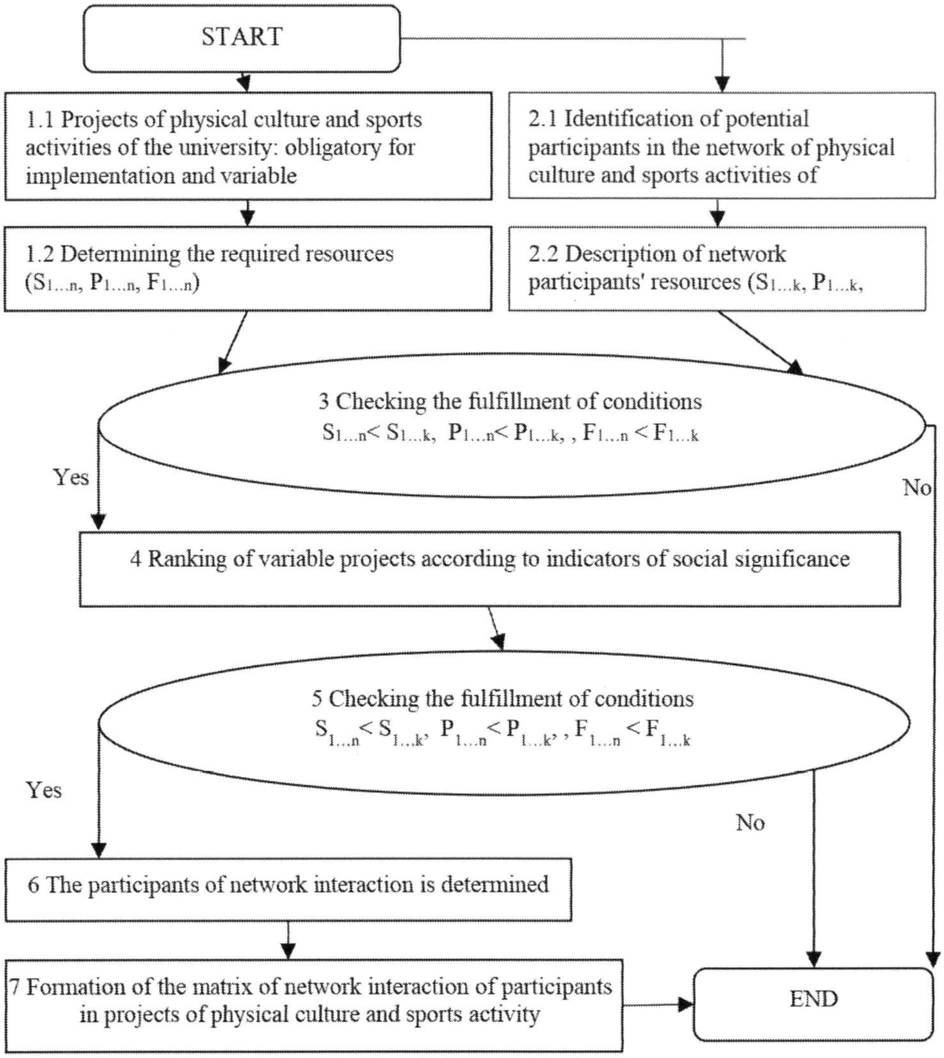

Fig. 1. Conceptual model of network management of physical culture and sports activities of the university within the framework of the project approach

Next, you should correlate the resources available to the university and the resources of partners. If it turns out that there are not enough resources, but at the same time there are network actors with these resources in a large (sufficient) volume, then you can proceed to the next step, which should assess the social significance of the projects planned for implementation. After ordering according to the degree of importance, it is possible to form a matrix of network interaction of the participants of the planned projects.

3 Results

On the basis of the proposed algorithm, a complex of projects of physical culture and sports activity of universities of the Sverdlovsk region was formed. Table 1 presents the characteristics of the projects obtained as a result of the implementation of stages 1.1 and 1.2 of the methodology.

Table 1. The list of projects of physical culture and sports activities implemented in the territory of the Sverdlovsk region and their resource requirements.

Name of projects of physical culture and sports activities of universities of the Sverdlovsk region			Resource requirements			
			Sports facilities	Sports equipment	Financial resources	Staff
Obligatory projects	Basic professional educational programs № 1-10	FSBEI HE "Ural State University of Economics"	4000	4000	4160	7644
	Basic professional educational programs № 1-3	ANO HE "Humanitarian University"	408	408	2104	2489
	Basic professional educational programs № 1-15	FSAEI HE "Russian State Vocational Pedagogical University"	1437	1437	7840	6578
	Basic professional educational programs № 1-45	FSAEI HE "Ural Federal University named after the first President of Russia B. N. Yeltsin"	18058	18058	30880	19133
	Basic professional educational program № 1	FSBEI HE "Yekaterinburg State Theater Institute"	71	71	1232	1213
	Basic professional educational program № 1	FSBEI HE "Ural State Conservatory named after M.P. Mussorgsky"	149	149	1760	1240
	Basic professional educational programs № 1-3	FSBEI HE "Ural State Medical University "Ministry of Health of the Russian Federation	829	829	4160	4178
	Basic professional educational programs № 1-7	FSBEI HE "Ural State Pedagogical University"	2862	2862	5320	3867
	Basic professional educational programs № 1-5	FSBEI HE "Ural State Law University"	1878	1878	5440	4000
	Basic professional educational programs № 1-5	FSBEI HE "Ural State Agrarian University"	608	608	1840	1333
	Basic professional educational programs № 1-3	FSBEI HE "Ural State University of Architecture and Art"	106	106	1040	711
	Basic professional educational programs № 1-12	FSBEI HE "Ural State Mining University"	3713	3713	6800	6133
	Basic professional educational programs № 1-12	FSBEI HE "Ural State Forest Engineering University"	2983	2983	6800	5756
	Basic professional educational programs № 1-7	FSBEI HE "Ural State University of Railway Transport"	3249	3249	5320	5511

https://doi.org/10.1051/shsconf/20208903005

Continuation of Table 1. The list of projects of physical culture and sports activities implemented in the territory of the Sverdlovsk region and their resource requirements.

Variable projects	Military sports games among university students	26	26	52	39
	All-Russian race "Ski track of Russia"	12	12	24	18
	All-Russian sambo team tournament among students in memory of the USSR master of sports	28	28	56	42
	Nation Cross	12	12	24	18
	Open championship of the Sverdlovsk region in sambo wrestling among senior schoolchildren for the "Cup of the Rector of Ural State University of Economics"	8	8	16	12
	Open badminton tournament	24	24	48	36
	Ural State University of Economics open rock climbing tournament "LIGHTNING" among junior schoolchildren	12	12	24	18
	Championship of Yekaterinburg in rock climbing	20	20	40	30
	Championship of the Sverdlovsk region in rock climbing	12	12	24	18
	Volleyball Championship of the Ural Federal District (2nd stage of the All-Russian Universiade)	26	26	52	39
	Regional Championship of the Association of Student Sports Clubs (chess, table tennis)	20	20	40	30
	Competition in Yekaterinburg in shooting from pneumatic weapons " Ural State University of Economics Open Tournament"	12	12	24	18
	Competition in shooting from pneumatic weapons "Victory Day"	12	12	24	18
	Spartakiad of Leninsky district	12	12	24	18
	Student regatta	12	12	24	18
	Universiade of the Sverdlovsk region	54	54	108	81
	"Ready for Labor and Defense" festival among correctional schools	24	24	48	36
	Fitness congress	24	24	48	36
	Championship of the Russian Student Sports Union in climbing, speed	24	24	48	36
	Relay "For the prize of the newspaper Vecherniy Yekaterinburg"	8	8	16	12
	Relay race of Leninsky district	24	24	48	36

The next step is to characterize the resource capabilities of each university from the standpoint of potential network interaction in the context of projects of physical culture and sports activities in accordance with paragraphs 2.1 and 2.2 of the developed methodology. Information on resource capabilities is presented in table 2.

Table 2. Resource characteristics of participants in network interaction

University name	Uni.No.	Availability of time for the use of sports facilities, hour / year	Availability of time for the use of inventory, hour / year	Availability of financial resources, thousand rubles	Time available for staff participation, hour / year
FSBEI HE "Ural State University of Economics"	1	3620	3620	5200	8600
ANO HE "Humanitarian University"	2	1230	1230	2630	3200
FSAEI HE "Russian State Vocational Pedagogical University"	3	4456	4456	9800	7400
FSAEI HE "Ural Federal University named after the first President of Russia B. N. Yeltsin"	4	18660	18660	38600	24600
FSBEI HE "Yekaterinburg State Theater Institute"	5	860	860	1760	1560
FSBEI HE "Ural State Conservatory named after M.P. Mussorgsky"	6	900	900	2200	1860
FSBEI HE "Ural State Medical University "Ministry of Health of the Russian Federation	7	2500	2500	5200	4700
FSBEI HE "Ural State Pedagogical University"	8	3700	3700	7600	5800
FSBEI HE "Ural State Law University"	9	3400	3400	6800	4500
FSBEI HE "Ural State Agrarian University"	10	1100	1100	2300	1500
FSBEI HE "Ural State University of Architecture and Art"	11	640	640	1300	800
FSBEI HE "Ural State Mining University"	12	5600	5600	8500	6900
FSBEI HE "Ural State Forest Engineering University"	13	4500	4500	8500	7400
FSBEI HE "Ural State University of Railway Transport"	14	4200	4200	7600	6200
Total		55366	55366	107990	85020

The analysis showed that the total time of possible use of sports infrastructure facilities and sports equipment of universities in the Sverdlovsk region is more than 55 thousand hours, the time of staff use is 85 thousand hours, and financial resources are estimated at more than 107 million rubles. Since these resources are unevenly distributed among universities, the optimization task, which is posed as part of the development of strategies for network interaction of universities, boils down to ensuring the most efficient use of resources, taking into account the preference of projects in terms of social indicators, as well as reaching the maximum possible number of participants. To solve this problem, we will consider the use of resources for the implementation, first of all, of projects of physical culture and sports activity that are obligatory for implementation in universities (point 3 of the developed methodology), table 3.

Table 3. The result of comparing the availability and use of resources of physical culture and sports activities of universities for obligatory projects

Obligatory project name	**University No.**	**Sports facilities**	**Sports equipment**	**Financial resources**	**Staff**
Basic professional educational programs № 1-10	1	-380	-380	1040	956
Basic professional educational programs № 1-3	2	822	822	526	711
Basic professional educational programs № 1-15	3	3019	3019	1960	822
Basic professional educational programs № 1-45	4	602	602	7720	5467
Basic professional educational program № 1	5	789	789	528	347
Basic professional educational program № 1	6	751	751	440	620
Basic professional educational programs № 1-3	7	1671	1671	1040	522
Basic professional educational programs № 1-7	8	838	838	2280	1933
Basic professional educational programs № 1-5	9	1522	1522	1360	500
Basic professional educational programs № 1-5	10	492	492	460	167
Basic professional educational programs № 1-3	11	534	534	260	89
Basic professional educational programs № 1-12	12	1887	1887	1700	767
Basic professional educational programs № 1-12	13	1517	1517	1700	1644
Basic professional educational programs № 1-7	14	951	951	2280	689

As can be seen from the table, for almost all participants in network interaction, there is no shortage of resources for the implementation of obligatory projects due to the fulfillment of the requirements of the main professional educational programs. At the same time, there is a shortage of space and inventory at FSBEI HE "Ural State University of Economics", so this university is forced to enter into network interaction with other universities in order to compensate for the emerging resource shortage of some of the sports facilities for the implementation of obligatory projects.

Next, we will perform the steps of stages 4 and 5 of the methodology and select participants for network interaction based on an optimization model, which includes the target function – maximizing the number of people participating in physical culture and sports activities, taking into account resource constraints: sports facilities, sports equipment, financial resources and staff. Let's consider the resource limitations of the formed matrix of the network model of physical culture and sports activity of universities of the Sverdlovsk region, table 4.

Table 4. Resource restrictions of the formed matrix of the network model of physical culture and sports activity of universities of the Sverdlovsk region

Resources	Indicator	University No.													
		1	2	3	4	5	6	7	8	9	10	11	12	13	14
Sports facilities	Allocated resources for networking projects	0	268	448	448	220	212	330	448	318	124	56	448	448	448
	Existing resources	-380	822	3019	602	789	751	1671	838	1522	492	534	1887	1517	951
	Resource allocation result	-380	554	2571	154	569	539	1341	390	1204	368	478	1439	1069	503
Sports equipment	Allocated resources for networking projects	0	268	422	422	220	212	330	422	318	124	56	422	422	422
	Existing resources	-380	822	3019	602	789	751	1671	838	1522	492	534	1887	1517	951
	Resource allocation result	-380	554	2597	180	569	539	1341	416	1204	368	478	1465	1095	529
Financial resources	Allocated resources for networking projects	0	520	828	828	424	408	644	828	620	248	112	828	828	828
	Existing resources	1040	526	1960	7720	528	440	1040	2280	1360	460	260	1700	1700	2280
	Resource allocation result	1040	6	1132	6892	104	32	396	1452	740	212	148	872	872	1452
Staff	Allocated resources for networking projects	0	414	645	645	342	330	507	645	489	186	84	645	645	645
	Existing resources	956	711	822	5467	347	620	522	1933	500	167	89	767	1644	689
	Resource allocation result	956	297	177	4822	5	290	15	1288	11	-19	5	122	999	44

As the analysis of the resources allocation of participants in network interaction shows (Table 4), the expansion of participation in projects of physical culture and sports activities of universities is mainly due to the provision of projects with the necessary additional personnel, and also, which turned out to be most typical for ANO HE "Humanitarian University" and FSBEI HE "Ural State Conservatory named after M.P. Mussorgsky", restrictions with financial support of projects were revealed.

The results obtained allow us to speak about the effectiveness of using the project approach in the strategy of network management of physical culture and sports activities of universities, since the number of participants involved can be significantly increased from 31 917 to 41 151 people with more effective use of existing resources. Further research can be aimed at a more detailed description of the project profiles of universities - participants in network interaction and analysis of the balance of the network strategy.

4 Conclusions

The results of the study allow us to confirm the provisions of the network theory and the resource approach in the strategic management of organizations regarding the achievement of higher indicators with the effective use of opportunities for the involvement and redistribution of resources within the network. The data obtained can be used within the framework of territorial locations to increase the effectiveness of the use of both sports infrastructure facilities and, in general, for social facilities.

References

1. A.A. Gresko, M. S. Rakhmanova, K. S. Solodukhin, Modern problems of science and education, **4** (2012)
2. K.S. Elenev, Yu.Ya. Eleneva, Modern problems of science and education, **1** (2013)
3. V.N. Lupanov, University Management: Practice and Analysis, **2** (2009)
4. A. Plakhin, M. Kocheryan, T. Volovik, Revista inclusions, **3** (2020)
5. M. Castells, *Formation of a society of network structures* (1999)
6. R Keast. et al. *Network structures: Working differently and changing expectations*, Public administration review, **3** (2004)
7. W. Mastenbrook, *Management of conflict situations and organization development: Translated from English* (1996)
8. J. Webster, The Journal of Strategic Information Systems, **1** (1995)

Technology for monitoring industrial enterprises by the level of economic security

Elena Karanina, *Nataliya* Maksimova*, and *Vyacheslav* Kazantsev

Vyatka State University, Moskovskaya St., 36, 610000 Kirov, Russia

Abstract. This article discusses the technology of monitoring industrial enterprises by the level of economic security. The problem of the financial crisis is a global problem. To get out of this state, it is necessary to monitor, review the management system and assess the level of economic security of the enterprise. The state and development of the industrial brunch is very important for the economy. In modern realities, the industrial enterprise is considered as a complex economic system, within which various processes are managed, among which an important place is occupied by innovative development. In the process of monitoring, it is necessary to pay sufficient attention to the target orientation and selection of evaluation indicators. To monitor light industry enterprises by the level of economic security, it is advisable to use economic and statistical methods. In the framework of this research the task is solved by the method of multidimensional statistical comparison. The reference values of the level of economic security have been determined and the rating of industrial enterprises of the North-West region has been determined.

1 Introduction

Today, the problem of the financial crisis is not only a national problem in Russia, but also has a global scale. In the process of emerging from the global financial crisis and overcoming economic instability, enterprises face many challenges. In this regard, there is an urgent need to monitor, review the management system and assess the level of economic security of the enterprise. The state and development of the industrial sector is very important for the economy as a whole. At all times, industry has been an economic engine. The level of security of the entire society with material goods directly depends on its state [1].

In modern realities, an industrial enterprise is considered as a complex economic system, within which various processes are managed, among which innovative development takes an important place. The efficiency of any enterprise always depends on financial and economic stability, flexibility of management decisions, orientation and possibility of innovative development [2].

Let's consider the concept of monitoring. In a broad sense, monitoring is defined as a complex system of monitoring of any object. In a narrower sense, carrying out of target monitoring at the industrial enterprises is carried out on a voluntary basis and is realized

*Corresponding author: info-mna@yandex.ru

according to the instruction of the management for definition of a vector of development by means of establishment of the best indicators of development, and also revealing of existing financial, economic and social problems which have developed at the moment of carrying out of research [3]. The result of the monitoring is to determine the rating of enterprises by certain criteria and identify the best ones. Those enterprises, which have shown the highest results, become the standard for the others.

Targeted monitoring can be applied at enterprises within the framework of state programs. Usually this is used to determine the attractiveness of an enterprise for investors [4].

In the process of monitoring, regardless of the industry, it is necessary to pay sufficient attention to the target focus and selection of evaluation indicators. Quantitative evaluation is always a higher priority than qualitative evaluation [5]. In the course of monitoring, a large number of different indicators are analyzed. Some of them are publicly available, while others have limited access.

Targeted monitoring provides an opportunity to optimize the innovation breakthrough in the industry under study - in particular, in the country as a whole [4].

Due to the large number of industry-specific features, it is unfortunately not possible at present to conduct monitoring in a uniform manner. Heads of enterprises make their own decisions about the need to conduct monitoring. It should be noted that information about plans and problems of a competitor's enterprise is a commercial secret, and management decisions cannot be made based on the analysis of undocumented information [5, 6].

2 Materials and methods

Theoretical and methodological concepts of building a system of indicators of economic security at the regional / business level were considered earlier in the works of authors: E. V. Karanina, D. A. Loginov, K. E. Kartavykh, E. S. Sapozhnikova, O. A. Ryazanova, L. Yurieva, N. Grischuk, A. N. Timin, L. P. Domracheva, M. Bortnikov, M. V. Palkina and others [7 - 17].

To monitor light industry enterprises on the level of economic security, in our opinion, it is advisable to use economic and statistical methods (the method of multivariate statistical comparison).

The analysis of scientific publications and the results of an expert survey of managers and leading specialists of industrial enterprises, as well as calculated indicators, allow us to justify the main indicators and threats to economic security.

So, we can distinguish the following indicators:

x_1 – the coefficient of update;

x_2 – the coefficient of absolute liquidity;

x_3 - – the coefficient of the borrowed and own funds of the enterprise;

x_4 – the coefficient of stability of staff;

x_5 –the coefficient of information armament;

x_6 – the coefficient of social benefits in the volume of net profit;

x_7 – the coefficient of environmental friendliness of products;

x_8 – the coefficient of export and import products;

x_9 – the coefficient of the current activity of the enterprise;

x_{10} – the coefficient of innovation potential;

x_{11} – the coefficient of educational level of lawyers at the enterprise.

Table 1 shows the initial data for calculations:

Table 1. Matrix of source data

Company	X_1	X_2	X_3	X_4	X_5	X_6	X_7	X_8	X_9	X_{10}	X_{11}
CJSC "Salyut"	0,12	0,74	0,15	0,60	1,05	0,03	0,23	0,20	1,10	0,89	0,67
JSC research and production enterprise "ANA"	0,16	0,72	0,14	0,74	3,11	0,02	0,18	0,15	1,20	0,98	1,00
"Severny Tekstil" LLC	0,10	0,70	0,10	0,90	1,24	0,04	0,24	0,10	1,01	0,92	0,67
JSC "Bolshevichka"	0,05	0,52	0,62	0,78	2,25	0,02	0,11	0,05	0,62	0,75	0,98
LLC "Sewing enterprise Galant"	0,01	0,11	0,79	0,27	2,18	0,01	0,08	0,02	0,36	0,38	0,97

For convenience of calculations, the matrix of source data is standardized as follows: the best value in the column is taken as 1, the remaining values are divided by the best. The calculation results are shown in table 2:

Table 2. Matrix of standardized indicators characterizing the economic security of the industrial enterprise

Company	X_1	X_2	X_3	X_4	X_5	X_6	X_7	X_8	X_9	X_{10}	X_{11}
CJSC "Salyut"	0,75	1	0,19	0,67	0,34	0,75	0,96	1	0,92	0,91	0,67
JSC research and production enterprise "ANA"	1	0,97	0,18	0,82	1	0,50	0,75	0,75	1	1	1
"Severny Tekstil" LLC	0,63	0,95	0,13	1	0,40	1	1	0,50	0,84	0,94	0,67
JSC "Bolshevichka"	0,31	0,70	0,78	0,87	0,72	0,50	0,46	0,25	0,52	0,77	0,98
LLC "Sewing enterprise Galant"	0,06	0,15	1	0,30	0,70	0,25	0,33	0,10	0,30	0,39	0,97

Then, according to the method, the values are squared, then the sum of the squares of the values for enterprises is determined, from which the square root is extracted. Based on the results of calculations, the rating of industrial enterprises was determined (table 3).

Table 3. Rating of industrial enterprises by the level of economic security

Company	X1	X2	X3	X4	X5	X6	X7	X8	X9	X10	X11	$\sum$	$\sqrt{}$	Rating
CJSC "Salyut"	0,56	1	0,04	0,45	0,12	0,56	0,92	1	0,85	0,83	0,45	6,78	2,60	2
JSC research and production enterprise "ANA"	1	0,94	0,03	0,67	1	0,25	0,56	0,56	1	1	1	8,01	2,83	1
"Severny Tekstil" LLC	0,40	0,90	0,01	1	0,16	1	1	0,25	0,71	0,88	0,45	6,77	2,59	3

Continuation of Table 3. Rating of industrial enterprises by the level of economic security

JSC "Bolshevichka"	0,10	0,49	0,61	0,76	0,52	0,25	0,21	0,06	0,27	0,59	0,96	4,82	2,20	4
LLC "Sewing enterprise Galant"	0	0,02	1	0,09	0,49	0,06	0,11	0,01	0,09	0,15	0,94	2,96	1,72	5

Based on the results shown in table 3, the reference ranges of values of integral coefficients for the level of economic security have been determined (table 4).

Table 4. Reference values of the level of economic security of light industry enterprises.

Level of economic security of enterprises	Range of values
1. High	2,6 – 3,0
2. Medium	2,2 – 2,6
3. The risk	< 2,2

3 Results and discussions

The definition of the economic security zone for the studied enterprises is presented in table 5.

Table 5. Zoning of light industry enterprises in accordance with the reference values.

Enterprise	Indicator	Zone
CJSC "Salyut"	2,60	
JSC research and production enterprise "ANA"	2,83	
"Severny Tekstil" LLC	2,59	
JSC "Bolshevichka"	2,20	
LLC "Sewing enterprise Galant"	1,72	

Based on the results of the study, the following conclusions can be drawn:

1. the Leaders in terms of economic security were two companies: CJSC "Salyut" and JSC research and production enterprise "ANA" (high level of economic security of the enterprise).

2. the Average level is reached by: LLC "Severny Tekstil" and JSC "Bolshevichka". These enterprises have the potential to increase the level of economic security to the highest values by eliminating existing problems and shortcomings, as well as by increasing the innovation potential.

3. the company "Sewing Enterprise Galant" was at risk due to its small scale and low level of implementation of innovative technologies.

4 Conclusions

It can be concluded that the development of an enterprise in the field of innovation is not only a quick reaction to the application of the latest developments, but also the results of this application, which will directly affect the company's profit and other economic indicators [5].

The ability of an enterprise to quickly implement various types of innovations determines its potential and the possibility of not only maintaining the existing level of economic security, but also its growth in the future. It is very important to stimulate new promising ideas and various kinds of inventions [18].

It should also be noted that monitoring of industrial enterprises by the level of economic security is largely based on the indicators of innovative activity of the enterprise and is of particular importance in the prospects for the development of the industry. The more attention the company management pays to the development of the innovation sphere, the more stable is the level of economic security and the more attractive the company is in the eyes of investors. Investment attractiveness is directly related to the allocation of subsidies to the company, which means its further development prospects.

5 Acknowledgements

We are very grateful to the organizers of the international scientific and practical conference "Transformation of corporate governance models under the new economic realities" for the opportunity to present and publish a report on topical issues of the impact of new technologies on the development and effectiveness of management systems.

References

1. E.V. Gorshenin, and N. Chumachenko, Russian entrepreneurship, 63 (2011)
2. E.N. Chizhov, V.Yu. Sorokin, O.O. Vesnina, E.V. Osychenko, Vestnik BSTU named after V. G. Shukhov, 156 (2018)
3. L.M. Putyatin, N.In. Tarasova, L.A. Lavrova, Bulletin Of The University, 49 (2016)
4. E.V Jamai., Y.P. Anisimov, S.A. Povekvechnykh, Finance. Economy. Strategy., 26 (2014)
5. N.In. Alekseeva, T.N. Autograph, O.V. Orlova, Bulletin Of The University, 69 (2018)
6. A.V. Masalov, Ekonominfo, 51 (2007)
7. E. Sapozhnikova, E. Karanina, O. Ryazanova, Svetlana Grin, Revista ESPACIOS, **39 (28)**, 14 (2018)
8. L. Yurieva, E. Dolzhenkova, E. Karanina, Espacios, **28**, 8 (2018)
9. E. Karanina, O. Ryazanova., N. Gritsuk, MATEC Web of Conferences: International Scientific Conference Environmental Science for Construction Industry, 193 (2018)
10. E.V. Karanina, O.A. Ryazanova, A.N. Timin, Advances in Intelligent Systems and Computing : 19th International scientific conference "Energy Managment of Municipal Transportation Facilities and Transport", 318 (2017)
11. E.V. Karanina, O.A. Ryazanova, A.N. Timin, L.P. Domracheva, E3S Web of Conferences 2019, 08002 (2019)
12. G. Snigireva, I. Sennikova, E. Karanina, A. Bakhtimov, L. Domracheva, E3S Web of Conferences 2019, 08063 (2019)
13. E. Karanina, K. Kartavyh, E3S Web of Conferences 2019, 02090 (2019)
14. D. Loginov, E. Karanina, E3S Web of Conferences 2019, 02021 (2019)
15. D.A. Loginov, E.V. Karanina, M.V. Palkina, *Growth Poles of the Global Economy: Emergence, Changes and Future Perspectives* (2020)
16. E. Karanina, M. Bortnikov, E3S Web of Conferences 164, 10011 (2020)
17. V. Bezdenezhnykh, E. Karanina, D. Arutiunian, E3S Web of Conferences 164, 09009 (2020)
18. Y.A. Vnuchkov, V.A. Moskovsky, E.I. Lukin, Scientific papers (Vestnik MATI), 190 (2011)

Transformation of corporate governance in the context of digitalization of the economy

Tatiana Kuzminova*, *Ilya* Marushcak, *Natalia* Onanko, and *Alexander* Yurasov

Moscow International University, Leningradsky Prospect, 17, 125040 Moscow, Russia

Abstract. The purpose of the work is to identify areas for improving the efficiency of corporate governance in the context of the digitalization of the economy. The characteristic features of a corporation as a system, the development of which is influenced by its subsystems and supersystem, have been highlighted and described. It has been shown that the quality of management impact on a corporation depends on how fully the experience of the corporation's development in the past is taken into account. Adequate forecasting of its possible states in the future is also important. Based on the study of the triad of the processes of development of socio-economic systems (ontogenesis, phylogenesis and technogenesis), it is shown that currently there is an imbalance caused by a sharp acceleration of technogenesis. The dialectical consideration of the triad of developmental processes led to the conclusion that the imbalance was caused primarily by the exacerbation of the contradiction between technogenesis and ontogenesis. As a point of relevance for resolving the revealed contradiction, it is proposed to shift the emphasis in corporate management to phylogenesis. Specific measures for the development of phylogonese in corporations are the introduction and development of new forms of joint activities, the transformation of organizational structures, the creation of an atmosphere of mutual understanding, justice, morality, ethics, and culture.

1 Introduction

The quality of control is determined by the depth and adequacy of the control subject's ideas about the controlled object. The holistic perception of the control object implies knowledge of its past, present and forecasting the future both at the level of the control object itself (modeled as a system of interconnected elements) and at the level of its subsystems and supersystems (Fig. 1). In this case, as a general rule, the goals of the system development lie in the supersystem, and the most effective are the impacts on the system from the supersystem. In the absence of an opportunity to influence from the supersystem, one can go down to the level of the subsystems. The properties of the elements (subsystems) of the system are determined by their interaction with other elements, therefore, a change in the structure of these interactions leads to a change in the system

* Corresponding author: tvk-07@mail.ru

itself. It is currently being actively used to implement change in corporations. A change in the organizational structure leads to changes in the characteristics of the corporation itself [1, 2].

At the corporate level, this manifests itself as three distinct models. There are Anglo-American, Japanese and German models of corporations. While maintaining the common elements characteristic of any corporate governance system (shareholders, directors, managers), different rules of interaction between participants lead to differences in the properties of corporations. For example, if in Japanese corporations, interaction is aimed at establishing long-term business ties, then in American corporations - at establishing a balance of power.

	Past	**Present**	**Future**
Supersystem	Subject of analysis for the purpose of making a management decision	Subject of analysis for the purpose of making a management decision	Management impacts that lead to changes in the state
System	Subject of analysis for the purpose of making a management decision	Real state of the control object	Desired state of the control object
Subsystems	Subject of analysis for the purpose of making a management decision	Subject of analysis for the purpose of making a management decision	Organizational changes that support changing the management object

Fig. 1. Management object as a complete system

Differences in the integral properties of corporations are also manifested in the transparency of their functioning, widely discussed by scientists and practitioners, which has both positive and negative effects on all participants in socio-economic relations [3]. The national composition of its members leaves its imprints on transnational corporations [4]. This reflects not only the interaction of elements in the system, but also the qualities of the elements that they manifest outside the system.

Possible options for the development of a controlled object under the influence of managerial influences depend on many factors located both in the internal environment of the control object (the level of the system and subsystems) and in the external (at the level of the supersystem) [5, 6]. It should be remembered here that the interaction of factors with each other also leads to a synergistic effect that distinguishes the final impact from the sum of the impacts of all factors as independent ones. At the same time, the level of uncertainty also rises, requiring special competencies from the manager in making decisions [7].

A distinctive feature of the modern stage of development of society is the rapid development of digital technologies, which led to the acceleration of the course of all socio-economic processes. “In 2017, the combined value of digital platforms with a market capitalization of more than $100 million exceeded an estimated $7 trillion, which is 67% more than in 2015" [8, P. 8]. "The dynamics of global Internet traffic in 2002 was 100 GB per second, in 2007 – 2000 GB, in 2017-46000 GB, by 2022 it is expected 150700 GB per second" [8, P. 2].

Management is increasingly faced with situations where previously well-functioning management methods lose their effectiveness in new conditions. The reason for this is the qualitative changes in interactions both within the corporation between the divisions and the corporation with external stakeholders as a result of a sharp increase in the amount of information that supports all socio-economic processes.

In these conditions, a further increase in the efficiency of corporate management systems is possible only with an emphasis on the processes occurring in the supersystems, identifying and taking into account the objective laws of the functioning of society in general and the economy in particular.

2 Methodological approaches and research methods

From the point of view of dialectics (ideal for G. Hegel and materialistic for K. Marx) development is a consequence of the manifestation of objective laws. Corporations change according to system laws. From the point of view of making managerial decisions, the most important of them is the unity and struggle of the opposites. At the same time, the formation and resolution of contradictions of the corporation is ensured by objectively different development goals: the corporation and its divisions (systems and its subsystems), different divisions of the corporation (different subsystems of the unified system), the corporation and the socio-economic system of the regions where the corporation's products are sold (systems and supersystems).

Another important dialectical law from the point of view of the issues under consideration is the transition from quantity to quality. The rapid growth in the amount of information, provoked by the intellectual activity of people, has led to a qualitative change in all spheres of society. This is one of the key reasons for corporations are undergoing changes that reduce the effectiveness of management practices that have proven to be reliable in the industrial age.

The fractality of the development of all life on Earth will make it possible to draw an analogy between the two extreme forms of power in society (dictatorship and democracy) [9, p. 208] and the priority of the participants in the management of the corporation (the board of directors and shareholders). At each moment in time, the corporate governance system is represented by a certain ratio of the power of the board of directors and shareholders. "The crisis of power comes when it is appropriate only for a very limited number of people or is not able to limit the dangerous individual or group egoism of people" [9, p. 209]. People in the corporation and its external environment are all its members and stakeholders. If earlier management considered customer satisfaction to be the most important element of its activities, then the modern management paradigm asserts that success is impossible without harmonious coordination of the interests of all stakeholders, individual and group ones.

3 Results and Discussion

The resolution of any contradiction must use a point of relevance that differs from the poles that entered the dialectical struggle. For this, it is necessary to move from a pair of opposites to a triad, as it happens, for example, in biological systems. Today, many models of objects of leadership in management are built by analogy with how biological systems manifest themselves. In this case, biological terms are borrowed. For example, it is widely known that corporations are divided according to the type of innovative behavior into violets, patents, exporters and commutators. Continuing this tradition, the modern researcher of business systems A.G. Teslinov proposed to consider the development of all socio-economic systems from the point of view of the simultaneous occurrence of three processes: two, similar to processes in all biosystems (phylogenesis and ontogeny) and the third, created by the creativity of people, characteristic only for human society - technogenesis [10]. With regard to a corporation developing in the context of digitalization, it looks like this (Figure 2).

The contradictions that ensure development are exacerbated at each specific moment, mainly between two processes, the development of the third serves as a point of relevance for its resolution.

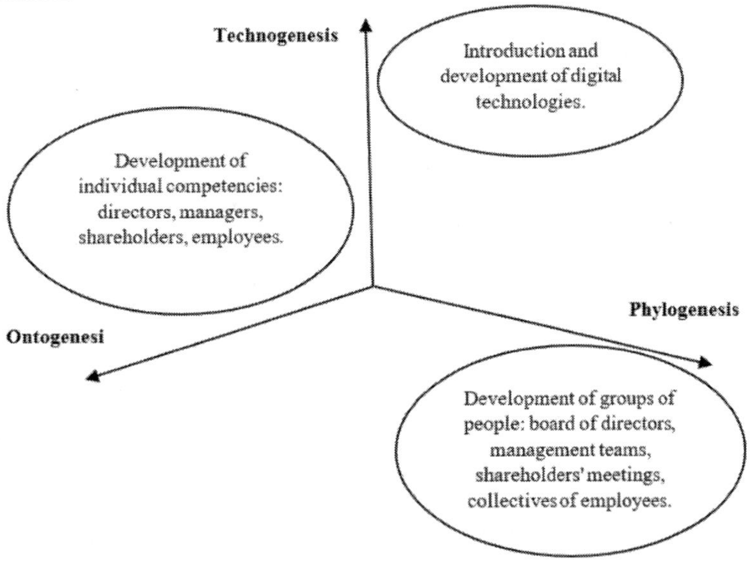

Fig. 2. The evolutionary whole of a corporation in the context of the digitalization of the economy

In our opinion, the contradiction between technogenesis, which is developing at an excessively fast pace, and ontogenesis, which is the basis of all democratic processes, has now sharpened, which has led to many ethical problems [11, 12], the resolution of which is possible by accelerating the development of phylogenesis. At the corporate level, "the development of phylogenesis is the emergence of new norms of joint activity, organizational structures, mutual understanding, justice, morality, ethics, culture" [10].

The triad of integrity is manifested at all levels of socio-economic systems. An attempt to look at the processes of globalization from the point of view of reducing costs [13] also reflects the manifestation of the development of the triad processes (Fig. 3).

The era of the Enlightenment and the New Age was preceded by the Renaissance, which returned human attention to the center, ontogenesis was accelerated, which led to geographical discoveries and many inventions that made it possible to reduce the cost of moving goods. This was the first stage of globalization.

"The second acceleration of globalization took place in the 1990s, when the ICT revolution drastically lowered the cost of moving ideas" [13]. Here the priority of the development of technogenesis is obvious, which is shown in Figure 3 and reflected by the central blocks.

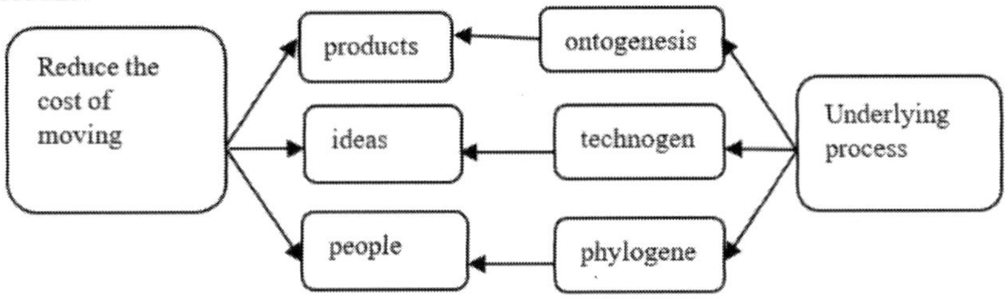

Fig. 3. Manifestation of the development triad in the cost of moving goods, ideas, and people

Further, R. Baldwin [13] predicts a third reduction in expenses, costs of moving people, which corresponds to the activation of phylogenesis and is designated by us as the direction of transformation of the corporate management system. Currently, this process manifests itself in the form of corporate social responsibility aimed at harmonizing the interests of all stakeholders of the corporation - both internal and external. In the future, the emphasis on these processes will be a condition for the effective management of corporations in the context of digitalization of all spheres of society.

4 Conclusions

1. The corporation exhibits properties that are common to all systems. Until recently, the effectiveness of corporate management depended on the ability to take a holistic view of it as a system.

2. A corporation, like any living system, develops according to dialectical laws, the main of which (from the point of view of making managerial decisions) is the formation and resolution of contradictions.

3. In any corporation, three processes occur simultaneously - phylogeny, ontogeny and technogenesis. For the long-term development of a corporation, a balanced development of all processes is necessary.

4. The digitalization of society has led to a significant bias of the development triad towards technogenesis, which has caused many ethical problems in society in general and at the level of corporations in particular.

5. Bringing the corporate development triad into balance requires a redistribution of management's attention to addressing phylogenetic issues, including the development of corporate ethics and culture.

Acknowledgements

The article has been prepared within the framework of the research project of ANOVO "Moscow International University" on the topic "Management in the digital economy: Russian and international experience."

References

1. S.P. Zemtsov, A.V. Chernov, Journal of the new economic association, *1(41)* (2019)
2. S.V. Shibaev, I.I. Marushchak, Economic Systems, **11(1)**, 60 (2018)
3. T. Rovinskaya, *Mirovaya ekonomika i mezhdunarodnye otnosheniya*, 64 (2020)
4. T. V. Kuzminova, V. A. Prokudin, Consolidation of intellectual resources as a Foundation for the development of modern science: collection of articles of the International scientific and practical conference (2020)
5. David Henriques, Ruben Pereira, Rafael Almeida, Miguel Mira da Silv, FORESIGHT and sti governance, **14** (2020)
6. T. V. Divina, N.A. Onanko, A. B. Yurasov, FES: Finance, Economy, Strategy, **15** (2018).
7. E. Vashurina, T. Bazarov, Organizational psychology, **10** (2020)
8. Digital economy report 2019: creating value and generating benefits: implications for developing countries, Review, 16 (2019)

 https://doi.org/10.1051/shsconf/20208903007

9. V.G. Maslennikov, Theory of change, Experience of combining ancient and modern knowledge, 251 (2000)
10. A. G. Teslinov, *The concept of development, The first stage of comprehension of development*, (2019) http://edu.dbaconcept.ru/
11. T. V. Kuzminova, Scientific achievements of higher education 2020: collection of articles of the International research competition (2020)
12. Ethics and "digital issues": Ethical problems of digital technologies, 207 (2020)
13. R. Baldwin, Journal of Economic Sociology, **18** (2017)

Using artificial intelligence for effective decision-making in corporate governance under conditions of deep uncertainty

Aleksandra Volosova[1,*], and *Ekaterina* Matiukhina[2]

[1]Faculty of Management, Automated control systems, Automobile and Road Construction State Technical University (MADI), Leningradsky prospect, 64, 125319 Moscow, Russia
[2]Russian Technological University (MIREA), Vernadsky Avenue, 78, 119454 Moscow, Russia

Abstract. The article deals with research related to the use of artificial intelligence technologies for effective decision-making in corporate governance under conditions of deep uncertainty. To process uncertainty, it is proposed to use the cognitive capabilities of artificial intelligence. Cognitivism can be used to implement intuitive, psychological and other components of the internal mental activity of a person when making decisions. These capabilities allow one to make informed decisions and predict the consequences of these decisions. To study the properties of deep uncertainty, the authors suggest using a tensor model. The tensor model of deep uncertainty makes it possible to study additional properties of uncertainty that are not available in traditional models, such as Bayesian formalism, Dempster-Shafer theory, fuzzy sets, a method based on certain factors (Stanford formalism), and others. The use of the tensor model allows one to study the spatial model of uncertainty, real and imaginary values of uncertainty, as well as uncertainty invariants with respect to various transformations of the coordinate system.

1 Introduction

Technological transformation and coronavirus has had a serious impact on the concept of management. To improve the efficiency of corporate governance, it is necessary to take into account new business formats. Corporate management of ecosystem, platform, and distributed companies is a highly computationally complex task. Artificial intelligence technologies are now being successfully used to solve such problems.

Tasks such as:

- data collection and analysis
- monitoring the organization's own activities
- monitoring the workforce problems
- representation of the company's brands , products and services in social networks, monitoring user views of this information
- risk management procedures

[*] Corresponding author: avolosova32@gmail.com

- market forecasts
- analysis of existing and forecasting new socio-cultural trends to monitor their impact on business
- monitoring the company's position on various social issues
- forming and optimizing the organization's own rating and analyzing the impact of the rating on the stock market, investors, consumers, and social networks.
- forecast the organization's investment in various areas in order to avoid the financial collapse of the organization
- assessment of risk factors
- data collection and analysis management decision-making, personnel decision-making, internal and external financial policy decision-making, disaster decision-making can be solved using artificial intelligence.

Now the penetration of AI into the daily lives of people and businesses is undeniable [3]. Obviously, artificial intelligence also has an impact on corporate governance. The complexity and rapid change of the world make making the right decisions a very difficult process [5][6][7][11]. And often the cost of mistakes is very high [1]. Of course, you can use automation to partially reduce risks. But it is better to use the cognitive capabilities of artificial intelligence. The cognitive capabilities of artificial intelligence can be used to implement intuitive, psychological and other components of the internal mental activity of a person when making decisions. The cognitive capabilities of artificial intelligence allow one to make informed decisions and predict the consequences of these decisions. Usually, the information that a person uses to make decisions contains noise that changes the measured values [2]. So we are dealing with uncertainty. And if decisions are made in dynamic interaction with the system, then there is such a thing as deep uncertainty. Decision-making under deep uncertainty is a complex process that which combines artificial intelligence technologies such as fuzzy knowledge processing and forecasting. Next, we consider possible approaches to handling deep uncertainty in the decision-making process. When we try to represent uncertainty in the knowledge base, we may encounter the following problems:

- the degree of uncertainty of information is so high that during formalization, an erroneous representation of this information inevitably occurs

- based on poorly presented knowledge, erroneous conclusions may be drawn.

In other words, the success of making a decision depends on how we present the uncertainty.

When using traditional tools for mathematical modeling of uncertainty, such as Bayesian formalism, Dempster-Shafer theory, fuzzy sets, the method based on certainty factors (Stanford formalism) and their various combinations and modifications, we cannot say that the resulting model fully reflects all the properties of uncertainty. This is due to the unavoidable use of heuristics in modeling uncertainty. We can't avoid using heuristics to formalize uncertainty, but we can try to minimize its impact on our research. To do this, consider the tensor basis and try to represent the uncertainty using this basis.

2 Tensor method of dual networks

The fuzzy set basis can be used to handle uncertainty [4]. To formalize uncertainty we use the a fuzzy quantity. A fuzzy quantity can be defined as an arbitrary fuzzy set:

$$Z = \{\zeta, \vartheta_Z(\zeta)\}, \zeta \in Z \quad (1)$$

This set ζ defined on the set of real numbers R and the membership function of a fuzzy quantity is a function of the form:

$$\vartheta_Z(\zeta) \to R[0,1] \quad (2)$$

Define $\Im$ as tensor:

$$\Im = \zeta \otimes (\vartheta_Z(\zeta))^T \quad (3)$$

$$\Im = \{\zeta_1, \zeta_2 \dots \zeta_n\} \otimes \begin{Bmatrix} \vartheta_Z(\zeta_1) \\ \vartheta_Z(\zeta_2) \\ \dots \\ \vartheta_Z(\zeta_n) \end{Bmatrix} = \begin{Bmatrix} \zeta_1 * \vartheta_Z(\zeta_1) & \zeta_2 * \vartheta_Z(\zeta_1) & \dots & \zeta_n * \vartheta_Z(\zeta_1) \\ \zeta_1 * \vartheta_Z(\zeta_2) & \zeta_2 * \vartheta_Z(\zeta_2) & \dots & \zeta_n * \vartheta_Z(\zeta_2) \\ \dots & \dots & \dots & \dots \\ \zeta_1 * \vartheta_Z(\zeta_n) & \zeta_2 * \vartheta_Z(\zeta_n) & \dots & \zeta_n * \vartheta_Z(\zeta_n) \end{Bmatrix} \quad (4)$$

So a fuzzy variable can be represented by a tensor.

Using the transformation, we can obtain from the dyad a symmetric tensor, an antisymmetric tensor, a sphere tensor, and a deviator. And we can determine additional properties of the fuzzy variable as a result of analyzing these tensors.

3 Results of applying the tensor basis for the system modeling of deep uncertainty

Let's consider the application of the tensor basis for modeling uncertainty in the problem of forecasting passenger traffic volumes in St. Petersburg and the Leningrad region for the period up to 2030[8][9][10][12]. The calculation is made for conservative and innovative options (See Tab.1, Tab. 2).

Table 1. Passenger traffic trends to 2030 (conservative option)

	2020	**2025**	**2030**
Passenger traffic (million people)	**2068,9**	**2130**	**2194,9**
Saint-Petersburg	1971,3	2027	2086,9
Leningrad region	97,6	103	108

Table 2. Passenger traffic trends to 2030 (innovative option)

	2020	**2025**	**2030**
Passenger traffic (million people)	**2139,1**	**2216,9**	**2322,3**
Saint-Petersburg	2040,7	2108,4	2202,4
Leningrad region	98,4	108,5	119,9

Passenger traffic for each year is a fuzzy value. Hence the following fuzzy variables can be defined for each calculation option (conservative and innovative) (See Tab. 3).

Table. 3. Fuzzy variables

Conservative option	Innovative option
ζ_{1c} - **fuzzy 2068.9 (2020)** ζ_{2c} - **fuzzy 2130 (2025)** ζ_{3c} - **fuzzy 2194.9 (2030)**	ζ_{1i} - **fuzzy 2139.1 (2020)** ζ_{2i} - **fuzzy 2216.9 (2025)** ζ_{3i} - **fuzzy 2322.3 (2030)**

Now we define universums and membership functions for fuzzy variables (See Tab. 4).

Table 4. Universums and membership functions for fuzzy variables

	conservative option	**innovative option**
2020	$\zeta_{1c}=[1906:26.84:2100]$ $\vartheta_Z(\zeta_{1c})=\text{trimf}(\zeta_{1c},[1907.86\ 2068.9\ 2081.12])$	$\zeta_{1i}=[1852.94:40.88:2220.86]$ $\vartheta_Z(\zeta_{1i})=\text{trimf}(\zeta_{1i},[1893.82\ 2139.1\ 2179.98])$
2025	$\zeta_{2c}=[2044.46:12.22:2154.44]$ $\vartheta_Z(\zeta_{2c})=\text{trimf}(\zeta_{2c},[2056.68\ 2130\ 2142.22])$	$\zeta_{2i}=[2108.04:12.98:2248.86]$ $\vartheta_Z(\zeta_{2i})=\text{trimf}(\zeta_{2i},[2123.54\ 2216.9\ 2232.46])$
2030	$\zeta_{3c}=[2114.04:12.98:2220.86]$ $\vartheta_Z(\zeta_{3c})=\text{trimf}(\zeta_{3c},[2127.02\ 2194.9\ 2207.88])$	$\zeta_{3i}=[2174.74:21.08:2364.46]$ $\vartheta_Z(\zeta_{3i})=\text{trimf}(\zeta_{3i},[2195.82\ 2322.3\ 2343.38])$

The Fig.1 (a-c) shows the fuzzy variables for the conservative and innovative options. So we can perform all the operations available to us from fuzzy set theory on variables. But each variable has a quantity behind it. In our case, this is passenger traffic. By studying the fuzzy set, we can draw conclusions about the fluctuations of this quantity within the boundaries of the fuzzy set. However, we can't say anything about this quantity beyond these boundaries. This limitation can be overcome by using a tensor basis. Applying formula 3 to the vector of values of each fuzzy variable and its membership function, we obtain tensors of fuzzy variables (see Fig. 1 d-j).

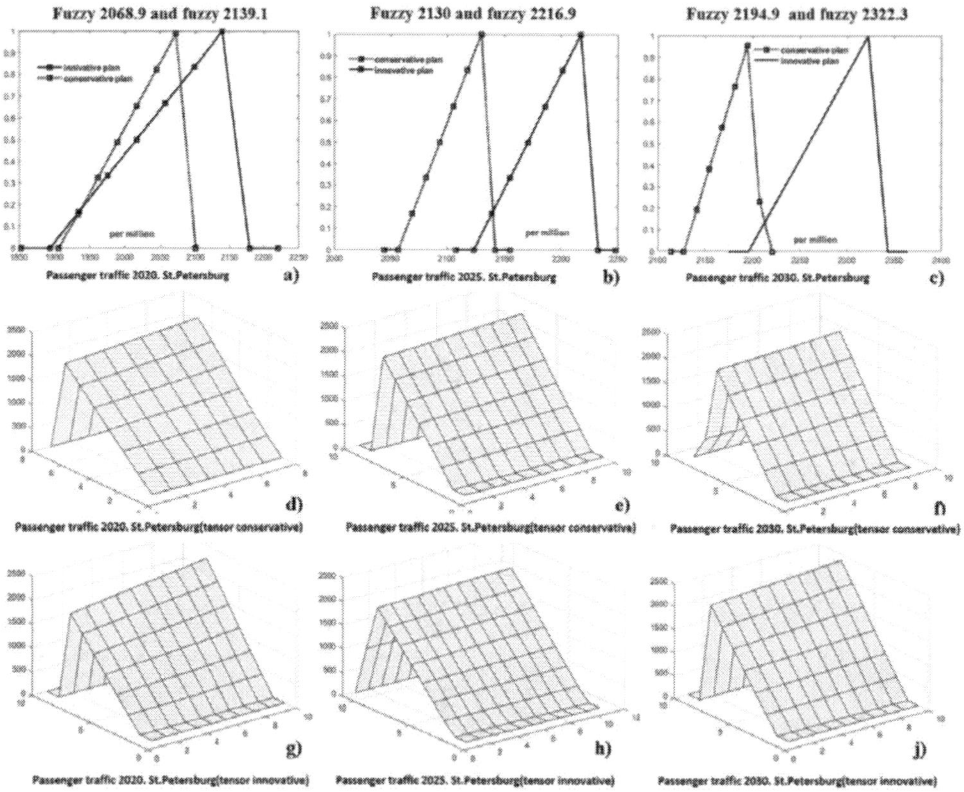

Fig. 1. The fuzzy sets and tensors

Hence instead of a set of values for a fuzzy variable we get a mathematical object and can now explore additional properties of uncertainty. Using the tools of tensor algebra, we can obtain additional tensors for each tensor of a fuzzy variable. As an example, we decompose the tensor for fuzzy 2068.9 into a sphere tensor and a deviator. The resulting tensors describe additional properties of this fuzzy variable – it's clear and fuzzy parts (see Fig. 2).

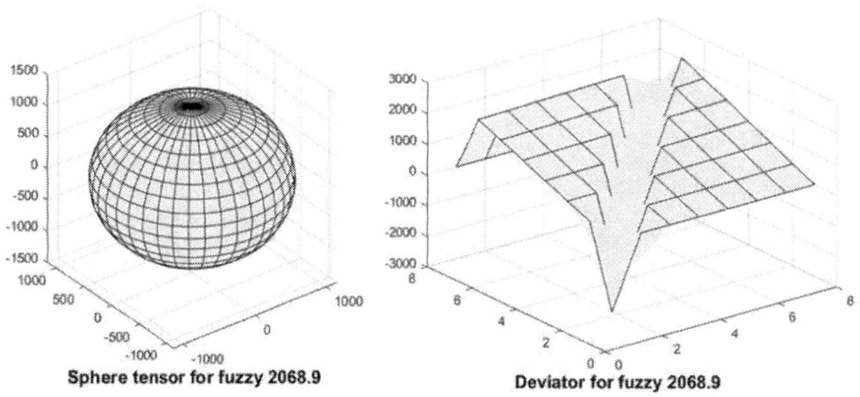

Fig. 2. The sphere tensor and deviator for fuzzy 2068.9

We can also decompose the tensor for fuzzy 2068.9 into symmetric and antisymmetric tensors (see Fig. 3).

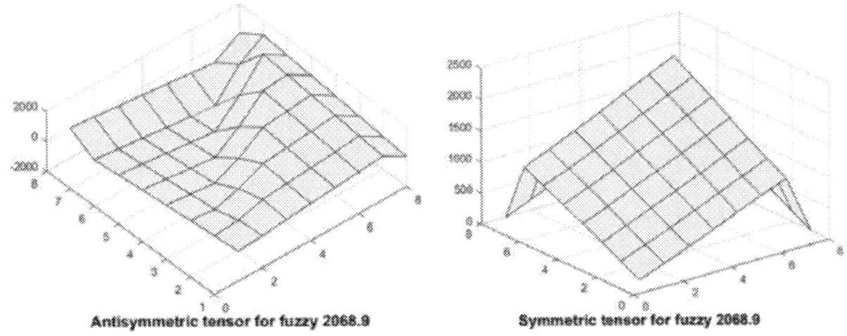

Fig. 3. The antisymmetric and symmetric tensors for fuzzy 2068.9

An antisymmetric tensor characterizes the transformation of a fuzzy tensor with respect to the base coordinate system. The antisymmetric tensor does not change when the coordinate axes are reflected and is constant in all coordinate systems where Newton's laws apply.

Finally, we can get the eigenvalues of the fuzzy variable tensor. The combination of these values forms the spectrum of the fuzzy variable tensor (See Fig. 4).

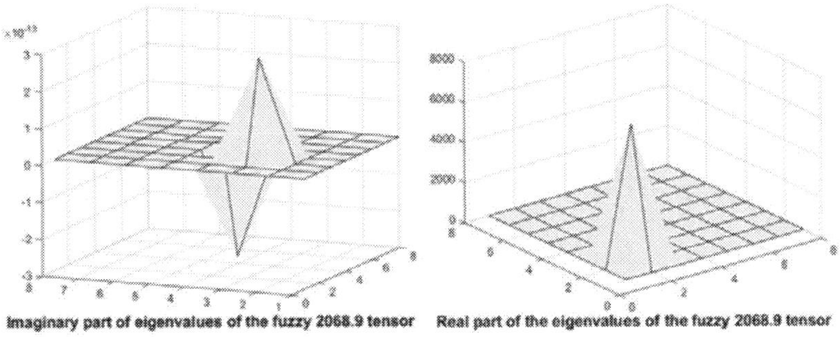

Fig. 4. The real and imaginary parts of the eigenvalues of the fuzzy 2068 tensor

So we see that the use of the tensor basis significantly expands the possibilities of studying the deep uncertainty that is an inevitable companion of artificial intelligence technologies. Especially when these technologies are used to solve complex corporate governance problems.

4 Conclusions

The research conducted in the article allows us to draw the following conclusions:

1. To improve the efficiency of corporate governance, it is necessary to take into account new business formats. This means using artificial intelligence technologies
2. The cognitive capabilities of artificial intelligence can be used to implement intuitive, psychological and other components of the internal mental activity of a person when making decisions. These capabilities allow one to make informed decisions and predict the consequences of these decisions.
3. Effective decisions in the corporate governance process are made under conditions of deep uncertainty, as these decisions are made in dynamic interaction with the management system
4. The tensor model of deep uncertainty allows you to explore additional properties of uncertainty that are not available when using traditional models, such as Bayesian

 https://doi.org/10.1051/shsconf/20208903008

formalism, Dempster-Shafer theory, fuzzy sets, and a method based on certain factors (Stanford formalism), etc. The tensor model is a spatial model of uncertainty; its real and imaginary values and invariants with respect to various transformations of the coordinate system.

References

1. O.I. Maksimychev, V.B. Golubkova, V.A. Vinogradov, Zav Zav Aung, *The Integration of a Set of Functions for the Product Life Cycle of a Virtual Enterprise, Advances in Business and Finance,* 2018 3rd BEM International Conference on Education, Sociology and Humanities (BEM-ESH 2018). August 14-16 (2018).
2. Kochenderfer, Mykel J., *Decision making under urcentainty: theory and application* (1980)
3. B.S. Subbotin and E.U. Ahmetzhanova, *Research on the integration of intelligent transport systems,* IOP Conference Series: Materials Science and Engineering, **832** (2019).
4. A.V. Volosova, Computational nanotechnology,1, (2019)
5. D. Akimov, P. Krug, A. Ostroukh, V. Ivchenko, T. Morozova, I. Sadykov, International Journal of Applied Engineering Research, **11(9)**, 6435 (2016)
6. E.Matyukhina, A. Ignatov, V. Ivchenko, P. Krug, T. Morozova, Applied Mechanics and Materials, **851**, 470 (2016)
7. L.V.Zhivoglyadova, V.G. Zhivoglyadov and V.V. Siliyanov, IOP Conference Series: Materials Science and Engineering, International Conference on Digital Solutions for Automotive Industry, Roadway Maintenance and Traffic Control (DS ART 2019) 1 November 2019, Cholpon-Ata, Kyrgyzstan, **832** (2019)
8. D.B. Efimenko, V.M. Vlasov, V.N. Bogumil, V.A. Demin and S. A. Filatov, IOP Conference Series: Materials Science and Engineering, International Conference on Digital Solutions for Automotive Industry, Roadway Maintenance and Traffic Control (DS ART 2019) 1 November 2019, Cholpon-Ata, Kyrgyzstan, **832** (2019)
9. N. Novikov, S.V. Eremin, A.G. Shevtsova, *Development of a methodology for optimizing the urban route in the metropolitan area,* IOP Conference Series: Materials Science and Engineering, International Conference on Digital Solutions for Automotive Industry, Roadway Maintenance and Traffic Control (DS ART 2019) 1 November 2019, Cholpon-Ata, Kyrgyzstan, **832** (2019)
10. E. Matyukhina, A. Ignatov, V. Ivchenko, P. Krug, S. Pavelyev, Applied Mechanics and Materials, **851**, 477 (2016)
11. E. Matyukhina, D. Akimov, A. Ostroukh, V. Ivchenko, P. Krug, ARPN Journal of Engineering and Applied Science, **12(11)**, 3433 (2017)
12. A comprehensive plan of transport services for the population of St. Petersburg and the Leningrad Region for the medium and long term (until 2030) in terms of suburban passenger transportation, approved by the order of the Government of the Leningrad Region dated 26.06.2019, 592 (2019)

Informal evaluation of corporate image based on text mining

Kostyantyn Malyshenko[1,*], *Vadim* Malyshenko[1], *Marina* Anashkina[2], and *Dmitriy* Anashkin[1]

[1]V.I. Vernadsky Crimean Federal University, Republic of Crimea, 295007 Simferopol, Russia
[2]Sevastopol State University, 299053 Sevastopol, Russia

Abstract. This article explores the possibilities of using text-mining technologies in order to determine the image of corporations based on data obtained from Twitter social network. The problem of low efficiency of traditional methods of consumer opinion research and the need to develop methods based on unsolicited data has been actualized. Consumer opinion is an indicator of the level of corporate image. Analysis of opinions allows you to develop an effective policy to improve it. The authors have developed a methodology for assessing the corporate image. The article analyzes the work of leading researchers. The features of the use of technologies when working with texts published in Russian have been analyzed. An index of customer (consumer) satisfaction has been developed, which is proposed as a basis for determining the level of corporate image. The obtained results of the study allow to make further adjustments to the corporation's policy in order to improve its image.

1 Introduction

In modern conditions of development, corporations play a key role. They are the largest economic entities and perform a number of socio-economic functions, among which are job creation, satisfaction of consumer demand, ensuring capital flow through financial channels, development of various sectors of the economy through investment, etc. The effective functioning of the corporation, its successful development in the changing economic, political and social conditions is the result of a competent policy, which is not possible without an effective analysis of external and internal factors. In the modern customer-oriented economy, the assessment of consumer opinion becomes a key factor in making appropriate management decisions and achieving maximum efficiency of the corporation's activities [1, 2].

A corporation's image is a key factor in its advancement in a competitive market. In this regard, the issue of searching for modern methods of analyzing the consumers' opinions regarding the quality of its work is relevant. Customer satisfaction with individual parameters of goods / works / services sold by the corporation forms the basis for developing specific solutions to eliminate identified problems and build further functioning policies.

* Corresponding author: docofecon@mail.ru

The choice of such methods, as a rule, should meet the following requirements

1. Information content, clarity and objectivity of the results.
2. Efficiency of obtaining results.
3. The relevance of the results obtained (must be based on actual data).
4. Possibility of forecasting.
5. Availability of incoming information.
6. Accessibility and ease of use.
7. Minimal financial investment.

The above requirements at the present stage of scientific and technological progress are fully consistent with big data analysis technologies, namely, text mining.

Krylov V.S. in his work "Digital Economy: Intelligent Text Analysis" [3] explores the possibility of using the software modules that allow extracting and visualizing the metadata components of text emotionality. The author focuses on the need to develop the methods of analysis and forecasting necessary for the development of intelligent information systems, their deep learning, neural networks, artificial intelligence, etc.

Chang Ts. [4] in the study "Product sales forecasting using macroeconomic indicators and online reviews: a method combining prospect theory and sentiment analysis" notes that macroeconomic conditions and user reviews have a significant impact on consumer purchasing decisions, and they can potentially be used to make more accurate forecasts of corporate development. The author has developed a sentiment analysis algorithm for converting text online reviews into numerical values.

The need to determine opinions in order to maintain the image of corporations and build an effective policy for their functioning is also considered in the works of El Alau I. [5], Moussa S. [6].

Traditional methods of determining the opinion of consumers today are becoming ineffective, since they are based on historical data. Most often, in order to determine consumer sentiment and satisfaction in Russian practice, statistical and survey methods are used. This, in particular, is reflected in the works of O.V. Takhanova. [7], Amirkhanova L.R. [8]. Pelaez J. The work "Products and services valuation through unsolicited information from social media" [9] raises the problem of the low reliability of the currently existing methods of collecting and evaluating consumer opinions about various goods and services. The author proposes the calculation of a special index based on the processing of unsolicited consumer information using text mining technologies.

In the work of M. Song "Forecasting economic indicators using a consumer sentiment index: Survey-based versus text-based data" [10], the effectiveness of modern methods of processing unsolicited data and their comparison with the traditional method of polling respondents has been investigated. The authors come to the conclusion that methods for analyzing big data have some limitations, but they allow obtaining better information.

Eskisi H. in the study "A text mining application on monthly price developments reports" [11] emphasizes that the use of text mining provides great opportunities for economic research. The paper proposes a methodology for using text mining technologies to measure statistical consistency with the annual consumer price index.

Meanwhile, both of these methods do not meet the requirements put forward for this indicator, which were listed above. So, using only statistical data as a basis for constructing allows you to get results for a period that is significantly remote from the moment of direct receipt of the final result. In this regard, further development of decisions on this basis may have low efficiency, especially in the case when it is necessary to take urgent measures to change the policy of the corporation in order to improve its image.

Thus, the study of the possibilities of using text-mining technologies in order to assess the corporate image of Russian companies is a topical issue. To date, the approaches to data processing by these technologies proposed by the world community are poorly developed.

It is necessary to study the peculiarities of their application in order to analyze the texts left by Russian-speaking users, and also to develop a methodology for using the resulting data in order to assess the corporate image.

The purpose of this work is to determine the possibilities and limitations of the use of text-mining technologies in order to determine the satisfaction of consumers (clients) of corporations and to develop a methodology for assessing the corporate image.

2 Materials and Methods

In this work, we used data on the two largest banks operating in the Russian Federation with a corporate structure. These are banks VTB and Sberbank. We have used the Orange software, which is a set of various big data analysis tools.

It also contains all the necessary set of text mining tools, which allows you to conduct research without hindrance. In particular, the software allows you to automate the data collection process, since it has a special component - "Twitter". Thus, it can be used to extract short user messages on any topic from Twitter social network.

The use of the data of this social network meets the requirements of the research - the need to obtain information in real time, or with a minimum gap in the time window between the collection and the direct receipt of the final analysis result. Tweets posted by users are selected according to a given topic (tag). Thus, filtering tweets by corporation name allows us to extract consumer feedback about it, presented in the form of user posts on the social network.

With the help of the software component, we collected data on the two above corporations using the tags "VTB" and "Sberbank" for November 2020. Their volume was 505 tweets. The study was carried out in several stages, shown in Figure 1.

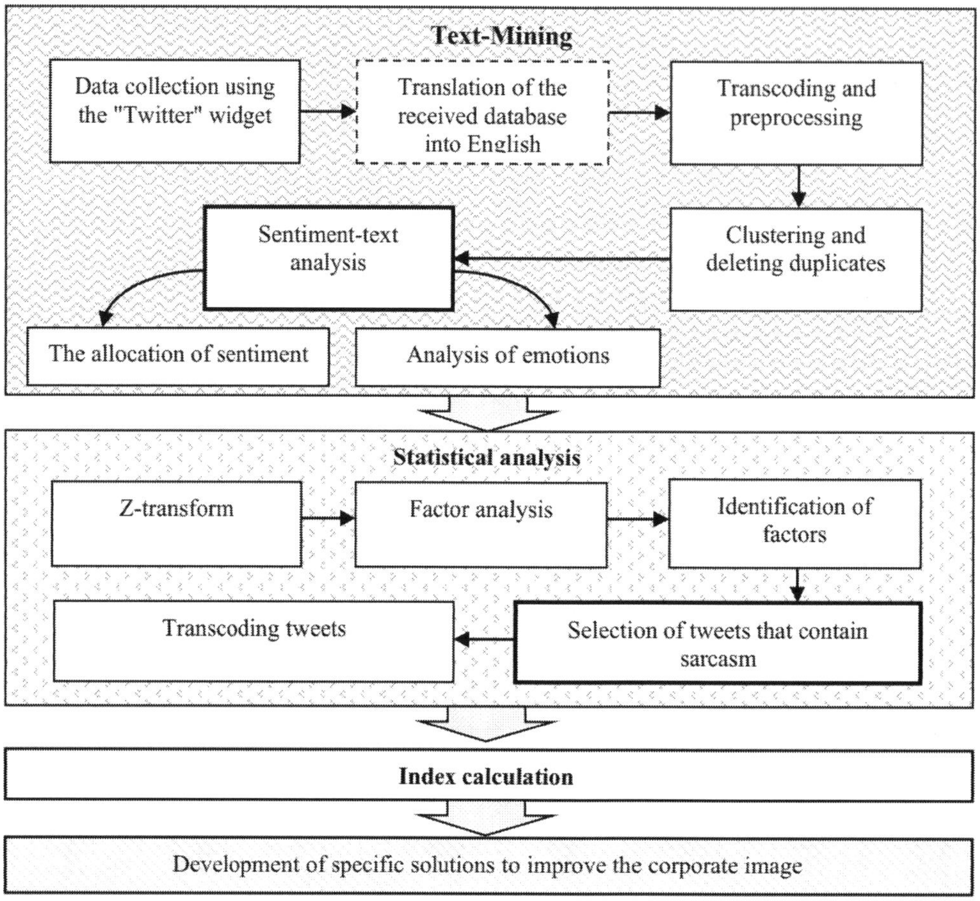

Fig. 1. Procedure for evaluating the corporate image.

Let's look at these stages in more detail.

3 Results and Discussion

As mentioned, the first step in research is to collect the necessary data. As a result, we obtained two databases of tweets with the tags "VTB" and "Sberbank", containing the texts published by users regarding these corporations. The software we use has a number of functions that allow automating the process of adding new tweets to the existing database. Thus, it is possible to obtain data in real time.

The main problem of using text mining technologies for analyzing texts left in Russian is the lack of necessary dictionaries in existing software for their further processing. In other words, the software does not recognize the meaning of such texts and, accordingly, does not allow calculating sentiment and extracting emotions from it. To solve this problem in this study, at the next stage, it is necessary to translate the obtained bases into English, since it is the language of the software developer.

At the next stage, the databases are loaded in English and prepared for further processing. For this, the software also has a special component. It performs conversion and preprocessing of the text - extra characters, punctuation marks, etc. are removed.

Next, you need to clear the text from duplicate tweets that are published by robots. This is necessary to obtain a more accurate result. Such tweets can affect the overall assessment of consumers' opinion on the functioning of the corporation, since it increases the share of certain emotions inherent in such a text through its numerous repetitions.

After that, sentiment and emotions are directly extracted according to the algorithms set by the program. We used the Vader algorithm for sentiment extraction, since it gives the most adequate assessment of the analyzed text. The Plutchik algorithm was used to highlight the emotions. It allows you to highlight all the groups of emotions necessary to define sarcasm in the text. The results obtained as a result of processing by text-ming technology require further statistical processing in order to clarify the belonging of individual tweets to one of three groups: positive, negative, neutral. This must be done in order to obtain the most accurate assessment of customer opinion, since the software does not recognize sarcasm.

For these purposes, we used SPSS statistical data processing software. The first step is to standardize the scores (Z-transform). This procedure is necessary because the obtained variables have different ranges and meanings in orders. They must be brought to a common denominator. Further, on this basis, we carried out a factor analysis, as a result of which the main factors in the selected groups of tweets corresponding to sarcasm were identified. As a result, it was revealed that sarcasm is a ratio of disgust, humor, sadness.

The next step is to define the criteria boundaries of these factors and identify the tweets containing sarcasm. Thus, tweets that are defined as positive and neutral, and contain sarcasm, should be categorized as negative ones. For this, we have carried out the appropriate conversion.

The resulting estimates are necessary to build an index of customer satisfaction. The higher the index, the higher the corporate image, and vice versa. We have formed three groups of tweets: positive, negative and neutral. To calculate the index, it is proposed to use the following formula developed by the authors:

$$Ind = (pos - neg)/(tot - neu) \quad (1)$$

In this formula, *pos* is the number of positive tweets, *neg* is the number of negative tweets, *tot* is the total number of tweets, *neu* is the number of neutral tweets.

The results of the frequency analysis are presented in Tables 1 and 2.

Table 1. Frequency analysis based on opinions about Sberbank.

If compound is greater than zero (positive tweet) - 1 if less than zero (negative tweet) - 2 if zero (neutral tweet) - 0					
		Frequency	Percentage	Percentage of valid	Accumulated percent
Valid	.00	159	28.7	31.5	31.5
	1.00	218	39.4	43.2	74.7
	2.00	128	23.1	25.3	100.0
	Total	505	91.2	100.0	
Missing	Systemic	49	8.8		
Total		554	100.0		

Table 2. Frequency analysis based on opinions about VTB Bank

If compound is greater than zero (positive tweet) - 1 if less than zero (negative tweet) - 2 if zero (neutral tweet) - 0					
		Frequency	Percentage	Percentage of valid	Accumulated percent
Valid	.00	148	29.4	29.4	29.4
	1.00	203	40.3	40.3	69.6
	2.00	153	30.4	30.4	100.0
	Total	504	100.0	100.0	

Let's calculate the indices according to the above formula:

$$Ind\ (Sberbank) = (218 - 128)\ /\ (505 - 159) = 0.26 \quad (2)$$

$$Ind\ (VTB) = (203 - 153)/(504 - 148) = 0,14 \quad (3)$$

We will interpret the index values according to the following table based on the Likert scale:

Table 3. Customer satisfaction scale.

Index value	Satisfaction level	Interpretation of the value	The level of the image of the Corporation
$1 \geq Ind > 0.5$	high	High level - the respondents' mood regarding the corporation is positive, the policy is effective, no revision is required	high
$0.5 \geq Ind > 0$	average	Average level - the general mood of the respondents towards the corporation is positive, the policy is effective, it needs to be continued	average
0 (± 0.01)	neutral	Neutral level - respondents' mood regarding the corporation is identified as both positive and negative, the policy is ineffective, it is necessary to take certain actions to improve the corporation's image	neutral
$0 > Ind \geq -0.5$	low	Low level - the general mood of respondents towards the corporation is negative, a revision of the existing policy is necessary	low
$-0.5 > Ind \geq -1$	crisis	Crisis level - respondents' mood regarding the corporation is negative, a complete change of the existing policy is required	crisis

Thus, the values of the indices obtained in this study for both corporations refer to the average level. That is, the customers are satisfied with the quality of services provided by the banks, dissatisfaction is generally insignificant. This is in line with the average corporate image score. The policy of their functioning should be continued, or it is possible to carry out a detailed analysis of customer dissatisfaction in order to eliminate the shortcomings and achieve a high level of image.

Conclusion

This article explores the possibilities of using text-mining technologies in order to determine the image of corporations based on data obtained from Twitter social network . We have developed a methodology for assessing the corporate image. The methodology was verified on the basis of data from VTB and Sberbank operating in Russia and having a corporate structure.

We have identified certain limitations regarding the possibility of using text-mining technologies when analyzing texts published in Russian. Existing software can analyze text mainly in the developer's language (English). At the same time, the training of the program is possible, but this requires the development of a Russian dictionary and its further introduction into the software.

To date, the application of the methodology proposed in the work requires the addition of an obligatory step - the translation of the resulting database into English. The results obtained do not have material distortions and allow further calculation of the index.

The proposed index can be interpreted according to the scale. Its results can be used as the basis for the development of specific management decisions aimed at changing the corporate image. This indicator has a number of advantages, in particular, it allows you to monitor the current state of the company's image, since it is based on current data. This allows you to timely respond to negative trends and take measures to prevent a fall in the image of corporations and a decrease in demand for their services.

Acknowledgment

The reported study was funded by RFBR and EISR, project number 20-011-31388 "Development of a methodology for using text-mining technologies to determine mood indices based on social networks".

References

1. K.A. Malyshenko, V.A. Malyshenko, M.V. Anashkina, Proceedings of the 32nd International Business Information Management Association Conference, IBIMA 2018 - Vision 2020: Sustainable Economic Development and Application of Innovation Management from Regional expansion to Global Growth, 1030, (2018)
2. K.A. Malyshenko, V.A. Fastynova, D.A. Mardar, M.A. Shostak, M.V. Anashkina, V.A. Malyshenko, Revista Inclusiones, **7**, 739 (2020)
3. V.S. Krylov, Scientific notes of the Crimean engineering and pedagogical University, **3(65)**, 120 (2019)
4. C.Zhang, Y.-X. Tian, Z.-P. Fan, Y. Liu, L.-W. Fan. Soft Computing, **24(9)**, 6213 (2020)
5. I. El Alaoui, Y. Gahi, R. Messoussi, Y. Chaabi, A. Todoskoff, A. Kobi, Journal of Big Data, **5**, 1 (2018)
6. S. Moussa, Journal of Retailing and Consumer Services, **49**, 86 (2019)
7. O.V. Tahanova, J.D. Tsibikdorzhieva, G.E. Alekseeva, A.R. Tsyrenov, Internet-journal "Science studies", 9, 6. (2017), https://naukovedenie.ru/
8. L.R. Amirkhanova, E.Yu. Bikmetov, Eurasian economic journal, **1(104)**, 349 (2017)
9. J.I. Peláez, E.A.Martínez, L.G.Vargas, Soft Computing, **24(3)**, 1775 (2020)
10. M. Song, K-S. Shin, Journal of forecasting, **38(6)**, 504 (2019)
11. H.B. Eskici, N.A. Kocak, Central bank review, **18(2)**, 51 (2018)

Developing a FinTech Ecosystem in Vietnam: Opportunities and Challenges for Startups

Van Thien Hao*

Quang Trung University, Dao Tan Str.,327, 590000 Quy Nhon city, Vietnam

Abstract. In the article, the author has analyzed the current situation of fintech market development and Vietnam's startup ecosystem in recent years. The research results show that, in the past 3 years, the number of fintech companies in Vietnam has nearly quadrupled, the market value in 2020 is estimated at 9 billion USD. With 72% of adults using smartphones, nearly 64% of the population using the Internet, along with the young generation of technology-savvy, the explosion of e-commerce and the support of the Government, Vietnam is a potential market for implementing tech startup models. Opportunities and challenges for fintech startups in Vietnam have been discussed by the author, from which solutions to develop fintech ecosystems are proposed in the upcoming time.

1 Introduction

Statista's research (2019) [11] shows that in the world there are more than 10,000 Fintech companies competing with traditional banks in everything from payments, fundraising, lending, forex trading and consulting. invest. In Vietnam, Fintech has gained great attention. They mainly focus on the areas of: payments, loans/deposits and other mixed services. The revenue of the Fintech sector in Vietnam is forecast to reach about 9 billion USD by 2020, when many products are strongly increasing in price, especially in mobile payment, Internet and e-wallets. However, compared to other countries in the region such as Malaysia, China, Singapore and other countries, Fintech in Vietnam is still very limited in terms of policy mechanisms, quantity, scope of activities and many other factors.

According Simon Kemp and Kepios [15], 72% of adults (population over 15 years old) in Vietnam use a smart phone. Meanwhile, according to Statistic, the number of smartphone users in Vietnam in 2018 & 2019 reached 32.43 and 43.7 million, respectively, and continues to increase rapidly year by year. Vietnam currently has 64 million Internet users (of which 61.73 million people access by mobile devices, accounting for 96% of Internet users and nearly 64% of Vietnam's population), ranking 6th in Asia-Pacific Ocean and 13th in the world (according to Internet World Statistics). In addition, with the young people's understanding of information technology, the explosion of e-commerce, the proportion of people with low bank accounts... are favorable factors to develop Fintech-based Finance - Banking services in Vietnam.

*Corresponding author: vthao@qtu.edu.vn

2 Fintech Ecosystem and Startups in Vietnam

Over the past few years, Vietnam is gradually closing the gap with the two leading startup hubs in the region: Indonesia and Singapore. One of Vietnam's advantages include its large population, which will reach 100 million people by 2030. The majority of the population is also young, tech-savvy and heavy mobile users. The potential for developing a technology startup model in this market is very lucrative.

According to a report on investments in Vietnam startups 2019 and the first six months of 2020 by the Do Ventures fund [2], 2019 marks a record high of venture capital value in startups. Vietnam with 861 million USD, an increase of 92% compared to 2018. This amount of capital poured into 123 investment deals, doubling over the same period. The main driver for the leap came from companies that were developing at a later stage such as Tiki, VNPay and Sendo.

Over the years, the development of technology has gradually changed the lives and behaviors of consumers. These structural changes have motivated Vietnamese startups to focus in technology investment and market expansion. The amount of investment capital and the number of domestic technology transactions have increased six-fold in just two years. Additionally, Vietnam is currently ranked second in Southeast Asia in terms of the startup ecosystem's growth rate according to Ernst & Young (2019) [4].

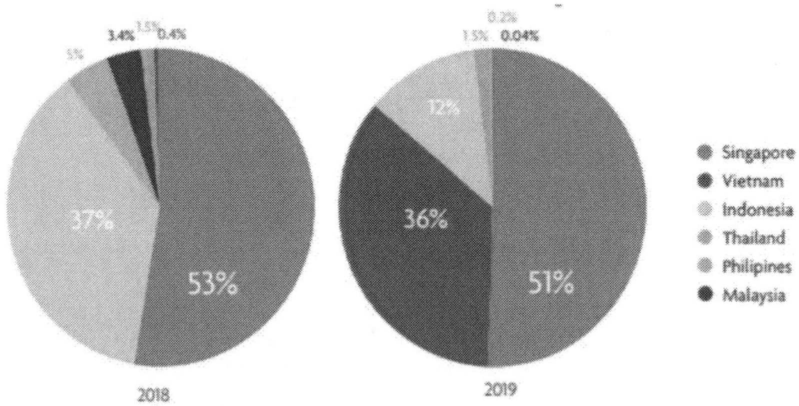

Fig. 1. The funding structure for the fintech market in Southeast Asia Do Ventures fund (2020) [2].

Besides, the impact of the digital revolution on the development of the banking industry is increasingly clear. Typical as the appearance of a variety of innovative banking products and services. New channels of distribution of banking services based on financial technology were born. In the first half of 2018, total investment in financial technology reached 31.7 billion USD for 450 successful investments [7]. This number tripled in value over the same period in 2017. In this context, the fintech segment in Vietnam is also proving its potential based on available resources. In the first half of 2019, 246 million USD was invested in startups in Vietnam, and 63% of that money went to gaming, fintech and e-commerce markets.

In recent years, many start-up incubator and acceleration programs have been opened to attract the attention of Vietnamese startups. Building a startup ecosystem is an essential need for the fintech market in Vietnam. Cooperation between the parties is important to develop the segment further. It helps create many innovations, promotes new solutions, attracts investment and qualified labor.

Fitch Rating, a financial rating agency, said that in the period of 2014 - 2018, the top 11 largest Vietnamese banks have expanded their portfolio with the compound annual growth rate of 28%, outpacing non-consumer credit's 15% [9]. It can be seen that Vietnam's consumer finance industry is entering a mature stage. This is a great opportunity for Vietnamese fintech companies to make disruption and catchup the trend of the fintech in the world.

Fig. 2. Fintech Vietnam Startup Map 2019 [19]

According to research by Solidiance, a financial consulting firm fintech Vietnam, in 2017, it reached 4.4 billion USD in 2017 and is expected to increase to 7.8 billion USD in 2020, equivalent to an increase of 77 % within 3 years. The development of Fintech in Vietnam is going strong, attracting 117 million USD in startup capital.

By the end of 2019, Vietnam ranked second in the ASEAN region in terms of capital to finance the Fintech market and attracted 36% of investment capital in the region, behind Singapore (51%) [2, 3, 4].

In 2019, the Fintech Vietnam market achieved a strong growth in funding thanks to two big deals for VNPay payment company (300 million USD) and the C series of MoMo Pay (500 million USD). These two deals are respectively the third and biggest deal in the area [5].

The Fintech Vietnam startup ecosystem currently has about 154 operating companies which divided into main segments:

- Payment
- P2P lending (peer to peer lending)
- Digital banking
- Wealth management
- Crowdfunding
- •Credit scoring
- Blockchain/crypto
- Comparison
- POS
- Insurtech.

The fintech sector has a very high growth potential in Vietnam and will continue to attract funding, especially from venture capital funds. Digital payments are the top segment today. P2P lending is the second largest investment segment according to CB Insights in Vietnam's Fintech market. Besides, other fintech fields such as credit scoring, asset management and personal finance are in the development stage with high growth potential.

Fintech in Vietnam mainly focuses on payment services with 47% company in the market, accounting for the highest percentage in the region. This segment always attracts many investors and becomes the general rule of countries with early developing Fintech markets. In Vietnam, companies in this segment also account for 98% of total funding for the Fintech market in 2019. There are 35 startups and platforms in the market. The notable companies and brands in this segment include M_Service, the developer of mobile payments app MoMo and one of the most well-funded fintech startups in Vietnam, Moca, a free mobile payment application for Vietnamese consumers, GrabPay, a mobile wallet integrated into Grab's app in Vietnam, and ZION, the company behind Zalo Pay, a service integrated with Vietnam's popular messaging platform Zalo that lets users link a payment card to make P2P payments, pay via NFC, QR codes, as well as purchase products and services online, mobile topups, and pay their utility bills.

P2P lending (or peer-to-peer lending) is an activity that allows individuals and businesses to lend and borrow money. With an efficient structure, P2P lending can offer low interest rates and improve lending processes between lenders and borrowers. A little bit different from banks is that these fintech companies are technically not involved in lending. It is simply the connection between the lender and the borrower and collects the user's fees. This is the second largest Fintech segment in Vietnam with more than 20 startups. The outstanding companies in this segment include: Timo, a consumer financial marketplace and P2P lending platform, Growth Wealth, a P2P lending platform for small and medium-sized enterprises (SMEs) in Vietnam, as well as TrustCircle, and VayMuon.

3 Opportunities and challenges for the fintech market in Vietnam

In three years, from about 40 Fintech companies, this number has quadrupled to more 150 companies today. The progress of Vietnam's fintech industry occurred due to:

- Vietnam provides the necessary infrastructure for fintech services, even in remote areas;
- Vietnam has a large proportion of the highest smartphone inSoutheast Asia [12];
- Besides, the income and consumption level of Vietnamese people has been increasing in recent years;
- Currently, the government has begun to pay attention to the development of fintech. In 2018, a Steering Committee was established to help improve the industry's ecosystem. Besides, the instructions to reduce cash transactions from now to 2020 to less than 10% is also an effort to promote the cashless economy [14].

"Vietnam is going through an important period. As the key components of the digital economy begin to form," according to joint research by ESP Capital and Cento Ventures [5]. The study also said that Vietnam's digital economy would enjoy from a young and tech-savvy demographic. There would be an extra 10 million users participating in the online consumer sector by 2023. These are the factors that create momentum for many consumer finance companies to partner with the fintech scene because they are digitizing heavily.

Vietnam hidden opportunity for fintech startups

Vietnam is a young country, capable of grasping trends quickly. Which is favored by the international market. Meanwhile, Fintech operates in many different fields. This creates a great opportunity for Financial Technology in Vietnam to have a chance to grow stronger.

The number of non-cash transactions in Vietnam is the lowest in Southeast Asia. According to a World Bank survey, the rate of non-cash transactions in Vietnam is 4.9%, which is much lower than Thailand (59.7%), Malaysia (89%) and China (26%). Therefore, Vietnam is a very potential market for startups in the fintech sector [9]. In order to promote the development of the fintech market in Vietnam, the Vietnamese government has many policies to encourage and support the market as well as startups in this field:

- In 2016, the government established the Nation Agency for Technology Entrepreneurship and Commercialization Development (NATEC) to provide, train, mentor, incubate businesses and provide financial support to tech startups
- Preferential policies for things for startups under certain conditions. Corporate income tax reduction for companies working in the high-tech sector or in high-tech zones (preferential tax rate: 10% for 15 years or 17% for 10 years compared to the normal tax rate of 20 %)
- Focus on the STEM (science, technology and math) education model by incorporating STEM education in high school extracurricular curricula. Universities have started training in data science.
- In addition, there are a number of accelerator program in the country to support startups.

However, currently in Vietnam, there are no large-scale Fintech companies, as well as no tax incentives for angel investors and venture capitalists like some Fintech centers in the world. This has created **many challenges for the growth of the fintech industry as well as for fintech startups.**

Part of the reason "constraining" the development of financial technology is due to the incomplete legal framework, some macro projects, and monotonous payment regulations.

In Vietnam, there is hardly any legal framework that clearly regulates the nature of products, services, standards of products/services; Operation model, legal status, establishment and operation conditions of Fintech company; consumer protection policy with financial products, personal information security.

The government agency does not have proper awareness of the role and functions of companies in the E-payment field. Fintech is considered as an extended arm of the bank that has not been operated independently as an "individual" in the financial system. E-wallets can only be used if linked to the consumer's existing bank account.

Although Vietnam is a potential market for companies in the fintech sector, the habit of using cash as well as technology infrastructure is still limited and has not met the technological foundation. It is difficult for companies in this field to expand their market.

Human resources with expertise in finance and banking are weak, and their capacity in informatics and foreign languages is limited. Senior managers often have a lot of expertise in finance but have limited technology knowledge.

In addition, the problems of business models, management skills and mid-long-term development strategy of enterprises are almost absent. So these businesses only operate in a short time and it is very difficult to grow up and continue to develop more. At the same time, for many people, Fintech is still a novel model. So has caused a lot of confusion with investors and people involved in Fintech financial activities.

4 Conclusions

It can be said that Vietnam has all the necessary elements for technology startups to operate and develop. However, to be able to take advantage of the great opportunities, the startups need to have a suitable investment direction to minimize the risks brought by the challenge.In addition, to develop and nurture the fintech ecosystem, it is necessary to maintain a close coordination between the three most important elements of an ecosystem, including:

- Business environment/accessibility market;
- Government/legal support;
- Access to capital.

The advantage of the Vietnamese market is that all three above factors have been formed. However, to remove the current barriers and difficulties, the development of the legal framework and supporting policies of the Government to develop the fintech ecosystem is very necessary. Proposed mechanisms to promote fintech start-up opportunities include:

1. Supplementing and completing mechanisms and policies for Fintech activities. The Government should study and consider building supportive policies and creating an investment environment for Fintech startups; It is necessary to soon have financing policies to support research and technology application in finance - banking and other fields; Early issue a legal framework, testing mechanism (Sandbox) for high-tech applications and especially Fintech technology in the financial and banking sectors.

2. To build national digital entrepreneurship centers and creative start-up support centers at universities. The National Digital Startup Center is a government funded and invested hub. This is a favorable environment for nurturing innovative ideas and start-ups. It also helps connect collaboration opportunities and technical support for young talent with creative ideas. There are also internship programs and exchanges to create conditions for young people to approach research practices; testing innovative solutions for technology, business models, products, services... Moreover, this is also a place to share experiences from organizations, enterprises, banks, big technology companies (bigtech) domestically and internationally for startups to make their ideas come true.

3. Center for Creative Startups at Universities. This is a place that provides facilities, infrastructure, and technology facilities for students to participate in promoting creative thinking, startup ideas. This center also helps connect creative ideas between students with faculty, successful alumni of the school, businesses, investors ... so that creative ideas become feasible. In addition, the center is also a place to support Creative Startups (set up a business; transfer technology; selling solutions and products to a business ...) for faculty and students to turn creative solutions into effective business solutions.

4. Build an ecosystem and diversify Fintech products. Fintech Ecosystem is a combination of Fintech Company with banks, finance companies, technology companies, state management agencies and client partners. Only when there is such a coordination between the parties will the conditions and the best environment for startups in the Fintech field be successful and thrive.

In addition to promoting ecosystem construction, the diversified development of Fintech products not only focuses on areas such as payment, money transfer, online lending, but Fintech needs to expand into potential applications such as financial management, online investment, personal financial consulting; fields of industry, agriculture, business, logistics, education, health care, tourism to meet the diverse needs of customers in the 4.0 revolution and the digital age.

5. Organizing digital startup forums, Fintech seminars and promoting communication activities. State management agencies need to coordinate with technology organizations and associations to regularly organize forums, seminars, seminars on entrepreneurship, fintech, digital technology to share experiences and introduce public technology, recommendations on state policies to promote digital start-ups, fintech and technology application.

The government should have a policy to support the Fintech community. The media also needs to promote activities to promote and disseminate knowledge about Fintech, introduce Fintech applications to the community of people and businesses to know the benefits that Fintech brings. At the same time, it is also important to disseminate preventive measures to limit risks in online transactions in the network environment.

References

1. Casey Hynes, *How Vietnam's Fintech Market Could Reach Nearly $8 Billion by 2020*, Forbes (2019)
2. Do Ventures fund, V*ietnam Tech Investment Report 2019-H1 2020* (2020)
3. Ernst & Young, *ASEAN FinTech Census 2018* (2019)
4. Ernst & Young, *Global FinTech Adoption Index 2019* (2019)
5. ESP Capital and Cento Ventures, *Vietnam Tech Investment Report* (2019)
6. The future of finance is emerging: new hubs, new landscapes. Global Fintech Hub Report (2018)
7. KPMG, *The Pulse of Fintech: Biannual global analysis of investment in fintech* (2019)
8. A. Patwardhan. K. Schmitz, K. Singleton, Financial Inclusion in the Digital Age. International Finance, **8** (2019)
9. Shaun Turton, Vietnam Fintech sector set for 'bloodbath' competition, Nikkei Asian review (2019)
10. A. Demirguc-Kunt, L. Klapper, D. Singer, S. Ansar, J. Hess, *The Global Findex Database 2017: Measuring Financial Inclusion and the Fintech Revolution* (2018)
11. Le Dat Chi, Tran Hoai Nam, Journal of Science and Technology of Vietnam, **5** (2019)
12. Finch Capital, MDI Ventures and Dealroom.com, *The Future of Fintech in Southeast Asia* (2020)
13. Álvaro Ortiz Vidal-Abarca, Le Xia, SumedhDeorukhkar, Tomasa Rodrigo and Alfonso Ugarte, *Fintech in Emerging ASEAN Trends and Prospects*(2017)
14. World Bank, *Vietnam's financial inclusion priorities: Expanding financial services and moving to a 'non-cash' economy* (2017)
15. United Overseas Bank and the Singapore FinTech Association,*FinTech in ASEAN: The Next Wave of Growth* (2019)
16. H. Gimpel, D. Rau, and M. Roglinger, Electronic Markets, **28** (2018)
17. In Lee, Yong Jae Shin, Fintech, Business Horizons, **61** (2018)
18. R. Alt, R. Beck, and M.T. Smits, Electronic Markets, **28** (2018)
19. Official website of Fintech news Singapore, https://fintechnews.sg/

SHS Web of Conferences **89**, 04002 (2020) https://doi.org/10.1051/shsconf/20208904002
Conf-Corp 2020

Fiscal and investment indicators for evaluating professional sports facilities

Mikhail Gordienko*, *Ksenia* Ekimova, *Nadezhda* Glubokova, and *Elena* Shuvalova

Plekhanov Russian University of Economics, Stremyanny Lane, 36, 117997 Moscow, Russia

Abstract. The article is devoted to the formation of a methodology for assessing the investment and fiscal attractiveness of professional sports facilities, for which there are many similar techniques, but no single universal one. The authors made an attempt to form such a methodology, based on the ecosystem approach, within which the attractiveness of a particular object is assessed in aggregate and taking into account the interrelationships of the object and the macroeconomic environment (state, territorial and sectoral regulation), as well as the microeconomic environment (the financial position of the organization and the characteristics of the object itself). The article gives examples of indicators, their variants of values and estimated values and presents a general algorithm for applying the technique and basic formulas for calculations.

1 Introduction

Today there are many different approaches to assessing the investment attractiveness of organizations. Various aspects of the definition of the term "investment attractiveness" are presented by D.A. Endovitsky, N.A. Baturin, V.A. Babushkina [1], E.I. Krylova, M.N. Kreinina [2], i.a. Issues of investment attractiveness and fiscal risks of professional sports facilities are considered by Gianluigi De Mare [3], S. Janaka Biyanwila [4], Hasnul Faizal Hushin Amri [5], A.L. Balduck [6], S.J. Biyanwila [7], i.a. Each approach has its own advantages and disadvantages, while there is no universal one. Thus, this study attempts to form a methodology for assessing the attractiveness of sports facilities, taking into account investment and fiscal evaluation criteria.

Investment attractiveness as a scientific category involves taking into account some nuances. First, despite the fact that investment attractiveness is an objective concept, its assessment by a potential investor is always subjective, since it is determined by his specific goals, which may be different for each investor. In this regard, it can be argued that, depending on the purpose of the investor, his assessment of the investment attractiveness of the selected object will always be unique. Secondly, the attractiveness for the investor is characterized not only by the current state of affairs at the facility, but also by future opportunities, therefore, objective research involves focusing not only on the results already achieved, but also on predicting the prospects for its financial condition.

* Corresponding author: gordienkomikhail@yandex.ru

It should be noted that each sports facility is characterized both by its own level of investment attractiveness and fiscal risks (internal factors), and depends on environmental factors. Therefore, it is important to determine the attractiveness at all levels of the environment, which together form the ecosystem, in order to compile a more holistic picture of investment potentials and fiscal risks of the assessed object. As a result, along with an analysis of the attractiveness and risks of an organization, we also propose to analyze the state of the country, region, industry and a specific organization to which the object of assessment belongs. Such an analysis will allow a potential investor to identify objects with great development potential, to more fully analyze the existing risks and prospects for making a profit. Each of the abovementioned levels has its own number of analytical indicators, due to which the most effective and comprehensive assessment is achieved. The significance of these valuation levels is different for different types of investors.

In the next section of the article, we will present the methodology of the author's approach to assessing the attractiveness of large-scale sports facilities and organizations. It should be noted that certain aspects of the methodology in each specific case need to be decomposed, refined and adjusted for the parameters of a specific sports facility. The technique requires constant improvement.

2 Materials and Methods

In practice, the analysis of investment attractiveness is often reduced to assessing its financial condition, which allows to analyze only the current position of the organization, but does not give the most complete picture of the prospects. In this relation, potential investor should pay attention to organizational and management structure, the level of management qualifications; the degree of uniqueness of the business concept, strategy for its development; potential for earning income from invested funds; the degree of financial transparency and openness; the level of competition in the market and the level of competitiveness; the degree of riskiness of investments; image, and other factors.

To make a decision, an investor needs a more detailed analysis of the enterprise and its attractiveness, taking into account not only quantitative but also qualitative indicators. A complex ecosystem, taking into account the interconnection of all spheres and links of the financial system, as well as factors of the internal and external environment of the object, has the following undoubted advantages:

- the object is assessed, taking into account the goals of different types of investors;
- an opportunity is created to assess organizations of different size and organizational and legal form due to the use of internal and external information about the object in the analysis;
- an approach is being universalized, which becomes possible to apply both to business owners and potential investors;
- a variety of factors are used in the assessment, which reduces risks;
- the qualitative characteristics of the enterprise and its activities are taken into account;
- a combination of several approaches allows you to compensate for the negative aspects and enhance the merits as much as possible.

Evaluation of the investment attractiveness of an individual object or organization of the professional sports industry should take into account both quantitative and qualitative characteristics, which will take into account the interests of the entire spectrum of potential users of this information. However, the potential availability of the required information must also be considered. While it is easy to obtain information about the financial performance of an enterprise, data on the qualitative characteristics of a company's activities are usually difficult to access and extremely subjective. Therefore, it is recommended to use only those qualitative indicators, the characteristics of which will not be difficult for the investor to verify.

Within the framework of the methodology, we consider it necessary to interpret the investment attractiveness as a relative indicator comparable to certain standard values. As a rule, the value of the weighting factors is determined by an expert judgment by the investor himself. However, in the proposed methodology, various options for calculating the integral indicator for each type of investors are possible, since the level of significance of individual indicators for them is often different. Thus, in the methodology, the assessment of the investment attractiveness of any object of professional sports will consist of five blocks and their characteristic estimated indicators (see Figure 1).

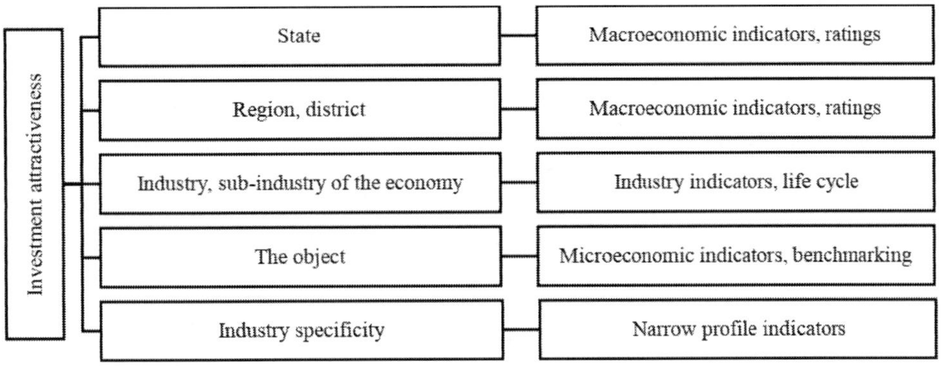

Fig. 1. The main blocks and the corresponding characteristics of the investment attractiveness of any object of professional sports.

Each indicator should be assessed by a set of factors, its components, the state of which is quantitatively determined using a point estimate, the variation of which will be in the range from 0 to 3. Thus firstly one need to consistently consider what factors characterize the investment attractiveness of the state, region, industry and sub-industry, as well as a specific object. This approach allows a comprehensive ecosystem-based consideration of the attractiveness of an enterprise at the level of the state, region, industry, organization or a specific investment object.

3 Results and Discussion

Since at present stage of development there is no standard international rating that most fully characterizes all aspects of the investment attractiveness of a state, this indicator will be analyzed using a set of ratings, as well as on the basis of an assessment of investment activity in the country. Table 1 shows as an example 1 indicator "Place in the Doing Business rating" with a breakdown by states and intervals of values.

Table 1. Indicators for assessing the investment attractiveness of the country

Indicators	Status and value ranges	Rating scale
Ranking place Doing Business	The country is in the top 30% of places in the ranking	3
	The country is in the second 30% of places in the ranking	2
	The country is in the third 30% of places in the rating	1
	The country is in the last 10% of places in the ranking	0

Source: compiled by the authors.

Since the volume of the publication is limited, let us list the rest of the indicators introduced by analogy and also estimated from 0 to 3 points: place in the Standard & Poors rating; place in the ranking of the global competitiveness index; place in the ranking of the index of weakness of states; dynamics of investment activity; the level of fiscal burden and the quality of tax administration, etc.

It is proposed to evaluate the attractiveness of the region in which the enterprise is located based on its place in the rating of investment attractiveness of the regions of the state. To do this, it is worth resorting to data from state statistics bodies or data from rating agencies. For Russia, the data of the rating agency Expert RA are representative, as one of the most complete and diverse studies, which is presented in Table 2.

Table 2. Indicators for assessing the investment attractiveness of the region

Indicators	Status and value ranges	Rating scale
Place in the rating of investment attractiveness of the regions Expert RA	Region belongs to category A	3
	Region belongs to category B	2
	Region belongs to category C	1
	Region belongs to category D	0

Source: compiled by the authors.

It is proposed to assess the attractiveness of the industry on the basis of an analysis of the qualitative and quantitative indicators of the organization of this industry, an assessment of the direction of investment flows in it. Each indicator, as in Tables 1 and 2, is expertly assigned intervals of values. Let's list the main indicators: the number of unprofitable enterprises in the industry; number of innovations; direction of investment flows in the industry; the degree of wear of the fixed assets of the industry; quick and absolute liquidity, financial dependence, financial leverage ratios; profitability of sales, assets; term of work in the market; market share; the degree of competition; the sensitivity of products to price volatility in the market and fluctuations in demand; business reputation; level of innovative activity; the level of tax burden on the industry.

To assess the investment attractiveness of an enterprise, an analysis of both quantitative indicators characterizing the financial condition of the enterprise and qualitative indicators describing the business reputation of the company, its position in the market, and the economic potential of the activity will be required. The financial position of an enterprise is proposed to be assessed by the indicators of 4 groups of coefficients: profitability, financial stability, liquidity and business activity. It is recommended to take no more than 2 indicators from each group, since the coefficients within the group have a high degree of correlation [7].

In addition to the presented macro- and microeconomic criteria, we consider it necessary to apply indicators in the methodology that take into account the industry-specific nature of organizations or specific investment projects implemented in the field of professional sports. Let us turn to the normative legal acts in the field of professional sports in order to find the most objective approaches to the classification of specialized objects. The most suitable, in our opinion, is the classifier approved by the order of the profile regulator […], which presents the most complete list of sports facilities and their detailed characteristics. The classifier takes into account the functional features and infrastructural parameters of a particular object, in accordance with which the object is assigned a certain point. We will give an example of classification for the "stadium" object intended for playing European football; we will not use other non-representative classification parameters from the order in this case. The rating scale remains the same (see table 3 below) [8, 9].

Table 3. Indicators for assessing the industry specificity of an organization or project

Indicators	Status and value ranges	Rating scale
Stadium	Indoor specialized with natural turf for football	3
	Indoor specialized with artificial turf for football	3
	Outdoor versatile natural turf for football and artificial turf for athletics	3
	Outdoor specialized with natural turf for football	2
	Outdoor versatile with natural turf for football and athletics	2
	Outdoor specialized with artificial turf for football	1
	Outdoor all-purpose artificial turf for football and athletics	1

Source: compiled by the authors.

Thus, the first stage of the methodology is to determine the values of the scoring factors of factors characterizing the investment attractiveness of a country, region, industry, enterprise, object – IR_S, IR_R, IR_O, IR_P, IR_{SO}, respectively. These indices represent an integral assessment of the lower-order attractiveness factors that are included in them, taking into account their weight for different types of investors.

The attractiveness of a country is estimated as a percentage using formula 1:

$$IR_S = \frac{\sum_{i=1}^{n} F_i,}{\sum_{i=1}^{n} F_{i\,max}} \times 100\% \quad (1)$$

where F_i – is the score of the i-th factor of the country's investment attractiveness; $F_{i\,max}$ – is the maximum possible score for the i-th factor of the country's investment attractiveness.

The attractiveness of a region is estimated as a percentage using formula 2:

$$IR_R = \frac{IR}{IR_{max}} \times 100\% \quad (2)$$

where IR – балis a point assessment of the region's place in the rating of investment attractiveness of the most authoritative rating agency; IR_{max} – is the maximum point estimate of the region's place in the investment attractiveness rating of an authoritative rating agency.

The attractiveness of the industry is estimated as a percentage using formula 3:

$$IR_o = \frac{\sum_{j=1}^{n} F_j,}{\sum_{j=1}^{n} F_{j\,max}} \times 100\% \quad (3)$$

где F_j – is the point estimate of the j-th factor of the investment attractiveness of the industry; $F_{j\,max}$ – is the maximum possible score for the i-th factor of the industry's investment attractiveness.

To measure the investment attractiveness of an enterprise, two groups of indicators are used: financial and non-financial.

The attractiveness of the company is estimated as a percentage using the formula 4:

$$IR_P = \frac{\sum_{k=1}^{n} F_k,}{\sum_{k=1}^{n} F_{k\,max}} \times 100\% \quad (4)$$

where F_k – is the point estimate of the kth factor of the investment attractiveness of the enterprise; $F_{k\,max}$ – is the maximum possible score for the k-th factor of the investment attractiveness of the enterprise.

Investment attractiveness also takes into account the industry specificity of the organization or the characteristics of a specific project in the field of professional sports, which is assessed as a percentage using formula 5:

$$\mathrm{IR}_{\mathrm{SO}} = \frac{\sum_{j=1}^{n} F_{\mathrm{p}}}{\sum_{j=1}^{n} F_{\mathrm{p}\,max}} \times 100\% \qquad (5)$$

where F_p – is the score of the j-th factor of the industry specificity of an organization or a specific project; $F_{p\,max}$ – the maximum possible score of the i-th factor, the industry specificity of an organization or a specific project.

The integral indicator of the methodology is the result of summing the values of investment attractiveness factors, which are obtained as a result of multiplying the assessment of these factors on a three-point scale by their weight in the overall assessment of investment attractiveness, which is assigned separately by each investor in accordance with his interests. The Integral Indicator (II) is calculated using the formula 6:

$$II = IR_S \times V_S + IR_R \times V_{IRr} + IR_o \times V_{IRo} + IR_P \times V_{IRp} + \mathrm{IR}_{\mathrm{SO}},\times V_{IRso} \qquad (6)$$

где IR_S, IR_R, IR_o, IR_P, IR_{SO} – is an assessment of indicators of the investment attractiveness of a country, region, industry and the enterprise itself, respectively, in percentage terms; V_{IRs}, V_{IRr}, V_{IRo}, V_{IRp}, V_{IRso} – the weight of the corresponding factors of attractiveness in the integral assessment, respectively.

The sequence of calculating the integral indicator of attractiveness:

1) each single indicator characterizing the factors of the attractiveness of a country, region, industry and enterprise (factors of the third level) is assessed on a scale from 0 to 3 points (a more complex scale is also possible);

2) by summing the scores of the attractiveness factors, the values of the integral indicators of the attractiveness of a country, region, industry, enterprise and a specific object are calculated - IR_S, IR_R, IR_o, IR_P, IR_{SO} - respectively. Then these estimates are presented in a comparable form, in percentage format;

3) the final integral indicator of attractiveness is calculated by summing the products of integral indicators of the attractiveness of a country, region, industry, enterprise and object, taking into account their weight for a particular investor (in this case, the sum of the weights should always be equal to 1);

4) the obtained value is compared with the maximum possible ideal (equal to 100%) and on the basis of this, a conclusion is made about the state of investment attractiveness and fiscal risks of the assessed object as a whole.

4 Conclusions

The use of the above-described methodology and the integral ecosystem indicator included in it in the analysis of the attractiveness of the object of analysis in the sphere of professional sports has the following advantages:

- bring together qualitative and quantitative indicators in one assessment;
- takes into account the significance of various factors, including industry-specific, for different types of investors in the final assessment (takes into account the typology of investors);
- makes it possible to compare the level of investment attractiveness and fiscal parameters of enterprises to select the best investment object;
- allows you to adjust the resulting factors to increase the attractiveness of the enterprise in order to further expand and develop.

Acknowledgements

The article was prepared within the framework of the research work "Development of scientifically grounded proposals for increasing the investment attractiveness of professional sports in the Russian Federation", with the support of the Federal State Budgetary Educational Institution of Higher Education « Plekhanov Russian University of Economics».

References

1. D. A. Endovitsky, V. A. Babushkin, N. A. Baturina, *Analysis of the investment attractiveness of the organization: scientific publication* (2012)
2. S.I. Krylov, *Financial analysis: textbook* (2012)
3. De Mare G., Granata M.F., F. Forte, CCSA 2015, Lecture Notes in Computer Science, Springer, Cham, **9157** (2015)
4. S.J. Biyanwila, Sports and The Global South, Global Culture and Sport Series, Palgrave Macmillan, Cham (2018)
5. H.F.H. Amri, S.A. Wahab, N. Salamuddin, M.T. Harun, ICoSSEET 2016, Springer, Singapore (2019)
6. A.L. Balduck, S. Lucidarme, M. Marlier, Voluntas 26, 2023 (2015)
7. E.V. Golubtsova, A.O. Zvereva, IBIMA 2019: Education Excellence and Innovation Management through Vision 2020 (2019)
8. M.E. Kosov, R.G. Akhmadeev, V.S. Osipov, Y.K. Kharakoz, I.I. Smotritskaya, Indian Journal of Science and Technology (2016)
9. R. Akhmadeev, T. Morozova, O. Voronkova, A. Sitnov, Entrepreneurship and Sustainability Issues (2019)

Legal regulation of the circular economy and ecology: current issues

Maria Mukhlynina[1,2*], *Natalia* Mishina[2], *Natalia* Vedysheva[3], and *Ilmira* Shikula[2,4]

[1]Institute of State and Law of the Russian Academy of Sciences, Znamenka str., 10, 119019 Moscow, Russia
[2]Moscow international University, Leningradsky Prospekt, 17, 125040 Moscow, Russia
[3]Kutafin Moscow State Law University, Sadovaya-Kudrinskaya str., 9, 125993 Moscow, Russia
[4]Moscow City University, 2nd Agricultural passage, 4, 1, 129226 Moscow, Russia

Abstract. The problem of legal regulation of the circular economy and ecology, sustainable management of municipal solid waste at the legal and organizational levels both in the Russian Federation and in the world is now coming to the fore and requires immediate solutions. Examining strategic planning documents and other legal acts, the authors attempted to examine current legal problems in the study area in the context of modern state of ecological policy of Russia and the implementation of circular economy, realizing the sustainable development goals of the UN in the environmental area. The authors suggest that a legal assessment of the possibility of transition of Russian industrial companies to a closed-loop economy is necessary, which is associated with an active approach to processing the materials of their products after the stage of use. In conclusion, a number of conclusions are drawn about the legal acts adopted in the Russian Federation and the organizational measures being implemented that contribute to solving various problems related to the sustainable development of the most important segment of the economy and the environment – the disposal and processing of waste, reducing its volume and the possibility of secondary use due to the development of the circular economy. Innovation is an important condition for this transition.

1 Introduction

The circular economy (CE), in contrast to the traditional linear development model, is a new regime that focuses on the reuse of resources to save money, protect the environment, and achieve economic development. In this regard, many countries have made the development of a closed-loop economy their strategy, as it has played an important role in adjusting the industrial structure, changing the way economic growth is conducted, building an ecological civilization, and finally promoting sustainable development.

Cyclicity, environmental friendliness and new technologies are the "beacons" that the economy of any country in the world needs to look up to at the present time. The outdated traditional model of the economy cannot adequately withstand the global challenges of our

* Corresponding author: muhlyninamm975@mail.ru

time. right before our eyes, it is creating a new resource-efficient model-the circular economy. The circular economy is a new model of economy based on the renewal of resources with the involvement of the latest scientific developments of scientists, innovative projects and technologies that make the process of using natural resources almost endless.

The reorientation to a new model of socio-economic development in Russia corresponds to the trend of the developed countries of the world, which, together with the member States of the United Nations, adopted the sustainable development Goals for the period up to 2030 on September 25, 2015. There are seventeen goals and one hundred and sixty-nine tasks to be completed in the same time frame. Focusing on this international agenda, President of the Russian Federation Vladimir Putin issued a Decree on July 21, 2020 N 474 "On national development goals of the Russian Federation for the period up to 2030" set an ambitious goal – to achieve breakthrough development of the Russian Federation, increase the country's population, and improve the standard of living of citizens. The focus on creating a high-performance export – oriented sector, primarily in two sectors-manufacturing and the agro-industrial complex, will undoubtedly put a heavy burden on the country's environment. The twelve key areas in which national projects have been developed in accordance with this decree relate to all major areas of the state's life, including the environment.The study of legal and organizational problems of legal regulation of the circular economy and ecology is reflected in the works of Vinogradova E.V., Mukhlynina M.M., Mukhlynin D.N., Solovyeva N.V., Lebedeva O.E. [1], Mukhlynina M. M., Vedyshev N.O., Solovyov N.V., Kozyrev A.A. [2], Mukhlynina M., Vedysheva N. [3], Pakhomova N.V., Richter K.K., Vetrova M.A. [4], Mukhlynin D.N. [5], Mishina N.V. [6]. However, the issues of comprehensive analysis of the legal regulation of the circular economy and ecology in the world and in the Russian Federation remain insufficiently developed.

2 Methods

Methodological basis of research was made following General scientific methods: analysis and synthesis to analyze the existing theoretical and methodological approaches, regulations and scientific research of circular economy and ecology; structural and logical methods of systematization of the factors influencing the implementation of circular economy and ecology. The method of factor analysis was used to determine the impact of various factors on the level of implementation of the state policy of the Russian Federation in the sphere of circular economy and ecology, as well as the degree of satisfaction of citizens with its results.

The information base of the research included legislative and regulatory acts, materials of state and local authorities, scientific publications of Russian and foreign researchers on the problems of improving the system of compensation for environmental damage [7].

The paper also uses the citation method as a method of argumentation, illustration, and reference load. The authors use the comparison method to compare the features, features, and characteristics of the studied objects in order to find similarities and differences between them and to describe them. The method of generalization in its particular manifestation as a method of systematization is also used by the authors in this study.

3 Results and discussion

As the governments of many countries (mostly developed) have become more interested in environmental issues, we have begun to meet more conditions on environmental conservation, resource management, sustainability, and the economic potential of a particular place (region). As a result of this policy, we often encounter terminology such as: sustainable development, global environmental trends, green economy, ecotourism, environmental protection, regional issues, technological impact on the environment (or a new version of industrial ecology), environmental issues, social responsibility, corporate responsibility, urban sustainability, sustainable tourism, circular economy, etc. In other words, the link between biodiversity, the economy, and technological progress exists in almost all socio-economic forms.

Economic regulation of nature protection devoted to Chapter 4 of the Federal law of the Russian Federation dated 10.01.2002 № 7-FZ "On environmental protection" (hereinafter - the Law on environmental protection), which is based on several constitutional principles, such as payment for use of natural resources and basic rights of every citizen to a healthy environment and to prevent violations of legislation in the field of environmental protection. The main emphasis in the articles of Chapter 4 of the law is placed on the legal regulation of payments for negative impact on the environment: goals, deadlines, rates, coefficients, control, state regulation, etc. It is gratifying that article 17 specifies options for state support in the implementation of a number of measures: the use of the best available technologies, the use of recycled water supply systems without effluents, and others. The legislator provides for the possibility of establishing other measures of state support for economic activities aimed at protecting the environment at the expense of budget funding based on the norms of Federal and regional laws. Environmental insurance is another economic and legal instrument designed, according to the Law on environmental protection, to protect the property interests of legal entities and individuals in the event of environmental risks.

The economic category of insurance protection of public production finds its material expression, according to M. M. Mukhlynina, N. O. Vedysheva, N. V. Solovyova and A. A. Kozyreva, in the insurance Fund, which is important as a measure of protection against the risk of loss of property, health and income of the population [2]. From an economic point of view, the purpose of insurance is to accumulate money paid by many policyholders and invest it in the economy.

Through economic measures, the state not only encourages increased production or other economic activities, but also their elimination, which is reflected in article 18.1 of the Law on environmental protection and concerns economic incentives to stop the production and use of ozone-depleting substances and products containing them.

One of the most pressing problems of humanity at the present time can be called overpopulation, which creates a serious threat to the future of the planet at the current level of production and consumption, as well as the existing attitude to the environment. In the 2015 UN report "world population growth Prospects", researchers defined the prospective population of the planet Earth as follows: if in 2015 – 7.4 billion, by 2030 – 8.3 billion, by 2050 – 9.7 billion. The UN Millennium development goals have significantly improved health, living standards and consumption. New challenges lie in ensuring a decent quality of life for all people in the world with a growing population, consumption and production, but at the same time – limited areas and dwindling reserves of natural resources. In may 2018, Russia hosted the St. Petersburg international economic forum, which included discussions on the circular economy, aimed at minimizing waste in addition to efficient use of resources.

Pakhomova N.V., Richter K.K., Vetrova M.A. [4], referring to the European investment Bank, among the reasons why it is necessary to make the transition to a circular economy, allocate resource, technological, socio-economic and environmental. At the same time, the latter is given priority. In particular, we are talking about taking into account negative externalities that lead to significant environmental damage, including in the socio-economic sphere, which should be internalized by the circular economy [8].

Closed-loop Economics is a very young discipline that deals with the efficiency of using scarce resources. Here we can observe the relationship between the rational use of natural resources in ensuring economic growth and job creation [9]. Circular Economics appears in various fields of research.

Recently, environmental Economics has become a very popular new discipline. According to Mele et al. [10], Long et al. [11], Patterson and Glavovich [12], ecological Economics combines both doctrines: on the one hand, it examines the connections between physical problems and biological spheres, and on the other hand, it examines the processes of monetary exchange at the micro-, meso-, and macro levels of the economy. This discipline, which is the link between Economics and ecology, arose from the need to emphasize their connection, given their social, environmental, political and sustainable significance at the present time.

Perhaps the approach recently proposed by Capra and Mattei [13] in their book "the Ecology of law" may provide better guidance in this regard. In their view, in order to have a functional economy linked to the extraction of natural resources, we need to transform "legal institutions from mining machines rooted in the mechanistic functioning of private property and state power to institutions based on ecological communities" [13]. Management and regulatory enforcement through such "ecological communities" that have appropriate education on natural mining restrictions is likely to be the best way to manage a circular economy.

Agrawal's concept of the "environment" [14] draws on Foucault's seminal work on knowledge and power [15] in governance beyond conventional political institutions. Here, the role of governments is to create conditions for the acquisition of knowledge in societies. In order to develop and develop optimal solutions, Governments should discuss possible contradictions in international forums. In our view, it will be necessary to conclude agreements on resource flows at the international level that would encourage environmentally and socially responsible waste management and more environmentally and cost-effective energy transfer across borders. Undoubtedly, to facilitate such processes, some reform of international relations in the field of ecology will be required. However, such reforms are likely and possible as a broader awareness and actualization of the circular economy gradually takes root in international Affairs in the field of sustainable development.

It should be noted that the acceleration of the transition from traditional to circular economy and ecology is possible with the introduction of artificial intelligence technologies. Artificial intelligence technologies can ensure sustainable development by optimizing business processes, design, and infrastructure of the economy. The ability of artificial intelligence to adapt to the environment using machine learning methods should be used to develop new models and knowledge, as well as to develop new decision-making rules. At the same time, it is necessary to take into account the possible risks of introducing artificial intelligence technologies into public relations, ensuring an appropriate legal regime for the development, implementation and functioning of artificial intelligence and a mandatory public control procedure at the stage of introducing artificial control systems into commercial operation [6]. The effectiveness of public control mechanisms in the transition to a circular economy and ecology will depend on the consistency and complexity of legal regulation in these areas.

4 Conclusions

Thus, in General, this concept is not yet widespread enough in Russia. We need a legal assessment of the possibility of transition of Russian industrial companies to a closed-cycle economy, which is associated with an active approach to processing the materials of their products after the stage of use. Innovation is an important precondition for this transition. The main problem of the existing "cradle to grave" system is the burial of valuable materials after one cycle of use. The circular economy works on a different principle - "from cradle to cradle". The company manages the material flow until the material returns to the beginning of the production process by itself or is delivered by other partners, which ensures a stable supply of materials at a fairly low cost. Environmental issues need to be addressed at the product design stage, not at the last stage. Russian practice shows, on the one hand, low involvement of industrial companies in the field of waste management, but, on the other hand, they are open to environmental innovations in the future. As such innovations, artificial intelligence technologies can be developed that will speed up the process of transition to a circular economy and ecology, optimizing the infrastructure, business models, and design of the economy and ecology.

References

1. E.V. Vinogradova, M.M. Mukhlynina, D.N. Mukhlynin, N.V. Solovyeva, O.E. Lebedeva, Journal of Environmental Management and Tourism, **1(9)**, 144-150 (2018).
2. M. M. Mukhlynina, N. O. Vedysheva, N. V. Solovyova, A. A. Kozyreva. *Insurance law in questions and answers:* textbook. Khimki: AGZ of EMERCOM of Russia, 120 p. (2018).
3. M. Mukhlynina, N. Vedysheva, APEEM 2020, 05003 (2020)
4. N.V. Pakhomova, K.K. Richter, M.A. Vetrova, Bulletin Of St. Petersburg State University Economy **33(2)**, 45 (2017)
5. D. N. Mukhlynin, Law and law, **6**, 97 (2018)
6. N.V. Mishina. *Features of legal regulation of artificial intelligence as a guarantee of sustainable development of society*, IFT (2020)
7. M. Mukhlynina, N. Vedysheva, APEEM 2020, 05003 (2020)
8. H.A. Saleem, Preprints 2016, 2016090040 (2016)
9. D.B. Vukovic, R. Shams, SUSTAINABILITY, **12(1)**, 179 (2020)
10. B.H. Mele, L.Russo, D. D'Alelio, Sustainability, **11**, 4393 (2019)
11. R.D. Long, A. Charles, R.L. Stephenson, Mar. Policy, 57, 53 (2015)
12. M.G. Patterson, B.C. Glavovic, Edward Elgar Publishing: Cheltenham (2008)
13. F. Capra, U. Mattei, *The Ecology of Law: Toward a Legal System in Tune with Nature and Community*, 1 (2015)
14. A. Agrawal, Environmentality: Technologies of Government and the Making of Subjects (2005)
15. M. Foucault, *The Government of Self and Others: Lectures at the Collège de France 1982-1983* (2010)

The assistance of a University graduate's career with using the ecosystem «Career Tech»

Oleg Zatepyakin, *Tatyana* Bobko, *Galina* Kazanceva, and *Elena* Ivanova

Siberian State Industrial University, Kirova St., 42, 654007 Novokuznetsk, Russia

Abstract. Today, universities are increasingly perceived as active participants in the processes of regional development and corporate relations, and the role of the University is not limited to educational activities or science and research jobs. Universities are urged to solve the problem of attracting young people to production in order to ensure the normal reproduction and sustainable development of business, and at the same time provide the opportunity to find the place in society for their graduates, to realize their abilities in present dynamic social-economic conditions. The importance of the problem of effective employment of University graduates is also due to the perception of young people in society as the most socially unstable ("explosive") group. It requires the creation of graduate employment ecosystems, especially in peripheral regions. The ecosystem model proposed in this article, aimed at promoting employment of graduates and creating conditions for their career development, can be used as an information, analytical and infrastructure basis for integrating young people into the labor sphere of life, for expanding the number of stakeholders and harmonizing their interests.

1 Introduction

In current conditions, universities practice an active policy of employment of graduates based on the development of their ecosystem. The University ecosystem is a flexible system with multiple connections that can quickly respond to external and internal changes. The essence of an active policy of employment of University graduates on the basis of the development of its ecosystem is that the University tasks to the graduate of employment at the initial stages of training and offering appropriate assistance using its multifunctional infrastructure, organizational and human resources (the University ecosystem). This approach to employment of graduates allows them to design professional self-determination as a process of forecasting and planning of their career throughout the entire period of training. This process is practically implemented in the goals, tasks, content and means of personal self-realization, and search and find meaning in the chosen professional activity.

Career planning in the active policy of University graduates employment can be perceived as integrative quality of the personality and consider as the readiness of the individual to be a subject of the labour market and professional self-determination in close connection with cognitive, activity, motivational-value and reflective-evaluative components through linking the "want", "must" and "can". In fact, we are talking about individual typological features of the individual, represented by temperament, character, motives, needs, attitudes, value orientations of the individual; about expectations, claims, ambitions of young people concerning the goals, content, objects, means, working conditions, its restrictions in society and the requirements of the Federal and regional labor markets, requests of specific employers, certain areas of production, trends in society, etc. [1].

The purpose of the article is to reveal the possibilities of a modern University for implementing an active employment policy for its graduates based on the application of its ecosystem.

Research objectives:

- to consider the employment of graduates on the basis on the University ecosystem;
- to offer a model of the University ecosystem for implementing an active employment policy for its graduates;
- to reveal the interaction of the University and the business environment in the proposed model of the University ecosystem to promote employment of graduates.

2 Materials and Methods

Research Background. One of the first researchers to apply the term "Ecosystem" to the economy was Michael Rothschild [2]. This concept was developed by Charles V. Wessner [3], Rucker Schaeffer Paola, Fischer Bruno, Chieroz serjou [4], G. V. Osipov [5] and others.

The position of young people in the domestic labor market in current conditions is analyzed in the papers of Yu. n. Dorozhkin [6], T. I. Zaslavskaya, M. A. Shabanova [7], I. K. Zolotova [8], O. A. Rakovskaya [9], V. I. Chuprov [10-12] et al.

The empirical base of the research is represented by the studies of domestic and foreign scientists, own analytical works of authors.

Approach and Methods of Analysis. The study used methods of analysis, synthesis, comparison, deduction, abstraction, system and dialectical approaches to substantiate the features of the proposed model of ecosystem of planning and developing a University graduate's career.

3 Results and Discussion

The digital transformation of society is changing the content and approaches to higher education and the role of universities in increasing the competitiveness of graduates in the labor market. Currently, there are two basic objectives for the higher education: 1) universities should work to develop students ' "skills of the future"; 2) universities should not just train personnel for the labor market, they should cultivate talents for the " knowledge economy".

In 2017, a direct interview method was used to study the popularity of the Siberian state industrial University in Novokuznetsk (SSIU). The 1st-year students of the University took part in interview. The highest satisfaction that influence the choice of University by entrants was noted by respondents, in the following indicators: the availability of distance learning, extracurricular activities, and the presence of additional education. The lowest satisfaction – on the criterion of the guarantee of employment after graduation [13]. Thus, creating an environment that promotes employment and career start for a future young specialist is a serious area of work for SSIU, which require the University to transform its activities. In addition to forming the student's competencies for the digital economy, the University began to create conditions for employment and the formation of the beginning of a graduate's career, which became the second significant function of the University after educational activities. Of course, the University paid attention to the issues of employment of graduates earlier [14,15]. But the proposed ecosystem model makes it possible to expand the number of stakeholders by using an integrated digital platform, to harmonize the interests of a larger number of stakeholders, ensuring the preservation of a temporary resource.

The designed ecosystem "Career Tech" with the help of computer tools and cloud services creates conditions at the University for early career formation of future applicants and students. The "Career Tech" Ecosystem model of SSIU is shown in figure 1.

Starting at school, prospective applicants have the opportunity to expand their social contacts by interacting with the University in a virtual environment. To do this, SSIU offers them career guidance programs; participation in University events: Olympiads for schoolchildren, scientific conferences, participation in joint projects, courses for passing exam, training for obtaining additional skills, for example, in nano-technologies, robotics, modeling, etc.

This approach allows school-children to get acquainted, actively interact and even move within the University's scientific and educational environment. Having access to communications with the University, pupils of lyceums and secondary schools can virtually attend test lectures, "walk" around the University, see its laboratories and equipment, get acquainted with the major stakeholders of the University and, thus, more consciously choose the University and the direction of training in it. In turn, interacting in a joint eco-environment with secondary educational institutions, SSIU can attract school graduates which have high results to enroll, especially important in the context of increasing regional migration.

After entering the University, the professional interest and motivation of students begin to increase due to offers to participate in scientific and educational projects, cultural and corporate events of the University.

During the "rotation" of students in the "Career Tech" information ecosystem, their best personal qualities are revealed, career guidance is provided, and as a result, the future graduates begin to consciously predict and plan their career. They have the opportunity to form their career by responding to invitations of University employee and employers: 1) to participate in scientific research, scientific publications and subsequent admission to master course and postgraduate programs; 2) to gain new knowledge while studying under additional professional education programs; 3) to participate in the activities of the trade Union Committee, student Council, construction teams, volunteer movement, etc., which contributes to the development of universal competencies, such as leadership, activity, initiative.

The "Student assessment Center" created at SSIU within the ecosystem should implement one of the main functions of managing the employment of graduates and the formation of their early career. The main objective of the "assessment Centre" will be the systematic assessment of growth of educational and professional potential of students; evaluate the effectiveness of forming of project teams, assessment of students, a reasonable estimate of their capacity for science leadership teams (heads of groups, heads of project teams, leaders of volunteer or construction teams), all letters of recommendation to employers, identify of needs for additional education, based not only on the interests of the University, but also individual preferences, ranking and preparation of materials for creation of a database of the characteristics of students which are considered when employers review candidates for employment.

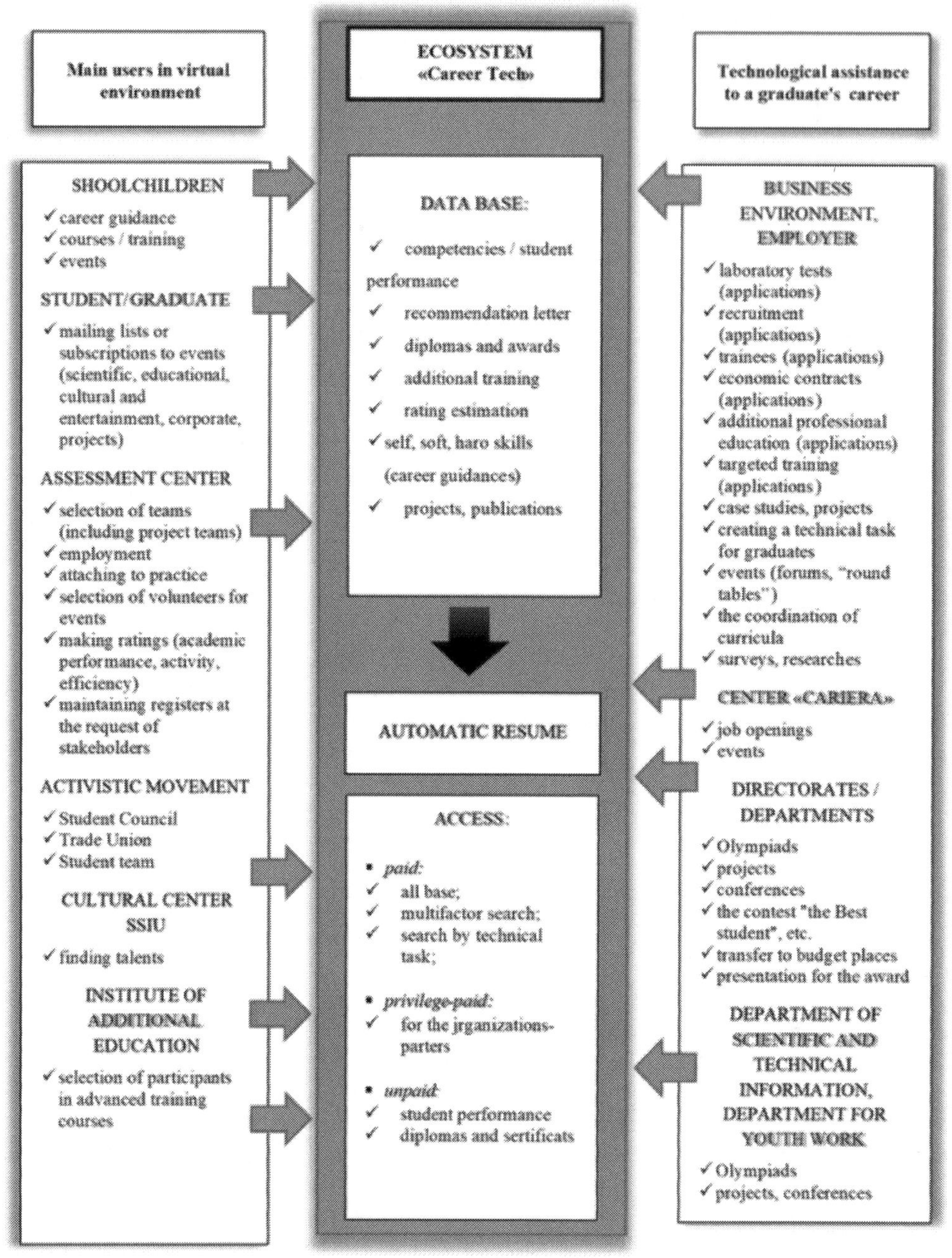

Fig. 1. The proposed model of SSIU ecosystem «Career Tech»

In 2020, major SSIU employers involved in the University ecosystem were interviewed. The study showed that business will be happy to use the y and its departments when applying for laboratory research, recruitment, contracts, etc; 2) ease eco-system because of: 1) the availability of services for distance interaction with the University of use (as a principle of "single window" in state institutions); 3) saving time for official circulation and formalize interaction with the University (faculties); 4) the availability of system information product with the insider information of the University about the generated skills of future young professionals. Due to this information system the employer can remotely interact with students to attract them to jobs, internships, events and employment.

Also, according to the results of the interview, some problems of the existing system of interaction were identified that restrain the achievement of high employment rates. For example: 1) lack of competence in the legal regulation of interaction between employers and educational organizations. As a result, there is an increase of mismatch in the expectations of employers and graduates during employment; 2) difficulty of interaction between SSIU and employers of medium and small businesses. Traditional interaction techniques exist in the format of "the University - a large organization." Such organizations have long-term plans for the development of production, stable incomes, a large staff, and, accordingly, job opportunity. Medium and small businesses, and especially individual entrepreneurs, have much fewer opportunities and resources, but they also have a deficit of personnel. Within the created SSIU ecosystem, tools and interaction technologies are emerging that provide access for various types of stakeholders.

4 Conclusion

The obtained results have a practical value: application the created ecosystem model "Career Tech" in the University expands the possibilities of attracting best schoolchildren for admission to the University. Also it creates a closed circulation of resources in the ecosystem (human, informational) that increases the efficiency of joint activities, builds mutually beneficial relationships which redistributes the potentials of the "knowledge and skills", increases quality control of training of students, raises the degree of congruence of stakeholder interests.

The created model has significant advantages over the existing ones, primarily focusing on promoting the career of graduates. It is also characterized by the ability to quickly adapt ecosystem elements to a changing institutional environment and scalability – the ability of the ecosystem to narrow and expand.

References

1. L.A. Piankova, A. V. Gromova, *Professional self-determination of young people as a process of career planning in modern socio-economic conditions: monograph* (2020)
2. M.L. Rothschild, *Bionomics: The inevitability of capitalism Hardcover* (1990)
3. C.W. Wessner, The Papers on Entrepreneurship, Growth and Public Policy (2004)
4. S.P. Rücker, B. Fischer, S. Queiroz, Foresight and STI Governance, **12**, 50 (2018)
5. G.V. Osipov, *Economics and sociology of knowledge: prakt. stipend* (2009)
6. L.M. Gaisina, Yu.N. Dorozhkin, et al., Espacios, **38**, 2 (2018)
7. T.I. Zaslavskaya, M.A. Shabanova, Social sciences and modernity, **5**, 5 (2012)
8. I. K. Zolotova, A. E. Zuev, Social sciences and modernity, **5**, 34 (1994)
9. O. A. Rakovskaya, *Social guidelines for young people: trends, problems, prospects* (1993)
10. Yu.A. Zubok, V.I. Chuprov, Social and humanitarian knowledge, **3**, 3 (2012)
11. Yu.A. Zubok, V.I. Chuprov, Social and humanitarian knowledge, **5 (373)**, 114 (2015)
12. Yu.A. Zubok, V.I. Chuprov, Economic and Social Changes: Facts, Trends, Forecast, **1**, 41 (2017)

13. T.V. Bobko, T.V. Petrova, Bulletin of the Siberian State Industrial University, **17**, 72(2016)
14. E.S. Bezikova, T.V. Bobko, Actual problems of economics and management of the XXI century, 97 (2017)
15 G.G. Kazanceva, T. V. Petrova, et al., Basic research, **7**, 131(2017)

Modern instruments for joint management of international projects in a pandemic

Elena Shatskaya[1,*], *Natalia* Kharchenko[1], *Mark* Popovichenko[1], and *Angelina* Lukyanova[2]

[1]North Caucasus Federal University, Stavropol, Russia
[2]Kazan Federal University- Naberezhnye Chelny Institute, Russia

Abstract. The article provides an overview of modern instruments for business structures in a pandemic and remote work. The most popular and relevant services that allow you to organize work on a project and maintain productivity in self-isolation have been summarized. The features and advantages of software products, which may be the future of project management, have been analyzed. A review, a critical analysis of the possibilities of the digital agenda of the studied countries made it possible to determine the importance of strengthening the interaction of government authorities and business structures in the framework of the implementation of social and economic digital projects for the development of international cooperation between the Republic of Belarus and the Russian Federation. A number of prerequisites were identified as a motivational toolkit for the involvement of business entities and society in the implementation and use of digital technologies on the territory of the union states - the Russian Federation and the Republic of Belarus. A complex of restraining factors of development in the post-pandemic period for the Republic of Belarus and the Russian Federation has been identified and substantiated, which makes it possible to recognize the problems of capacity building in states and the disunity of interests of the subjects of project management.

1 Introduction

The Russian Federation and the Republic of Belarus are member states of the Eurasian Economic Union (EAEU). Joint international cooperation is gaining a breakthrough character due to the new opportunities of the digitalization trend. Moreover, increasing the competitiveness and economic growth of countries in modern conditions is impossible without the transition to a digital economy.

Joint cooperation of the countries through strengthening the interaction of authorities and business structures in the framework of the implementation of socio-economic digital projects, contributed to the strengthening of international cooperation between the Republic of Belarus and the Russian Federation. The successful practice of domestic business in project management emphasizes the success of developing and maintaining their own sites,

* Corresponding author: elena.sh.05@mail.ru

announcing online projects, recording feedback in the post-support format. The accumulated experience is also in demand in Russian-Belarusian projects, especially in the context of a pandemic stagnation, with the use of automatic tools for setting tasks and planning and the use of CRM systems.

An analysis of Russian and foreign practice, as well as theoretical studies, made it possible to conclude that the development of a project economy in the context of digitalization declares the following challenges to public administration in Russia, specific to the Russian administrative and regulatory context: compliance with the rate of change, management flexibility, updated reporting forms, digital environment of trust, introduction of information and digital technologies into the structures of public administration [4, 5]. To solve the identified problems, it is envisaged as follows: the creation of an information and communication technological basis for the development of the economy and social sphere and, as a result, the improvement of administrative procedures, the digitalization of public services. By 2024, as part of the implementation of the federal project "Digital Public Administration", public services that are most in demand among the population should be contactlessly available. That is, to eliminate personal visits by citizens to state (municipal) authorities, 90% of intradepartmental and interdepartmental document flow will be transferred to digital format. The digital format of documentation and the presence of an electronic digital signature (EDS) will also affect the interaction of commercial organizations and authorities, increasing the workflow up to 70%. Business and the state should have an interest in working together to support digitalization, the state of which directly affects not only the growth of economic indicators, but also the intensity of economic development.

The policy of radical economic transformations in Belarus is carried out on the basis of digital technologies, as in other EAEU member states. One of the priority directions of the strategy of the Republic of Belarus is the digitalization of the country. For the transformation of socio-economic spheres, a legislative and regulatory environment is being created, international infrastructure projects aimed at globalization are being implemented, and a crypto exchange has begun to function. As well as on the territory of the Russian Federation, the issues of development and implementation of IT-tools are being actualized in the Republic of Belarus, conditions are being created for three-way electronic interaction: citizens - business community - state (municipal) authorities. Close partnership of a number of Russian regions with Belarusian regions is based on the implementation of international cooperation agreements. Most of the Russian-Belarusian projects are being implemented under the patronage of the Chamber of Commerce and Industry, research centers, technology parks, and individual organizations. Direct investments of domestic businesses imply an injection into Belarusian transport and manufacturing industry. The Republic of Belarus actively cooperates with partners from Moscow and St. Petersburg. One of the brightest examples of international cooperation between Belarus and Russia is an investment project to create a modern agro-town in the Sakhalin Region.

However, despite the existing reserves and the readiness of the business community for further cooperation and mutually beneficial projects, the unresolved issues of a number of administrative and legislative problems may complicate the practical implementation of the integration declared in the EAEU format. The transition to effective models of interaction between business structures in the context of digital globalization of Belarus and Russia is difficult at the organizational, regulatory, methodological and economic levels without appropriate training.

The modern world is dictating its terms and challenges faced by an increasing number of organizations. The pandemic is limiting face-to-face meetings and office work, employees are moving to remote work, and this affects the productivity of companies that were not prepared for such conditions. In addition, time and space restrictions are erased, the capabilities of project management participants are expanded, and a new competitive environment is formed. Due to the high level of mobility of the business environment, information becomes available to a wide range of participants in international projects. It should be noted that with the growing role of information and the IT sphere, these processes contribute to the emergence of obstacles to the management of international projects and stimulate monopolization in a pandemic.

"The practice of innovative design in the field of high technologies shows that the traditional form of project management slows down the development of a project, often making it impossible to implement it due to the difficulties in obtaining information necessary for the implementation of the project, lack of experience in implementing such projects, lack of developed technologies, competitive advantages of foreign analogues" [1].

Therefore, in the sector of corporate information products, more and more services began to appear that simplify joint project work, including at the international level. Information about these services will allow the management of companies to choose the most suitable ones for themselves and, as a result, increase the efficiency of project management.

2 Materials and methods

The study used the methods of detailing, abstraction, comparative theoretical analysis of economic and managerial literature, a critical review and generalization of practical experience, the study of regulatory documents on the organization of strategic planning of socio-economic and scientific-technological development of the studied countries, methods of logical modeling. At the final stage, the methods of statistical data analysis were applied, their graphic and semantic interpretation was carried out.

3 Results and discussion

Analysis of the practice of international cooperation between the Republic of Belarus and the Russian Federation in the framework of the implementation of socio-economic digital projects made it possible to substantiate that the countries historically have a long history of integration interaction. Currently, we can talk about new opportunities for the development of interaction, especially as EAEU member countries. Among the main instruments, the following services can be distinguished, as the most popular:

- Trello;
- Notion;
- Miro;
- MindMeister;
- Slack;
- Figma;
- Zoom;
- Google online services for working with documents.

Trello is an online project management tool. "Individual resources in Trello are a highly customizable analogue of a Kanban board, consisting of columns in which you can place individual cards, which, in turn, contain additional information. Thus, a three-level hierarchy of information is realized "[2].

Trello gives the project manager the ability to distribute tasks, assign performers and set deadlines. This service is based on the Agile methodology (which allows you to build work on the principles of "just in time" and "the project is always ready"), namely, on the Scrum technology. "Scrum is based on empirical processes, that is, knowledge comes only with experience, and decisions can be made based on known actions. Scrum uses an iterative-incremental approach to optimize predictability and risk management "[3].

The principles of "just in time" and "project always ready" mentioned above are that the implementation of the project is carried out through the structural decomposition of work at iteration. These iterations are called the "sprint". Upon completion of the iteration, the project manager can always demonstrate the result to the customer and inform that this version is a draft and will be further refined.

Such a methodology, together with an open Trello board, allows you to build a transparent interaction with the customer, despite the fact that counterparties can be located anywhere in the world and track the process of work both from a computer and from a smartphone or tablet.

An important advantage of Trello is that this board is easily integrated into the workflow, employees quickly master the technology and complete tasks with greater motivation, since this will allow you to move the card from the "To Do" column to the "Done" column.

Notion is an advanced analogue of Trello and has more popularity in the international environment, while Trello is a domestic startup. At the same time, the expanded functionality of Notion allows you to store all incoming content in a special table "warehouse", use databases to work with information and manage tasks in different ways (task lists and calendars).

With the help of Notion, you can manage a project, maintain a content plan, share documents, take notes and attach accesses to other platforms, which makes it an absolute favorite among similar tools. However, for employees, the transition to Notion can seem difficult and time consuming. Therefore, when introducing this platform into business processes, it is important to conduct a little familiarization with the platform, which in fact is not as complex as it seems at first glance.

Another domestic collaboration product is Miro. This project resembles a blackboard that hangs in most offices and contains important notes and tasks. The advantage of this service is its flexibility, simplicity and convenience. Here you can also attach the stickers and leave notes, draw tables and columns with tasks. This tool is common among creative teams and is used for generating ideas, brainstorming and posting additional information.

MindMeister is an association mapping platform. Association or mind maps can be used in business for a variety of purposes:

- creation of a business plan;
- positioning;
- SWOT-analysis;
- branding;
- naming.

MindMeister allows you to build an easy-to-interact association map that can accommodate an unlimited number of project ideas and customize to generate new ones. This product will help you structure ideas, build a hierarchy, prioritize ideas and find new solutions to current and strategic problems in the company.

Slack is a popular corporate messenger. In it, the management can build channels of communication with employees for various purposes: personal correspondence, correspondence on the project, informal communication, discussion of ideas, interaction with the contractors. Employees can communicate with management and among themselves, customize the platform for themselves and receive notifications about new tasks or management ideas. In turn, management can give access to for certain chats and conversations to certain employees.

Figma is a development tool for designers, but has great value for leadership. This tool allows you to include both the management and other employees of related specialties in the development process. The platform offers a wide range of functionality for branding, website and application development. At the same time, the manager can open the project at any time and see at what stage the designer is. A programmer can get a ready-made design project and work with it most efficiently, since Figma has a separate tab for developers, where information about styles is stored in the form of ready-made code. The project manager can leave comments and make small adjustments at any time. So, using this tool, leadership of creative teams can significantly save the development time.

Zoom is an online conferencing platform. This service allows you to create and conduct online meetings, conferences, business negotiations, interviews, brainstorming sessions, etc. Meetings can be private, and unauthorized users cannot connect to them. Any conference participant can record it, if necessary. The host can appoint co-organizers to effectively manage the meetings. In addition, the participants can be divided into groups and pairs on this platform. Meeting attendees can use a virtual background that can be a corporate one. Also an important function is the ability to demonstrate your screen for project presentations in front of the customer or manager.

Google online services include a large suite of software products:

– Google Documents;
– Google Slides;
– Google Sheets;
– -Google Forms, etc.

The main advantage of these Google services is the interoperability. Management can collaborate with employees to create documents in Google Docs, design presentations in Google Slides and work in spreadsheets in Google Sheets. At the same time, these platforms can be used for different purposes:

– development of investment documents;
– creation of a business plan;
– conducting marketing research;
– creating a content plan.

In addition to these services, the Google Forms service can play an important role in project work. It can be used to conduct staff questionnaires, customer briefing and Cust Dev target groups of a startup.

In the union state of the Republic of Belarus, preference in project management is given to special services that allow you to organize a business process without leaving your home. Among the most popular solutions are Microsoft 365 cloud applications, Microsoft Teams video conferencing and teamwork service, remote workplaces and cloud storage. The centralized Microsoft Teams system is represented by several spaces.

For group work on a project, to-do lists are created. The repository of communication exchange are message boards, each of which is dedicated to a specific topic. Statistical information is professionally presented in the form of graphs that reflect the appointed dates of upcoming tasks and events in the work on the project. A separate place in the Microsoft Teams system is occupied by the cloud space that accumulates documents, files and images. Real-time mode allows you to support group chat, reflecting the responsiveness in the "ask a question - get an answer" section. Thanks to Check-in questions, questions concerning all participants are concentrated in one place and replace numerous chats and constant meetings. The indisputable advantage of this system is the convenience of aggregating everything you need in one place.

4 Conclusion

A review, a critical analysis of the possibilities of the digital agenda of the studied countries made it possible to determine the importance of strengthening the interaction of government authorities and business structures in the framework of the implementation of social and economic digital projects for the development of international cooperation between the Republic of Belarus and the Russian Federation. Analysis of the experience in the development and implementation of digital projects indicates the presence of both evolutionary and artificial components that arise with the direct participation of public authorities. A number of prerequisites were identified as a motivational toolkit for the involvement of business entities and society in the implementation and use of digital technologies in the territory of the Russian Federation and the Republic of Belarus.

1. State background. From year to year, in the countries of the Russian-Belarusian commonwealth, the amount of financial support to small and medium-sized businesses is increasing, the main goal of which is to assist small and medium-sized businesses. Today, the use of end-to-end digital technologies is one of the strategic priorities for the development of the economies of these countries. The IT sphere plays a decisive role in ensuring the technological breakthrough of countries and ensuring their competitiveness in the world market.
2. Regional background. The Russian national project "Small and Medium Enterprises and Support for Individual Entrepreneurial Initiatives" and the State Program for Supporting Small and Medium Enterprises in the Republic of Belarus provide for various support measures at all stages of the entrepreneurs' life cycle.
3. Cognitive background. In order to meet the needs for specialists in information training areas, effective interaction with universities and colleges that train specialists for the ICT sector of the economy.

The accumulation of the efforts of interaction between the authorities and business structures will make it possible to more quickly adjust the implementation of socio-economic digital projects based on the modern requirements of regulatory legal acts and support measures at the regional level, as well as contribute to the development of international projects. Thus, the development of such interaction is impossible without the development of project management tools at the level of business structures.

In post-pandemic conditions, the information and IT-environment of Russia has acquired a different scenario of formation, namely, as part of the globalization of the information society. The spheres of industry, agriculture and transport, medicine, education are permeated with information projects. Even at the international level, there is no industry that does not have IT blocks or electronic services. Modern IT companies are forced to compete for the fact of being ahead of the creation and launch of digital services. The domestic public sector turned out to be no exception, and in order to increase the availability of social services and the provision of public services, it successfully eliminates the problems of digitalization of public administration. Of course, this provides for the full participation of the Russian Federation in the processes of information and economic integration of regions, countries and peoples, as well as the creation and development of information and knowledge as factorial project management. Thanks to the implementation of the national program "Digital Economy of the Russian Federation" dated June 4, 2019, it is planned to solve the following significant tasks:

- development of the information society;
- raising the standard of living of the population;
- development of the socio-economic and cultural spheres of society;
- increasing the level of competitiveness of domestic goods in the field of digital technologies.
- The results of the implementation of the program by 2024 provide for:
- increase in the number of citizens using information in digital form, up to 70%;
- ensuring joint activities of the state, citizens and entrepreneurs based on the use of digital technologies;
- increasing the level of labor productivity through the introduction of digital technologies;
- creating conditions for the integration of the Russian Federation into the world information society;
- carrying out legal and legally significant actions in digital form;
- increasing the level of Internet users to at least 95% of citizens.

In the period of development and digital transformation, it is impossible to overestimate the role of international projects and their management processes. The main tasks are to substantiate the advisability of choosing certain management tools and technologies to achieve the goals of the socio-economic system that make up the national economy of Russia. This is especially important for the use of advanced project management technologies and tools for digitalizing management decisions at the international level.

The process of transition to new conditions of interaction between the authorities, society and business structures in the international market has been launched, numerous social and economic digital projects are being implemented, international cooperation between the Republic of Belarus and the Russian Federation is becoming more transparent. The investigated complex of services and platforms can fully satisfy the needs of business structures in joint project management. The global crisis associated with the epidemiological situation has led to a digital breakthrough in the field of corporate information products, digital services, which have made it possible to maintain the efficiency of companies and, in some cases, save operating costs.

Acknowledgements

The study has been carried out with the financial support of the RFBR within the framework of the research project "Development of the concept of forming the effective models of interaction between the authorities and business structures at the regional level within the framework of digital integration of the Republic of Belarus and the Russian Federation", project No. 20-510-00025.

References

1. O.V. Serbskaya, Materials of Afanasiev readings, **2(15)** (2016)
2. V. G. Zaitsev, A. A. Zheltova, E. V. Tibirkova, Higher education in Russia, **12** (2016)
3. Ken Schwaber, Jeff. Sutherland, The definitive Guide to Scrum: The Rules of the Game, 23 (2016)
4. National Program "Digital Economy of the Russian Federation", http://government.ru/

5. Official website of the Eurasian Economic Commission, http://www.eurasiancommission.org/
6. E. Yu Shatskaya, O. A. Kozlova, E. V. Sukhostav, N. A. Anashkina, O. N. Tkachenko, Perspectives on the Use of New Information and Communication Technology (ICT) in the Modern Economy, Advances in Intelligent Systems and Computing (2019)
7. V.N. Parakhina, E.G. Popkova, The Future of the Global Financial System: Downfall or Harmony, 939 (2019)
8. S. Zenchenko, Y. Radyukova, A. Zaitsev, 32nd International Business Information Management Association Conference (IBIMA): «Vision 2020: Sustainable Economic Development and Application of Innovation Management from Regional expansion to Global Growth», 3042 (2018)
9. Nadezhda V. Miroshnichenko, Svetlana V. Nedvizhay, Elena A. Ponomareva, Elizaveta V. Skiperskaya, Natalia P. Kharchenko, Perspectives on the Use of New Information and Communication Technology (ICT) in the Modern Economy, 378 (2018)
10. E.Yu. Shatskaya, N.P. Kharchenko, (2365) Economy and entrepreneurship, **8(121)**, 1336, 250 (2020)

Assessment of the environmental aspect of corporate sustainability of companies

E.S. Blinova*, *and T.V.* Ponomarenko

St. Petersburg Mining University, Vasilievsky Island, 21 line 2, 199106 St. Petersburg, Russia

Abstract. Strengthening international environmental and climate policies dictates a reorientation of companies towards corporate sustainability management. The article discusses the issues of assessing the environmental performance of companies in the context of corporate sustainability. The main methods of assessing the environmental aspect of corporate sustainability have been analyzed. The classification of the main methods for assessing the corporate sustainability has been presented. The application of methods for assessing the dynamics of environmental indicators of corporate sustainability of companies, including mining companies, has been substantiated. Mining companies have been proven to cause significant damage to the surrounding ecosystems. The use of indicators of economic impact and environmental and economic assessment has been substantiated.

1 Introduction

The development of Russian business as a whole corresponds to the global trend of business transformation in the interests of sustainable development and the formation of models of corporate sustainability [1].

Global trends in the development of international communities show that the management of companies is aimed at increasing the corporate sustainability. The corporate sustainability is inextricably linked to both sustainable development and corporate social responsibility. In this regard, when managing the corporate sustainability, such spheres of the company's functioning as ecology, economy and society are singled out and evaluated [2-3].

To date, there is still no single approach to the definition of "corporate sustainability". This fact predetermines the fact that there are no uniform criteria that reflect the development of the company in the field of ecology, economy and society. This article will consider approaches to assessing the environmental aspect of the corporate sustainability, since the tightening of environmental and climate legislation will actualize environmental requirements for activities even more.

Analysis of methods for assessing the environmental aspect of the corporate sustainability has shown that all existing assessment methods can be conditionally divided into several groups:

- individual indicators of environmental impact,

*Corresponding author: ekblinova@icloud.com

– economic and environmental indicators,
– environmental indexes.

The importance of assessing the corporate sustainability for mining companies is determined by the specifics and strength of their impact on the environment, including: greenhouse gas emissions, pollution of natural water bodies, deforestation, soil pollution, dehydration and salinization of the soil layer. Therefore, the companies are faced with the need to implement the environmental projects that create social capital. It is understood as the development and implementation of the strategies for environmental safety, allowing to increase confidence on the part of counterparties and stakeholders of the company [4-5].

The environmental assessment of corporate sustainability has several objectives:

1) assessment of the dynamics of environmental indicators to identify the effectiveness of the adopted management decisions;

2) benchmarking to compare the companies with similar business models to identify the best technologies and their possible use.

The aim of the study is to substantiate the selection of the method for assessing the environmental aspect for mining companies.

The research objectives are:

– review of existing assessment methods,
– systematization and analysis of the methods,
– analysis of the applicability of valuation methods for mining companies

2 Methodology

The article used the method of analysis and systematization of existing methods for assessing the environmental aspect of corporate sustainability.

The analysis of the scientific literature related to the concept of corporate sustainability has shown a different understanding of this category. In this regard, the need to assess the environmental aspect is also understood differently. Three approaches to the definition of the corporate sustainability have been identified.

The first approach equates the sustainable development with the corporate sustainability. An assessment of the environmental aspect in this approach is necessary; without increasing the environmental sustainability, it is impossible to increase the corporate sustainability. But this approach does not define the environmental aspect as the main one, since the enterprise should equally increase economic sustainability, production, marketing, and organizational and develop comprehensively in all industries [6-8].

The second approach to the definition of the corporate sustainability equates it with the corporate social responsibility. And in this approach, environmental development is key, but not the main one [9-10].

The third approach defines the corporate sustainability in terms of competitiveness and profit growth. The preservation of ecosystems, reduction of the impact on climate change is of secondary importance.

None of the approaches emphasizes the need to assess the environmental aspect, since all the approaches assume the presence of many factors that determine the corporate sustainability [11-12].

Identifying the individual indicators of environmental health and results is the simplest method for determining the dynamics of the corporate sustainability. It is easy for the company's management to identify key indicators of the environmental development, taking into account the specifics of production. The same indicators are reflected in the non-financial statements of companies. For the industries of the mineral resource sector (MRS), among the individual indicators of interaction with the environment, one can single out [13]:

- indicators that characterize water consumption,
- indicators that characterize energy consumption,
- indicators that characterize the use of waste,
- indicators that characterize the reduction of emissions and other indicators that characterize resource consumption.

In the companies of the mineral resource sector (MRS), these indicators are widespread and are used as representative indicators of the dynamics of the environmental sustainability. At the same time, MKS companies have different scales of activity, and with an increase in capacity, the absolute values of pollution, emissions and waste will grow. All this creates a contradiction according to which only small companies in the industry will be more environmentally sustainable. This factor makes the assessment of the environmental aspect of the corporate sustainability of MKS companies by the methods of the first group in terms of final indicators not fully objective, in terms of specific indicators it is also difficult to compare.

It is easy to measure and monitor the dynamics of individual environmental indicators, therefore, it is possible to assess the effectiveness and efficiency of measures aimed at achieving the values of the set indicators.

The complexity is added by the need to separate indicators of production efficiency (they were discussed above) and indicators of management efficiency.

Some authors highlight the following criteria for achieving the corporate sustainability in the field of ecology:

- environmental policy,
- environmental (social, integrated) reporting,
- company's management system (environmental management) [13].

Without the development of management performance indicators, the development of production is impossible, since the company needs a regulatory framework to implement any projects. There are examples of evaluating managerial efficiency, for example, in Gazprom company. At the same time, they are largely of an evaluative nature, which complicates the possibility of their correct application. It is impossible to independently assess the indicators of management efficiency, because the final indicators are the indicators of production efficiency. In addition, the mere presence of functional environmental structures in companies does not allow assessing the quality of the environmental policy being implemented.

The second group of methods for assessing the environmental aspect of the corporate sustainability includes economic and environmental methods.

The essence of the methods lies in the fact that in addition to the positive effect of the introduction of environmental protection measures, the company receives an increase in the economic efficiency of the business entity. Thus, project appraisal influences the company's environmental sustainability assessment.

Based on the essence of these methods, the corporate sustainability of a company can be determined by the following criteria [14]:

- presence/absence of economic effect from the implementation of technologies in the field of energy and resource conservation,
- presence/absence of functional structures in the company that implement the environmental policy,
- dynamics of development/implementation of projects to protect the environment.

Within the framework of the 3 listed criteria, 10 indicators have been distinguished with weights established by an expert, namely:

- economic effect from the introduction of energy and resource saving technologies,
- coefficient of environmental pollution,

– effectiveness of environmental protection measures,
– utilization rate of secondary raw materials,
– waste utilization rate,
– availability and functioning of functional structures for environmental management in companies,
– reduction amount of the environmental fees,
– development and implementation of R&D projects in the field of the environment,
– implementation of separate collection of industrial waste,
– organization of safe storage of industrial waste[14].

All values of the indicators are multiplied by their weight coefficients for each parameter, after summing up, a complex indicator of sustainable environmental development is obtained. This coefficient is intended to assess the dynamics of the environmental aspect of the corporate sustainability within a single enterprise. The advantages of the index include the simplicity of calculation; the calculation of the coefficient for the enterprises of the industry will be able to distribute enterprises according to the levels of environmental development. The disadvantage of the method is its subjective nature, represented by the expert definition of the authors of the method of weight values and a set of indicators [15-17].

The proposed methodology raises doubts, since the criteria, indicators for assessing the ecological state, and their weight values are not regulated. There are also environmental and economic indicators that are not comprehensive, but reflect the economic effect of the introduction of the environmental protection measures. These assessment methods include the dynamics of the magnitude of the environmental impact. The indicator is the sum of the average values of payments for the negative impact on the environment (including possible penalties), taxes on the use of minerals, the costs of maintaining the operation of dust-collecting and treatment plants, waste disposal, depending on the type of environmental impact.

The third group of indicators that assess the environmental aspect of corporate sustainability include the environmental indices. Basically, all indices assess the corporate sustainability in general, that is, they can be considered broad-based indices focused on assessing the environment, economy and society. For example, such indices include: Dow Jones Sustainability World Enlarged Index, MSCIEAFEESG Index, FTSE4Good Europe Benchmark Index, NASDAQ OMX CRD, etc. It is not possible to evaluate exclusively the ecological aspect according to the listed indices [18-19].

The environmental aspect of the sustainable development is assessed only by a few indices, among which the NERAX-Eco and the FTSE CDP Carbon Strategy 350 Index can be distinguished.

The NERAX-Eco Index was developed by the Independent Environmental Rating Agency (NERA). The assessment is based on a number of indicators that assess the resource and energy efficiency of companies, as well as their fundamental and technological components [18-19]. The developers do not provide an evaluation mechanism in detail, since this information is intellectual property, the principal evaluation mechanism is as follows.

There are two groups of companies on the Russian market:
– the best companies in the field of environmental production efficiency,
– the best of the company's progress to reduce the impact on the environment.

Among the most liquid issuers, stocks of companies included in the rating of social and economic responsibility are selected, then 20% of companies are selected from this list, which are the leaders in the two directions presented above. An additional condition for getting into the list of leading companies is the provision of transparent social and environmental reporting.

Next, the average price level is calculated (using capitalization as the basis for the weighting factors). The resulting value is divided by a fixed coefficient, which is selected in such a way as to get 100% at the initial moment of calculating the index. Companies that fall into 2 categories at the same time have double weight in the index.

Such an assessment system allows one to determine the internal problems of the company's management, since the environmental efficiency of production and the rate of its change reflect the success or failure of technical and managerial modernization.

Although the NERA methodology assesses the environmental component of production, it nevertheless links it to the economic activity of the company, and ultimately serves as an indicator for assessing the investment risks. According to the authors of the methodology, environmental indices are more objective in determining the corporate sustainability, since, unlike financial statements, they cannot be adjusted [20]. There are examples when environmental indices showed negative company dynamics, while companies were stable in terms of financial performance. For example, YUKOS, assessed by NERA ratings, had internal problems from 2000-2003. This was unexpected for foreign partners when they had to deal with technical and environmental problems associated with the acquisition of the company.

FTSECDP Carbon Strategy 350 Index assesses the climate change and greenhouse gas emissions. The research methodology is not explained, since it is intellectual property, in principle, the mechanism is as follows.

The company sends out questionnaires to the world's largest corporations (today the pool includes about 3,700 companies). The content of the questionnaire is focused on collecting information on climate change, the level of emissions in companies, the risks and opportunities of companies to manage climate change, as well as the environmental management system. After processing the questionnaires, companies are assigned the points for good environmental strategy management [18-19].

The dynamics of this index shows the extent to which companies are ready to implement measures to reduce the emissions, to comply with the tightening international climate policy, to reduce the risks of climate change.

This index can be considered the most relevant for companies that emit greenhouse gases, primarily for the companies in the fuel and energy complex. For the companies in other industries, the FTSECDP Carbon Strategy 350 Index will be unrepresentative.

3 Results and discussion

The review of assessment methods allows you to systematize the methods that are designed to determine the level of environmental development of the companies within the framework of the corporate sustainability (Figure 1).

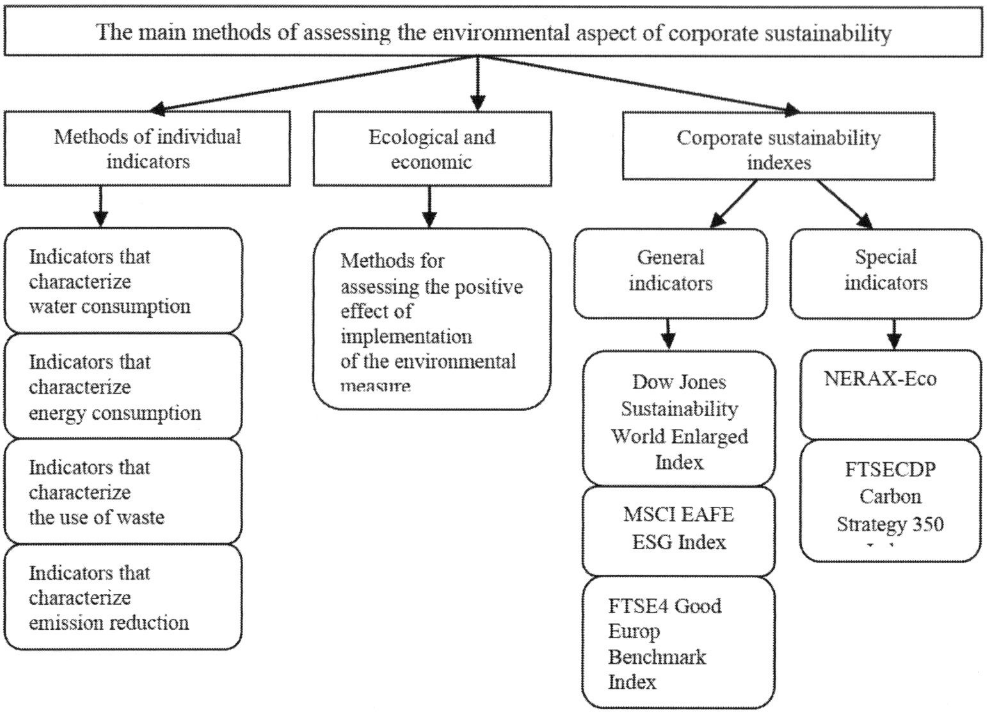

Fig. 1. Systematization of the main methods for assessing the environmental aspect in the corporate sustainability.

Based on the analysis, Figure 2 was compiled, which identifies the main advantages and disadvantages of assessment methods.

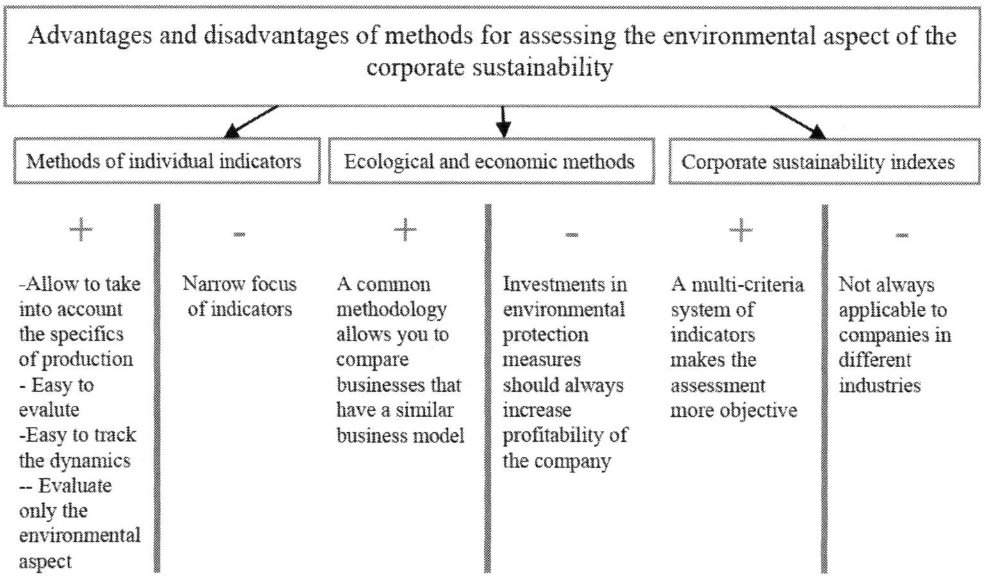

Fig.2. Advantages and disadvantages of methods for assessing the environmental aspect of the corporate sustainability.

In our opinion, among the methods considered, there is no assessment system that would give an objective understanding of the level of the environmental sustainability of the companies in the mineral resource complex. In this regard, a combination of methods is necessary.

The methods of the first group should be used primarily for the development of performance indicators of the environmental strategies and the environmental programs when evaluating the environmental protection projects in mining companies. The developed system of indicators should be based on an integrated approach to assessing the environmental sustainability. Firstly, to take into account the production indicators for the environmental performance, and secondly, to assess the quality of the environmental management in the companies.

Comparison of the companies with similar business models in terms of the indicators of the first group is permissible, but does not take into account the specifics of the mining and geological conditions of the deposits, as well as the mining technical factors of operation. For example, a deep ore body will increase the overall level of production waste compared to mineral deposits at shallow depths.

The existing methods for assessing the environmental aspect of the corporate sustainability using the environmental and economic methods are also partially applicable for MRS companies. Among them, only methods that take into account the indirect economic effect from the introduction of the environmental projects are applicable, since their implementation should not be aimed at increasing the production efficiency.

The second group of indicators also does not take into account the peculiarities of mineral deposits. For example, there is a difference in the gas content of coal mines even within the same field. Accordingly, this group of indicators is applicable only to assess the dynamics of the environmental sustainability within one company, and sometimes even within the same field.

It is impossible to provide an objective assessment of the methods of the third group, since no methodology is disclosed in detail. These methods are focused on the analysis of the largest companies. Small and medium-sized businesses will not be included in the sample, and their cumulative impact on the environment, including climate change, may be large and need to be assessed. In this group of methods, the specific characteristics of the mining company's mineral deposits are not taken into account.

4 Conclusion

The main conclusions of the study include:

1. To date, there is no unified approach to assessing the environmental aspect of the corporate sustainability of the companies in any type of economic activity, including the mineral resources sector.

2. Assessment of the environmental aspect of the corporate sustainability in almost all analyzed methods is linked with the financial results of the company or with the economic effect of the introduction of the environmental protection measures.

3. Each assessment method has advantages and disadvantages. Among the main disadvantage of all methods, one can single out the lack of a universal method used for different industries and scales. In this regard, the management of the company must independently reasonably choose a methodology to identify the dynamics of the assessment of the environmental aspect within the framework of the corporate sustainability, taking into account the specifics of the company.

4. For the companies of the mineral resource complex, it is most expedient to use the methods of individual indicators, as well as some of the environmental and economic methods. However, neither the first nor the second group of methods takes into account the specifics of the characteristics of the companies' fields. In this regard, the use of

benchmarking is impossible, the development and analysis of indicators for assessing the environmental aspect of the corporate sustainability is necessary for management.

Acknowledgment

The study has been carried out with the financial support of the Russian Foundation for Basic Research and the Ministry of Culture, Education, Science and Sports of Mongolia in the framework of the scientific project No. 19-510-44013\19.

References

1. K.Yu. Belousov, Problems of modern economy, 120 (2012)
2. T. A. Pikalova., N.V. Smirnova, *Albena: International Multidisciplinary Scientific Geoconference*, The development of methods for assessing the level of sustainable development of mining company, 3, 629 (2015)
3. T. V. Ponomarenko.; M. A. Nevskaya, O.A. Marinina, *Sustainability (Switzerland)*, 12, 5582(2020)
4. V. A. Roerich, Science and education today, 3, 38 (2019)
5. Website of the international consulting company Pricewaterhouse Coopers, https://www.pwc.ru
6. E.N. Kucherova, OSU Bulletin, 9, 76 (2007)
7. L. Lankoski, Journal of Cleaner Production,139, 847 (2016)
8. C.Villiers, Journal of Cleaner Production, 136, 78(2016)
9. W. Hediger, Journal Socio-Economics,39, 518 (2010)
10. I. Montiel, Organizational Environment,21, 245 (2008)
11. S.A. Panova, Russian entrepreneurship,8, 78 (2003)
12. R.Lozano, Corporate Social Responsibility Environmental Management, 22,.32 (2015)
13. I.Montiel, J.Delgado-Ceballos, Organization and Environment, 2, 1 (2014)
14. E.L. Zimina, D.K. Pankevich, S.M. Goryacheva, Bulletin of VSTU, 1, 128 (2019)
15. O. A. Marinina, IOP Conference Series: Earth and Environmental Science, 302, 1 (2019)
16. M.A. Nevskaya, O.A. Marinina, Research Journal of Pharmaceutical, Biological and Chemical Sciences, 7, 2352 (2016)
17. T.V Ponomarenko., R.P.Volnik, O.A.Marinina, *Notes of the Mining Institute*, 222, 882(2016)
18. A.V. Khoroshavin, *Implementation of the sustainable development goals: European and Russian experience: collection of scientific articles based on conference materials*, 184 (2019)
19. A.V. Khoroshavin, *North Palmyra: collection of scientific papers of the IX Youth environmental conference. Saint Petersburg,* 134 (2018)
20. V.V. Artyukhov, A.S. Martynov, *Metodics of assessing the environmental and energy efficiency of the Russian economy* (2010)

Development of 'smart'manufacturing in Russia: Empirical region-specific study of system technologies

Evgeniy Starikov, *Marina* Evseeva[1], and *Irina* Tkachenko

Ural State University of Economics, 620144 Ekaterinburg, Russia

Abstract. The purpose of the paper is to consider the concept of Industry 4.0 through the prism of 'Factories of the Future' program as a system of integrated technological solutions for creating 'smart' industries.The paper analyzes the TechNet road map proposed by the National Technological Initiative, a program aiming at enhancing the global competitiveness of Russian industry by 2035. Aregion-specific empirical study focuses on system technologies that contribute to the development of 'smart' manufacturing in four industrial regions of Russia: Sverdlovskaya, Chelyabinskaya, Kurganskaya and Tumenskaya Oblasts. The conclusion is made about the upward trendin thedevelopment of 'smart' industries in Sverdlovskaya and Chelyabinskaya Oblasts.

1 Introduction

A territorial production system at any level can be viewedasa complex of technologies. All technologies can be conventionallycategorizedintobasic, component, and system technologies [1-4]. Basic technologies aretechnologies for the production of materials and other substances (e.g., alloy used for manufacturingaircraft engine turbine blades). Component technologiesare technologies for the production of components and subassemblies that are not products of final consumption (e.g., an aircraft turbine engine).System technology integrates component technologies to produce thefinal product for end users by developing highly engineered mechanisms, machines, and apparatuses(e.g., a passenger aircraft) [5].

Thus, system technologies and related component technologiesare carriers of the most advanced technical solutions; accumulate theintellectual labor of thousands of people, research and development in the field of AdvancedManufacturing Technologies (AMT) whichaccumulate multidisciplinary knowledge and intellectual know-how [6].

A key feature of the current stage of technological development ofterritorial production systems is an overall change by digitalization and automation.The concept of Industry 4.0 as a set of technological changes in manufacturing was coined by the German government in 2011. The main paradigm for automation and data exchange in manufacturing technologies includes the use of the Cyber-Physical Systems whichenable to connect the physical things with virtual models [7]. In Russia, the concept of digital transformation ofthe production

[1] Corresponding author: evseevamv@usue.ru

sector through the development and implementation of AMT is applied within the framework of the National TechnologyInitiative (NTI), a program aiming at enhancingthe global competitiveness of Russian industry by2035 [8].

In this regard, we may hypothesize that a variety of system technologies andlocalization of system integrators in a region maytestify that the processes of digital transformation of the domesticproduction sector and the development of SmartManufacturing are taking place in Russian regions.

System technologies determine the production specialization and competitiveness of the region, since they offer opportunities to manufacture high-tech industrial products with highadded value. At the same time, to ensure the implementation of systemtechnologies,component technologies develop with a multiplicative effect. Component technologies may also contribute to gaining territorial competitive advantage by establishingnew component manufacturing with a high level of employment based on theintroduction of new engineering and manufacturing solutions and the development of newproducts. The development of basic technologies is largely due to the presence andcomposition of local mineral resource base. The growth of systemtechnologies used in the region prompts the correspondinggrowth of component technologies. Thus, varioustechnologies and localization of system integrators is a factor that maintains stability and competitiveness of regional industries. This results from possible engineering integration along global technological chains and visual effect of diversification: the territory retains its economicstability even if a product manufactured with the use of a particular technology is not in demand. On the contrary, the dominance of a technology (evenif there are many system integrators) is the factor thatmakes regional economy more sensitive to market fluctuations.

The NTI TechNet roadmap introduces the concept of FactoriesoftheFuture(FoF), FoF are understood as systems of integratedtechnological solutions (in fact, integrated technologicalchains) that enable to design and manufacture anew generation of globally competitive productsin the shortest possible time. The TechNetroadmap suggests a classification of FoFon the basis of leading groups of technologies: DigitalFactories, SmartFactories and Virtual Factories. DigitalFactoriesare charged with the task of designing manufactured products by using software, digital design and modeling tools. SmartFactoriesrefer to flexible production facilities with awide application of Process Automation Systems (PAS) and planning systems based on robotic processes.The combination of these technologies is aimed at increasing an overall level ofintellectualization and production efficiency, increasing faulttolerance, and creating the basis for further development of new markets forinnovative manufacturing solutions. Finally, VirtualFactories is anaggregator thatconnectsall parties into a single logistics chain: consumers of high-tech products, manufacturers, designers andprogrammers of production software (Digital Factories), facilities for components production; assembly and final product manufacturing using themost advanced technologies (Smart Factories), as well as suppliers of materials. Creation and scale application of technological solutions proposedin the TechNet roadmap are aimedat developing and increasing the export potential of the domestic industry, aswell as at resolving the issue of substituting imported high-tech products through the use of advanced production technologies [9].

It has been witnessed that some sectors of the Russian economy conform to international standards with regard to the degree of digitalization (e.g., finance, state and municipal services, education). At the same time, the manufacturing sector cannot meet theinternational standards of digitalizationand application of advanced technologies. Nevertheless, in recent years many regions have startedmodernizing the existing production facilities and building new ‘smart’ capacities which integrate computer high-precision and information components with high-performance labor and create cooperative chains, territorialmanufacturing and technological clusters. Surveys show that 78% of Russian large companies’ management considerdigital transformation to be their priority for the next 3-5 years, whichcertainly gives grounds for optimistic forecasts of the manufacturing sector growth in Russia in Industry4.0 [10].

2 Methods and results

An empirical study of system technologies was conducted for four legal entities within the Ural Federal Okrug: Sverdlovskaya Oblast, Chelyabinskaya Oblast,KurganskayaOblestand Tyumenskaya Oblast(autonomous okrugs excluded). The choice of the manufacturing sectoras the subject of study is determined bythe pronounced industrial type of these regions' economies: a significant share of thegross regional product is formed by processing industries.

To identify the totality of technologies implemented in the regions, disaggregation of the Russian National Classifier of Types of Economic Activity (OKVED-2) codes has been applied. We understand a type of economic activity as a combination of similar technologies, which results in a final product or a product for intermediate consumption.Given this provision, we have detailed the OKVED-2 codes to the levelof a single technology.For example, production of air and space craftand relatedequipment (OKVED-2 code 30.3) includes system technologies (manufacturing of aircraft; helicopters; unmanned systems and flight hardware;launch test vehicles; artificial satellites,planetary probes, orbital stations, shuttles etc.) and component technologies (production of components for aerial vehicles; fuselages, wings, doors, pitch attitude control fins; control wheels; chassis, fuel tanks; open cockpits;,aircraft motors and engines; components for turbojet and turbopropaircraft; equipment for launch complexes for rockets and spaceship).

The sample includes three types of companies: system technologyintegrators; component technology integrators; and basic technology integrators. Technological affiliation of the company was determined by its core economic activity.SPARK-Interfax was used to verify the study sample (Table 1). The observation periodis 12 years, from 2007 to 2019.

Table 1. The study sample (2019)

Factor	Kurganskaya Oblast	Tumenskaya Oblast*	Sverdlovskaya Oblast	Chelyabinskaya Oblast
General totality (*N*)	290	541	1673	1468
Sample companies (*N*)	144	255	865	859
Aggregated market share by profit (%)	85.5	83.6	86.0	86.3
Identified basic technologies (*n*)	8	12	37	31
Identified component technologies (*n*)	26	38	101	86
Identified system technologies (*n*)	8	11	72	24

*Autonomous okrug excluded

The next step was to calculate Shannon andSimpson indices for each region, which are used to measure diversity ofobjects in a limited population (Table 2).Shannon Index (Eq. 1) characterizes the diversity and alignment of objects in thepopulation. Simpson Index (Eq. 2) indicates the dominance of certain objectsin the aggregate. The combination of these indices is believedto most fullycharacterize the measure of diversity[11].

$$H = - (p_i/p \cdot ln (p^i/p)) \quad (1)$$

$$D = (\Sigma p_i (p_i-1)) / p(p-1) \quad (2)$$

wherep_i=i technology integrators;
p=total companies;
n= total technologies by type

The obtained data can beinterpreted as follows:

1)The more technologies in the region's industrial sector and the less variable the number of the companies integrating these technologies, the higher Shannon Index. With a small diversity of technologies, Shannon Index tendsto zero.

2)The more pronounced the dominance of a technology (see number of technology integrators), the higher Simpson Index. With a uniform distribution of technology integrators, Simpson Index tends to zero.

Table 2. Calculated Shannon and Simpson Indices

Shannon Index				
	2007	2011	2015	2019
Sverdlovskaya Oblast	5.01	6.12	5.78	5.81
Tumenskaya Oblast*	3.36	3.40	3.67	4.32
Chelyabinskaya Oblast	5.36	5.61	5.62	5.69
Kurganskaya Oblast	3.58	3.50	3.51	3.33
Simpson Index				
Sverdlovskaya Oblast	0.112	0.110	0.091	0.084
Tumenskaya Oblast*	0.43	0.41	0.32	0.36
Chelyabinskaya Oblast	0.093	0.092	0.097	0.081
Kurganskaya Oblast	0.21	0.22	0.27	0.28

*Autonomous okrug excluded

According to the 2019 data, most system technologies were used in Kurganskaya Oblast (19% of the total), followed by Tumenskaya Oblast (18%) and Chelyabinskaya Oblast(17%). The share of system technologies implemented in Sverdlovskaya Oblast was the lowest and accounted for 14%. However, the number of implemented system technologies has been increasing steadily in Sverdlovskayaand Chelyabinskaya Oblasts, as can be seen in Fig. 1.

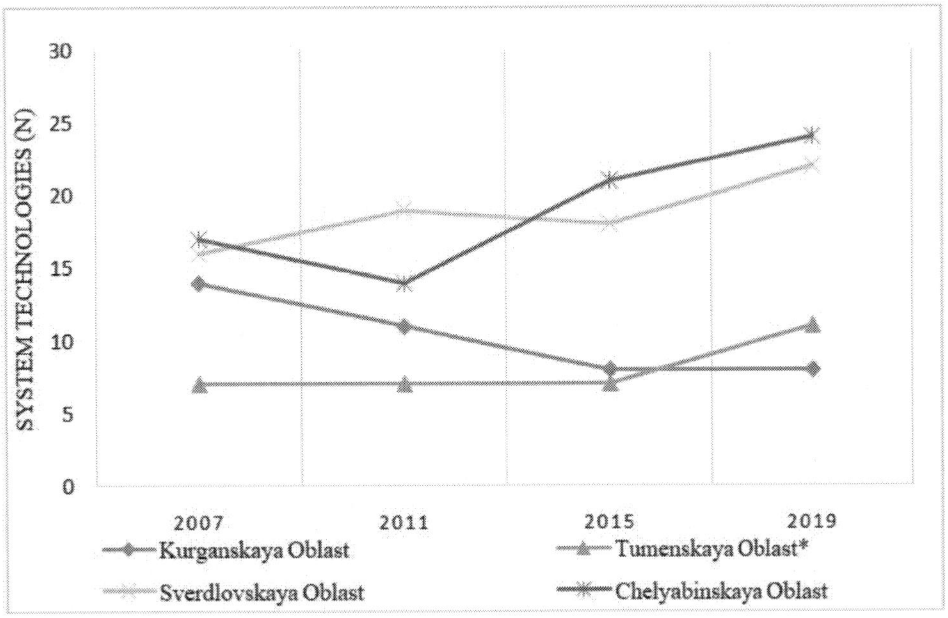

Fig. 1. System technologies implemented in the studied regions.

For each group of companies Total Factor Productivity was found. It measures productivity growth and has become a proxy for technological progress.A state-of-the-art measure of TFP is given by the following formula (Eq.3):

$$TFP = (Y/K)\alpha /(1- \alpha) \quad (3)$$

where Y = total output;
K = capital input;
L = labout input;
A =total factor productivity;
α = perspective share of output(= 0.3).

3 Discussion

By and large, diversity indices for Sverdlovskaya and Chelyabinskaya Oblasts arecomparable in terms of value and show a similar trend toward an increasing diversity of technologies.A number of system integrators in these two regions is growing. Currently,144 system integrators are located in Sverdlovskaya Oblast (Fig. 2) and 132 in Chelyabinskaya Oblast. However, at all technological levels we observe the smallest variance of TFP, which indicates a lowfragmented inter-firm environment. A free flow of resourcescontributes to more even distribution and localization of technologies in the mostproductive companies.

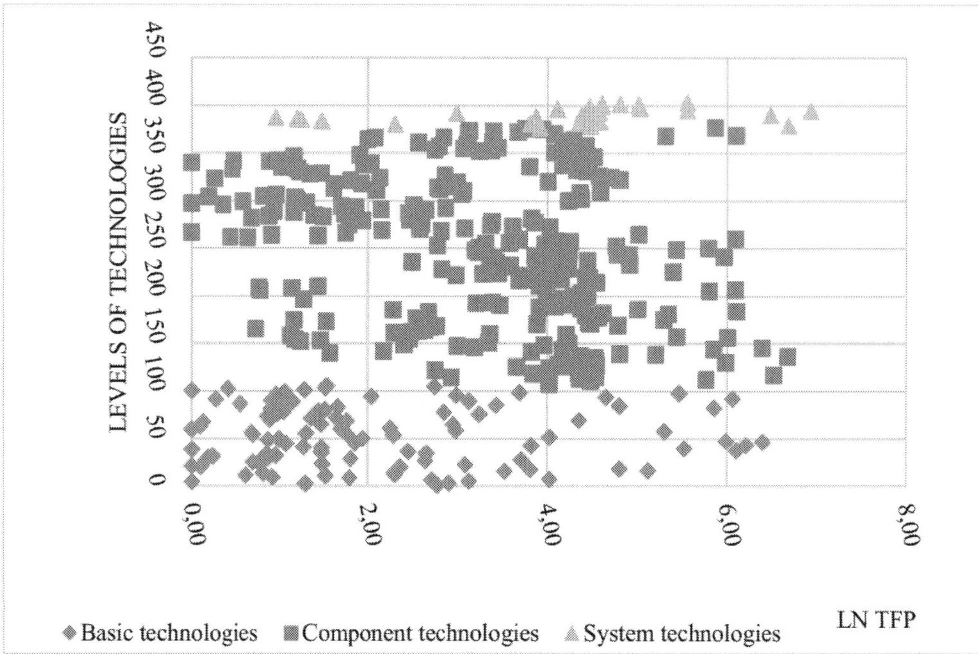

Fig. 2.TFP of firms located in Sverdlovasksys Oblast (2019)

As it was expected, the manufacturing sector of Tumenskaya Oblast has the highest Simpson index, which proves the presence of the dominant technology that forms a vector of development of the regional manufacturing sector and reduces itsresistance to market fluctuations. The component integrator groupincludes the firms producing constitutive elements for the dominantsystem technology. However, we can observe a downward change in Simpson Index, whichindicates that other system and component technologies emerge in the region. It is worth noting that during the entire observation period,system integrators in Tumenskaya Oblast showed the highest TFP.

Kurganskaya Oblast is also characterized by high Simpson Index and lowvalues of Shannon Index. This proves a small variety oftechnologies and the presence of a dominant technology. However, on the one hand, a downward trend is observed since the range of technologies is narrowing. On the other hand, component technologies do not match the profile of systemtechnology. This exacerbates areduction in gross value added,generated by system integrators.

For Tumenskaya, Sverdlovskaya and Chelyabinskaya Oblasts TFP in the group of system integrators is higher than inthe groups of basic technology and component integrators. TFP levels in Kurganskaya Oblast are comparable along all three groups and are similar to basic technology productivity in Chelyabinskaya and Sverdlovskaya Oblasts. In general, system technologies create higher added value.

However, we argue that the more system technologiesimplemented in the region, the greater the gap inTFP levels in the system andcomponent technologies. The variety of technologies in the system and component groupscontributes to higher gross value added generated bysystem integrators.

4 Conclusions

Thus, the analysis of technological development of the manufacturing sectors and the establishment of system integrators in four regions of Russia shows that the most dynamicprocesses of creating 'smart' industries based on system technologies take place in Sverdlovskaya and Chelyabinskaya Oblasts. Localization of system integrators in these regions, a significant variety of system technologies and, as a result, the formation of supporting groups of component technologies, provide multiplicative integrated development of the industrial sector in these regions based on scaling up modern 'smart' industries. This smart manufacturing can generate high added value, which, in turn, might attract additional funds and high-skilledhuman resources. This will ensure economic stability and a high level of regional competitiveness.

References

1. C.W.I. Pistorius, J.M. Utterback, *Research Policy,* **26 (1)**, 67 (1997)
2. N. Meade, T. Islam, *Management Science,* **44(8),** 1115 (1998)
3. S.J. Lee, D.J. Lee, H.S.Oh, *Technological Forecastingand Social Change,* **72 (8),** 1044 (2005)
4. G. Zhang, D.A. McAdams, V. Shankar, M.M. Darani, *Technological Forecasting and SocialChange*, **125,** 116(2017)
5. G. Zhang, D. Allaire, V. Shankar, D.A. McAdams, *PLoS ONE,* **14(6),** e0218370 (2019)
6. N.V. Vasilenko, K.V. Kudryavtseva, Digital transformation of the economy and industry: problems and prospects, 67 (2017)
7. S.I. Tay, T.C. Lee, N.A.A. Hamid, A.N.A. Ahmad. *Jour of Adv Research in Dynamical & Control Systems*, **10** (14), 1379 (2018)
8. A.V. Babkin, D.D. Burkaltseva, D.G. Kosten, Yu.N. Vorobiev. *Scientific and technical statements of the St. Petersburg State Polytechnic University.Economic sciences.***3 (10),** 9 (2017)
9. I.A. Gunina, E.V. Shkarupeta, V.V. Reshetov, Innovative clusters of the digital economy (2018)
10. I.N. Tkachenko, E.N. Starikov, *Izvestiya of Saratov University. New series. Series: Economics.Management. Law*, **3(20),** 244 (2020)
11. M.V. Evseev*a, Journal of New Economy*, **3(21),** 132 (2020)

Successful implementation of corporate strategy in the context of digital transformation

Daria Shalina, *Natalia* Stepanova, and *Pavel* Bochkov*

Ural Federal University (UrFU), Mira, 19, Ekaterinburg, Russia

Abstract. In our research, we cover the problem of strategic planning in the modern period within a separate organization. Long-term fixed strategies are losing their relevance in an unstable, constantly modernizing world. During the CAVID-19 crisis, companies realized the need for changes in the direction of digital transformation. Features of digitalization cause the need to form a new strategy of the organization, the implementation of which requires certain actions. The purpose of the study was to justify these actions. After studying the material that makes up the new corporate strategy, we identified the requirements for its implementation with maximum results. The research substantiates the concepts of corporate strategy, corporate culture, digitalization and its features. The results of the research will become recommendations for modern companies in the formation and implementation of corporate strategy.

1 Introduction

In the modern world, the role of information data is rapidly increasing [1]. Every year, digitalization is transformed from a simple way to improve performance to a real boost to the development of the global community [2]. Along with this, people's activities are also being transformed. We are beginning to introduce elements of digitalization in various areas of activity everywhere in order to increase efficiency.

The trigger for digitalization in 2020 was the COVID-19 crisis. The pandemic was the reason for the dynamic transition of organizations to the digital space, which is still incomprehensible and new to us today. Some companies have moved their operations online in the shortest possible time. Now their interaction with external stakeholders is carried out through digital technologies. So digital technologies have become a Central element of society [3].

However, there are companies that have not been able to cope with the digital transformation. The main reason for this was their unstable position in the market, which is provided by the internal organization of the company, namely the corporate culture [4]. Employees were simply not ready to adapt quickly and flexibly to the new conditions of existence due to the optimized corporate culture.

* Corresponding author: bochkov05@mail.ru

1.1 Relevance and practical significance

Before the COVID-19 crisis, digitalization was an element of the future for many. However, under unfavorable epidemiological conditions, the situation has changed. The company's success over the years is ensured by its ability to adapt dynamically to changes. The willingness to show flexibility in the process of transformation is becoming more relevant and necessary quality every year [5].

Companies now face the challenge of maintaining sustainability in an unstable environment. They need to properly implement technologies that are relevant to their business so that they maintain close relationships with employees, customers, and suppliers. And the better they do it, the less losses companies will suffer during the crisis and the more success they will expect after it [3].

At the heart of any company is a strategy that is set for several years. Attempts to adapt to the external situation may seriously contradict the company's plan. Hence, there is an urgent need for new requirements for the formation and implementation of the company's strategy. Knowing these needs, modern organizations can change their vision for strategic planning and will consider the new conditions of existence.

1.2 Goals and objectives

The purpose of our research is to compile and justify a list of requirements for the formation and implementation of a corporate strategy in the context of digital transformation.

To achieve this goal, the following tasks were identified:

- Study the corporate strategy of a modern enterprise;
- Identify the main components of the corporate strategy;
- Analyze the most significant areas of development;
- Identify and justify requirements when forming and implementing a corporate strategy.

In the study, the digitalization is estimated as a positive factor of influence on modern company. However, negative features are also highlighted and ways to prevent them are suggested.

1.3 Scientific novelty

Determination of the need to transform existing models of corporate strategies in the new economic conditions. Identify several factors that need to be considered when planning and organizing a corporate strategy. Their justification. Finally, the introduction of new strategic trends formed in the context of digitalization and the Covid-19 crisis into the corporate strategic planning of modern organizations. In this way, we suggest that businesses change their traditional understanding of corporate strategy to a new digital mindset.

2 Materials and methods

2.1 Effectiveness of the corporate strategy

The company's strategy is an important and integral element of any modern company. Regardless of size or type, companies try to achieve the desired state in the future. And this is the achievement of the set goal of the state of development of the company. The most important strategy is corporate. This is a General strategy for the development of the entire organization and all its elements. The corporate strategy contains the main guidelines and development directions that the company must adhere to in order to quickly and effectively achieve the desired result [6].

In today's ever-changing environment, companies must take a detailed approach to strategic planning. One of the factors of a well-designed strategy is to optimize the use of resources. For the most effective activity, managers must rationally allocate resources among the components of the business. Hence, another important factor is ensuring the company's stable position despite external unstable conditions [6, 7].

The effectiveness of the corporate strategy is the result of a correct assessment of the conditions and opportunities for the company's development, which include current trends and development directions [7]. The most significant trend is digitalization. Today, most companies need to improve production and economic processes, as it is necessary to quickly and accurately meet the new needs of consumers. Therefore, information and communication technologies are being introduced that can quickly and accurately determine the needs and interests of consumers, as well as improve the company's efficiency and corporate culture [2].

2.2 Corporate culture as a component of corporate strategy

Corporate culture is a set of internal elements of the company, on the organization of which the work of the entire company depends. It creates a unique image of the entire company, which distinguishes it from others. Therefore, the success and effectiveness of corporate strategy implementation depends on corporate culture [4, 8].

Corporate culture contains six main elements: corporate identity, mission, values, traditions, rituals, and corporate mythology. Each of them is aimed at improving the quality of work of the company's employees (see Table 1).

Table 1. Main elements of corporate culture [4]

Element	Description
Corporate identity	It involves combining employees based on common principles (from clothing style to documentation)
Company mission	It will convey the meaning of the company's functioning and reflect the internal direction of its development
Corporate value	They form the desired behavior of employees. The management of a modern company focuses on such a list of values for its employees that they are motivated to perform work with maximum results
Traditions and rituals	They are needed to unite the team. With the help of these elements strengthens the current rules in the company. They ensure the company's sustainability in a time of changing conditions and innovation
Corporate mythology	Corporate mythology helps to convey to employees the meaning of all elements of the company's mission. It is the Foundation of the corporate culture. Employees perceive myths as the main postulates of the company's development

Elements of corporate culture provide modern organizations with a stable position in the market. The company responds to changes by adapting its corporate culture to new operating conditions. Employees themselves must be prepared for changes, constantly update their knowledge and analyze market changes [8].

However, digitalization, by introducing its own templates, destroys the uniqueness of the company. It is lost against the background of competitors. In this regard, along with the modernization of the entire company's processes, we should not forget about improving the corporate culture.

Internal employee interactions affect customer loyalty. If a friendly family atmosphere prevails in the company, then customers will feel comfortable. To create such interaction, it is necessary to increase the trust of employees in each other, since cohesion and teamwork are the key to successful operation of the enterprise. The corporate culture should contain a positive attitude, trust, responsibility, creativity and cohesive interaction [9].

Modern companies are forced to adapt their business model to new conditions, and therefore modernize their corporate strategy, including the development of corporate culture and elements of digitalization [2, 8].

2.3 Digitalization of the company

2.3.1 General characteristics of digitalization

The concept of "digitalization" has many formulations. First of all, digitalization is technologies that process large amounts of data by converting information into digital form or into digital data sets. Therefore, costs are reduced, and new opportunities appear. In this context, we are talking about digitalization of a separate element of the company's activities [1, 2, 5].

Digitalization covers production, business, science and the social sphere in a broader concept, where digitalization is interpreted as a modern direction of development of the world economy and society. And this is already helping to increase the efficiency of the economy and improve the quality of life.

Global digitalization involves effective use of its results by specialists and ordinary people. Therefore, there is a need for skills in working with digital information [2].

In General, digitalization is a tool in the form of technologies that people can use to optimize and improve their activities.

2.3.2 Advantages of having digitalization in a modern company

The spread of digitalization is characterized by high rates. This is due to a large range of positive characteristics of digitalization, which help to improve the process of activity and its result [2].

Digitalization contributes to overall improvement, but in different areas of activity, this improvement can be presented in different ways. So, at the level of an individual company, there are certain features of improvement through digitalization (see Table 2) [1, 5].

Table 2. Digital features of companies. Sources [1, 2, 5].

Feature	Description	Practical using
The creation of new Internet technologies	The Internet space gives its users unlimited opportunities. Each company finds appropriate tools in it and uses them to improve internal processes [1, 5]	Exclusion of intermediaries. Manufacturers can create a website where they will communicate directly with their potential customers. The sales process will speed up [2]
The use of the corporate platform	At the company level, you can use your own platform, technology environment, where certain tasks can be performed. This accelerates business processes and reduces communication time [2]	Or cost optimization. The Internet provides a large amount of information, you can use it to promote products, sign contracts, and much more. Free or saved time can be devoted to the development of personal and professional skills of employees [1, 2]

Continuation of Table 2. Digital features of companies. Sources [1, 2, 5].

Quick response to market changes	In modern conditions, successful companies respond quickly and flexibly to changes. Thanks to digitalization, large amounts of information are automatically processed and synchronized [2]	Employees can make changes to projects without notifying other employees due to automatic updates, distribute information without losing its accuracy and clarity
Strengthening interdisciplina ry collaboration	Digitalization has a strong impact on the company's internal system. Based on digital technologies, all employees are United by one idea and are ready to implement the company's strategy together [5]	The needs of consumers are constantly updated, which requires considerable time to analyze and create a new product or service. Companies through digitalization better understand their potential customers and can accurately meet their existing and new needs [2]

Thus, the use of such a tool as digitalization increases the productivity and efficiency of the company. Internal processes are almost completely digitized. At the same time, digitalization technologies help to predict changes with greater accuracy and build the necessary strategies for the future successful development of the company [1, 2, 5].

2.3.3 Disadvantages of digitalization

Digitalization helps to improve the performance of modern companies and solve many problems with optimizing the company's resources and sustainability. But along with the positive effect of digital technologies, other problems may appear in the company. For example, indifference to living people appears and increases, people's dependence on technology increases, their individuality disappears, everyone is like each other, new technical opportunities for violating privacy appear, and so on [1, 2].

Negative features of digital technologies contribute to the emergence of complex slowing down situations. Distrust of new technologies can slow down the introduction of digitalization elements. Natural rejection of the new occurs for reasons of lack of necessary knowledge and lack of preparation. Even the most advanced company will not be able to work effectively based on digital transformation without appropriate training of employees and jobs [2, 3, 10].

Elements of digitalization provide many positive opportunities that can lead to misperceptions of new technologies. Digitalization is just a tool for achieving results faster and more effectively. When basing its activities and corporate strategy on it, the company should remember that it uses digital technologies to its advantage, and not Vice versa [2, 5].

When making corporate decisions, management is based on experience, intelligence, and intuition. The person decides in which direction the company will move, what results need to be achieved and what tools will be used. It is necessary to differentiate the areas of human activity and use digitalization technologies in the company to prevent corporate disputes and problems [1, 10].

Thus, digitalization is changing social society and modern companies. It is only possible to use the digitalization tool to increase the company's efficiency if this phenomenon is properly understood and perceived. It is necessary to master the skills and knowledge of digital technologies [1, 2].

3 Results and discussion

Before the introduction of digital technologies, the company has potential risks that can lead it out of the state of sustainability and lead to a decrease in efficiency. With the introduction of digitalization, new additional risks appear. The company's lack of preparation can lead to the implementation of risks that were previously suppressed, and new risks [10]. The problems that have arisen due to the introduction of digitalization elements make companies perceive this phenomenon as a complex tool that requires preparatory operations.

First of all, the company must clearly understand its current situation and potential problems. It requires a clear analysis of the market structure, the organization's capabilities at this level of development, as well as an analysis of the prospects for implementing a particular technology. It will be able to really assess the opportunities from the introduction of digital technologies and will use them as tools for solving these problems. This is where the digital corporate strategy of a modern company begins to take shape.

We found out that digitalization cannot exist separately from a person, so the second preparatory action is to analyze and improve the abilities of employees. Digitalization requires certain knowledge that is constantly changing and updated. The same concept should be followed when training employees. You should constantly conduct new trainings or courses.

Digitalization creates a new intellectual potential of the company based on the corporate culture. A new trend of strategic planning is emerging, which makes the company much more efficient even when working remotely (see Figure 1).

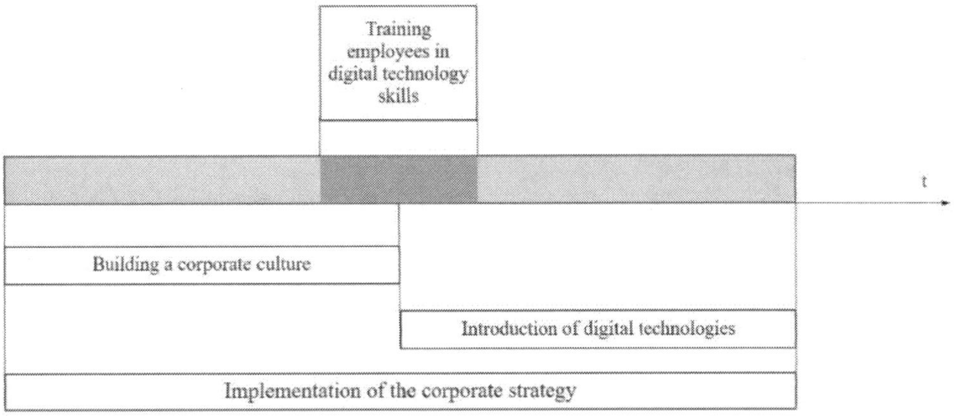

Fig. 1. The trend of improving the intellectual potential for the implementation of corporate strategy.

Digitalization requires skilled workers. The management of a successful modern company is aware that employees need skills and knowledge to work with digital technologies. And at the same time, human activity becomes valuable, important and unique.

This makes strategic planning more detailed, operational, and flexible. The combination of the current modern environment, corporate culture and digital technologies defines new strategic directions for the development of enterprises[11, 12]:

1. *Focus on the people.* Along with digitalization, the value of human labor is growing. People provide human potential and use digital tools to optimize their activities.
2. *Independence from location.* The digital world does not require linking to a specific place of work. However, it requires appropriate technologies to function effectively.
3. *Stable service delivery.* The ability to adapt quickly will ensure the organization's stable position in the industry. It is relevant in the modern world with unstable conditions.

Each of these areas can be divided into strategic technological trends of the present and future, which are presented in Table 3.

Table 3. Strategic technology trends.

Trend	Description	Reason
Focus on people		
Internet behavior	Chipping employees through gadgets or sensor implants	Tracking people's behavior through a digital footprint in order to influence them
Cumulative experience	Combining the experience of customers, employees, users, and a multi-channel environment	The excess of competitors on key indicators of satisfaction
Calculations that increase privacy	Ensuring the protection of the data used, maintaining secrecy and privacy	Need to process data in untested environments and perform multi-party data analytics
Independence from location		
Distributed cloud	Using shared cloud services that are not handled by providers	The solution of technical problems due to the location of clouds anywhere
Ubiquitous operations	An operating system that supports customers, employees, and services from any location	Optimization of mutual contact and communication between clients and employees
Network cyber security	Secure access to a digital item from anywhere in the world	Ensuring the security and availability of digital services
Stable service delivery		
Intelligent composition business	Forecasting business processes based on the analysis of their data	Optimization of the company's activities, including the decision-making process
AI engineering(AI-artificial intelligence)	Optimize the production process by presenting it in a single clear way	Increase productivity, scale, reliability, and return on investment
Widespread automation	Use of effective advanced technologies and tools	High-quality automation of business processes, complement of human potential
RESULT: Synergistic effect of digitalization on the company's strategic corporate interests to achieve sustainable development		
Intellectual potential that unites employees through digitalization	Planning, organizing and managing employees ' work with digital technologies based on corporate culture	Improving the efficiency of the company, increasing the value of the enterprise, accelerating the achievement of the corporate strategy of the company, a new attitude to employees

Following these trends will help companies organize a plastic position for development within 5–10 years.The Internet creates an opportunity for" add-ons " of neighborhood interactions to work in teams on projects, successfully implementing a corporate strategy in the context of digital transformation.

Visually, trends can be represented as the formation of corporate efficiency based on a community of practices of people United by common interests through virtual and real space (see Figure 2).

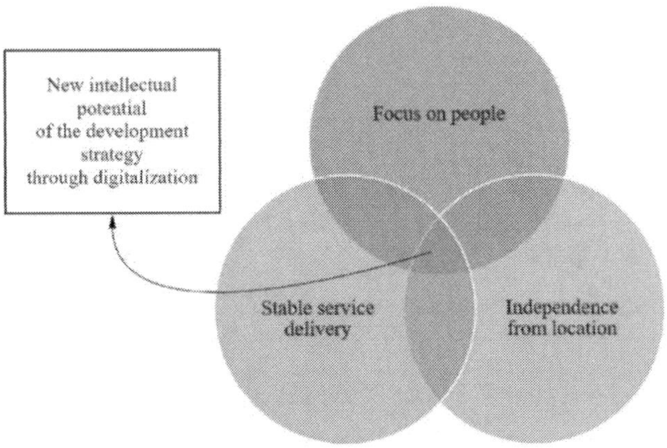

Fig. 2. Current trends in the implementation of corporate strategy in the context of digitalization.

Thus, the successful implementation of a corporate strategy depends on the ability to implement and use all current trends in strategic development as a combination of the power of the Internet and people's intelligence.

4 Conclusions

Today, modern companies have the opportunity to develop dynamically through digitalization. The availability of digital technologies allows companies to better meet the needs of consumers, adapt to changes, develop the creative potential of employees and be ahead of the competition.

The corporate strategy directs the organization to the desired result. The desired result changes – the strategy changes. Now the main trigger for changing the desired result is digitalization. Therefore, when implementing digital technologies, companies must train employees in the knowledge and skills of working with digital technologies and improve the corporate culture for effective interaction of technological and human labor. This will allow companies to form corporate strategies for a short period of time, be more flexible to external changes, and increase their competitive advantages.

References

1. A. G. Chernyshov, *Strategy and philosophy of digitalization,* Power, **5** (2018)
2. V. G. Khalin, G. V. Chernova, *Digitalization and Its impact on the Russian economy and society: advantages, challenges, threats and risks,* Management consulting, **10** (2018)
3. S. Blackburn, L. LaBerge, C. O'Toole, J. Schneider, *Digital strategy in a time of crisis*, McKinseyDigital (2020).
4. E. Y. Dusharina, *The Main elements of corporate culture,* Economics and business: theory and practice, **3** (2020)
5. M. Rachinger, R. Rauter, C. Müller, W. Vorraber, E. Schirgi, Journal of Manufacturing Technology Management (2018)
6. E. E. Kraus, A. A. Kirakosyan, V. A. Poltavtsev, E. E. Golova, Electronic scientific and methodological journal of Omsk state university, **1** (2018)
7. E. R. Feldman, Strategic Management Review, **1** (2020)
8. N. A. Dosalieva, Problems of modern science and education, **16** (2016)

9. M. Hitka, M. Vetrakova, Z. Balazova, Z. Danihelova, Business Economics and Management 2015 Conference, 34 (2015)
10. S. Ziyadin, S. Suieubayeva, A. Utegenova, *Digital Transformation in Business,* Digital Age: Chances, Challenges and Future (2019)
11. Gartner named 9 major technology trends in 2021. Inc. (2020). https://incrussia.ru/
12. K. Panetta, *Distributed cloud, AI engineering, cybersecurity mesh and composable business drive some of the top trends for 2021.* Gartner (2020). https://www.gartner.com/

Assessment of the impact of digitalization on the formation of strategic corporate interests of the company using the apparatus of fuzzy logic

Viktoriia Kokhanova*
Private educational institution of higher education «SOUTHERN UNIVERSITY (IMBL)», M.Nagibina av., 33A/47, 344068Rostov-on-Don, Russia

Abstract. Faced with the need to implement digital tools / processes, the company is forced to independently calculate not only costs, but also methods and mechanisms for assessing the effectiveness of these innovations. In this situation, it is necessary to choose, among other things, methods and assessment tools, and the use of the apparatus of fuzzy logic can be used as one of the tools. In this article, we give an example of using the apparatus of fuzzy logic, indicate the criteria that can be used as criteria for building a system for assessing the effectiveness of digitalization in order to achieve the strategic corporate interests of the company, and also provide an overview of the available developments. Within the framework of the article, we have also identified the directions of modification of classical assessment methods that are possible using the apparatus of fuzzy logic.

1 Introduction

The development of the digital economy does not give companies the opportunity to stay on the sidelines; companies increasingly have to think about the choice of technology or methodology for introducing digital technologies into their activities. In this situation, the assessment of the economic feasibility of these technologies comes to the fore.

In our opinion, it is necessary to provide for the following main directions digital transformation: creation and development of new business models; formation of a new approach to data management; digital modelingimplementation of information technologies and online systems; creation of a digital ecosystem of the company.

As part of the subject matter of this article, we will dwell in more detail on the issue of creating a digital ecosystem and assessing its effectiveness from the standpoint of the company's strategic corporate goals.

*Corresponding author: kohanovavs@yandex.ru

2 Methods for assessing the economic efficiency of digitalization using fuzzy logic tools

To assess the feasibility, you can use a variety of tools, including tools of fuzzy logic.
The following ideas of fuzzy logic are most in demand. According to the basic idea of fuzzy logic, all levels of economic parameters can be measured not only quantitatively, but also qualitatively, which is important for us within the framework of the task at hand. To do this, it is necessary to define the linguistic variable "Level of parameter X", the carrier of which is the domain of definition of the parameter X, in this situation X can take on values of the subset "very low level, low level, medium level, high level, very high level".
For pentascals, it is necessary to construct a system of membership functions of the carrier X to the corresponding fuzzy subsets. The simplest way to set is a system of trapezoidal fuzzy numbers [1]. This idea corresponds to the use of a system of fuzzy-logical conclusions - standard fuzzy multi-level [0,1] - classifiers.

The pentascal is optimal in most cases [2], but if necessary, you can use the simplest case of a binary scale such as High, Low or Bad, Good. Depending on the statement of the problem, the number of terms can vary: for example, in accordance with the established classification of assessing the state of systems, three, four or even ten terms can be used [3-6]. If the number of terms is not definitively determined, it is possible to use not trapezoidal, but sigmoid membership functions possessing smoothness sufficient for mathematical rigor and defined by the formulas:

$$\mu_k = \exp\left(-\left(x - \frac{k-1}{4}\right)^{2n} / \sigma^{2n}\right), \quad k = 1,2,3,4,5. \tag{1}$$

Here $n \in N$ is a natural number characterizing the steepness of the graph of the membership function. The nodes of the classifier, respectively, are located at the points:

$$\overline{g_k} = \frac{k-1}{4}, \; k = 1,2,3,4,5. \tag{2}$$

The parameter σ is selected based on the completeness of the constructed system of functions. Indeed, we require that at the intersection points of two adjacent graphs the following relationship holds:

$$\mu_k\left(\frac{k-1}{4} + \frac{1}{8}\right) = \mu_{k+1}\left(\frac{k-1}{4} + \frac{1}{8}\right) = \frac{1}{2}.$$

Final form of membership functions has the form:

$$\mu_k = \exp\left(-\left(x - \frac{k-1}{4}\right)^{2n} 8^{2n} \ln 2\right), \quad k = 1,2,3,4,5. \tag{3}$$

The given system (3) is characterized by completeness, therefore, it can be used to develop a five-level classifier. Thus, the above formulas can be connected as follows:

$$\mu_k = \exp\left(-\left(x - \frac{k-1}{4}\right)^{2n} 2^{2n}(K-1)^{2n} \ln 2\right), \quad \overline{g_k} = \frac{k-1}{K-1}, \; k = 1,2,\ldots,K \tag{4}$$

https://doi.org/10.1051/shsconf/20208905003

3 Tools for analyzing the main data sources to assess the effectiveness of digitalization in terms of the strategic corporate interests of the company

Anticipating the assessment of the effectiveness of digitalization of the company, it is necessary to outline the basis for analysis. We need to select in the sample companies that can be designated as financially stable, actively introducing digital technologies, and on the other hand, the average, so to speak, enterprises whose experience can be broadcasted to all similar enterprises. That is, the question of the representativeness of the sample is quite acute.

To assess the impact of digitalization on the formation of strategic corporate interests of the company, we used data on companies from the rating of 600 largest companies in Russia RAEX-600, the Sustainable Development Vector index of the RUIE.

To assess, we need to determine not only the sources of information, but also the criteria for the assessment. We suggest using the following KPIs as criteria:

1. Financial strategy:
1.1. Increase in assets,
1.2. Capital gains,
1.3. Loan growth, (share of borrowed capital)
1.4. Asset turnover,
1.5. Increase in profitability,
1.6. Financial stability assessment
1.7. Dynamics of investment in innovation
2. Production strategy:
2.1 Dynamics of sales
2.2 Dynamics of production volume
2.3 Labor productivity dynamics
2.4 Marriage rate
3. Branding strategy:
3.1 Goodwill
3.2 Increase in the market value of the company
3.3 Availability of an innovative development program
4. Organizational strategy:
4.1 Dynamics of staff turnover
4.2 Personnel engagement rate
4.3 Innovation Index,
4.4 Personnel Creativity Index
4.5 Employment satisfaction index
5. Marketin gstrategy:
5.1 Marketing expenses.

Note that the growth of not only the information flows themselves, but also the quantity and quality of methods for analyzing this information act as the main driving forces for the modification of management and forecasting models in corporations.

Digital technologies act as a trigger for constant changes in the formats of functioning within the business, contribute to the formation of new opportunities, due to which new forms of management objects, new forms of business, etc. appear. In parallel with this, digital technologies pose challenges to the existing management system.

In order to be able to assess the effectiveness of the digital ecosystem, in our opinion, it is necessary to introduce the following variables:

d - coefficient of influence of digitalization on the formation of strategic corporate interests of the company

t - time moment
F is some operator that integrates the action of the entire set of factors in the i-th area of interest, while considering n areas
W_1 - a set of some external disturbances
W_2 - a set of some internal disturbances
e - random factor

$$d(t) = \sum_{i=1}^{n} F_i(t, W_1, W_2, e) \quad (5)$$

A set of some external disturbances (W_1) may include indicators: the level of investment in the company as a whole, the share of investment in innovation, the increase in the company's market value.

A set of some external disturbances (W_2) may include indicators: dynamics of sales, dynamics of production volume, labor productivity dynamics, marriage rate, capital gains, loan growth, (share of borrowed capital), asset turnover, increase in profitability, financial stability assessment, dynamics of investment in innovation, the presence of an innovative development program and others.

Analysis of the literature shows that the possibility of these modifications has already been considered by experts in a number of studies of the stability of economic systems.

YachmenevaV.M. [5] offers an assessment of the level of economic sustainability of an enterprise based on fuzzy logic. To assess the ability of an enterprise to reflect external threats, maintain and ensure its financial stability, carry out its current activities fully, Yachmeneva suggests introducing a fuzzy number X from the range. The higher the X value, the more stable the enterprise.

Karpova N.A. [6] developed a methodology for assessing the sustainability of consolidated groups of companies based on a combination of fuzzy-multiple and cognitive approaches to modeling. Algorithms developed by O.A. Nedosekin to assess the risk of company bankruptcy [7]. In another article, Karpova N.A. it is noted that the adoption of managerial decisions is aimed at maintaining or changing the existing situation [8], which means that the control system built on the principles of fuzzy logic will be dynamic, supporting all changes in the enterprise, taking into account their predicted values. This conclusion is confirmed by E.Yu. Khrustalev, O.E. Khrustalev [9], the work shows that to understand the functioning of complex systems, the most effective is the use of fuzzy cognitive models, since they also allow predicting the course of development of the system.

Currently, fuzzy logic is spreading its influence as a tool for analysis, is used in the development of various methods of data mining, widely used in the study of economic processes. An example of this is:

- fuzzy neural networks;
- adaptive fuzzy systems;
- fuzzy queries to databases (fuzzy queries);
- fuzzy associative rules;
- fuzzy cognitive maps.

Compared to "pure" methods of data mining, fuzzy-multiple modifications have wider opportunities to take into account the uncertainty of external conditions and expert assessments, greater compactness, clarity and ease of use.

4 Conclusions

The choice of a parameter and its justification is a separate, rather developed area of research. The corresponding methods for constructing quantitative assessments of the stability of economic systems, with their undoubted practical value and relevance, have a

https://doi.org/10.1051/shsconf/20208905003

number of disadvantages. This is typical of classical economic models and is irresistible within the framework of the traditionally used mathematical apparatus. At the same time, they can be overcome with the help of a fuzzy logic apparatus, including fuzzy logic inference systems. Currently, active work is underway to modify traditional mathematical methods based on fuzzy logic, which will make it possible in the foreseeable future to achieve significant success in creating a modern mathematical apparatus that meets the challenges of the digital economy.

References

1. L. K. Konysheva, D. M. Nazarov, Fundamentals of the theory of fuzzy sets: a training manual, 192 (2011)
2. N. G. Vovchenko, M.B. Stryukov, L.V. Sakharova, O.V. Domokur, Advances in Intelligent Systems and Computing, **896**, 709 (2019)
3. E. A. Arapova, G. V. Lukyanova, L. V. Sakharova, G. I. Akperov, Advances in Intelligent Systems and Computing, 896, 643
4. L.V. Sakharova, T.V. Alekseychik, T.V. Bogachev, E.A. Arapova, Bulletin of the Rostov State University of Economics, **3(63)** (2018)
5. V.M. Yachmeneva, Economics and Management, **4-5**, 107 (2007)
6. N.A. Karpova, Bulletin of Eurasian science, **5(30)**(2015)
7. A.O. Nedosekin, Mathematical foundations of modeling financial activities using fuzzy-plural descriptions: dis doct. econ. Sciences, 280 (2003)
8. N.A. Karpova, Economy, Business, Banks, **2(3)**, 118 (2013)
9. E.Yu. Khrustalev, O.E. Khrustalev, Economic analysis: theory and practice, **10(313)**, 2 (2013)

Achieve synergy of the effects of innovation in the information business

Svetlana Kirilchuk*, and *Ekaterina* Nalivaychenko

V.I. Vernadsky Crimean Federal University, Sevastopolskaja Str., 21/4, 295017 Simferopol, Russia

Abstract. In the article there are researched the main directions to achieve synergy effects of innovations in the information business. There is shown the method consolidation as a necessary condition for increasing the profitability of companies of the information field, and нейтрализации its negative factors. There are considered problems of intellectual property in the information economic. Based on statistical, studies it was concluded, that specific of software products is their biggest insecurity as intellectual property. So, often the solution of this problem is drawing up terms of patenting and licensing of this activity by the state.

1 Introduction

Globalization processes are deeply associated with the spread of scientific and technological achievements in the world. In today's environment, the application of results of scientific and technical activities in a particular area of the world is impossible. All parties of development of the international community is organically linked, so natural is the interpenetration of scientific achievements between different regions of the Earth. Inventors who provide information, interested in this process through established mechanism of receiving payment for use of the invention in the form of royalties. In another country, on another continent, in another market, or in another company, these gains can be used with greater advantage than the transmitting side.

2 Methods and types of the Earth's remote sensing

New forms of STP are also in the joint establishment and use of the products of intellectual property by international companies, in developing and performing of international scientific and technical programs. In this regard, the notion of technoglobalizm was introduced in science, which is a planetary process of internationalization of the creation, development, production and commercial use, transfer, distribution of technologies [1, p.208].

The commercialization of technologies most facilitates to their spreading. Technology commercialization requires three conditions – extensive base of scientific and technological

*Corresponding author: skir12@yandex.ru

research, venture capital ready to be attached to the new product, talent manager, ready to manage such capital. According to Dr. Sal Kassys'ye (USA) [2], even in developed countries these conditions are not always fulfilled and this leads to the necessity of intervention in this process both the state and international organizations.

3 Results and discussion

On balance and dissemination of the availability of the international system of property is directed activity of the World Intellectual Property Organization (WIPO) – a specialized agency of the United Nations. WIPO was established by WIPO's Convention in 1967 and received a mandate from its members to promote the protection of intellectual property throughout the world through cooperation among States in cooperation with other international organizations. Strategic direction, budget and activities of WIPO are determined by its 191 member states who sit in the assemblies, committees and other legislative bodies [3]. Now the intellectual property system plays a critical role in shaping the digital world, network space. WIPO began implementing a far-reaching program of introduction of digital regulation that reflects the impact of the Internet and digital technologies on the intellectual property system. This program touched on the use of intellectual property laws to the rules of trade and financial transactions over the Internet, copyright compliance in a global network of trade marks, patents, etc. [4].

The development of information technology resulted in the creation of new products and services such as distance learning, buying electronic goods in stores, remote management of bank accounts, entertainment and so on. Underlying many of these goods and services are works protected by copyright and works that are objects of related rights. In this connection it became necessary to make new laws by technologically advanced countries. Thus, in the U.S. there is passed the law of copyright protection in the electronic environment, regulating the use of media types in the Internet, and also prohibits the creation and sale of technologies intended for hacking devices, controlling copyright. In analyzing the dynamics of the patent in the world (fig. 1) we see that the total number of patents increases.

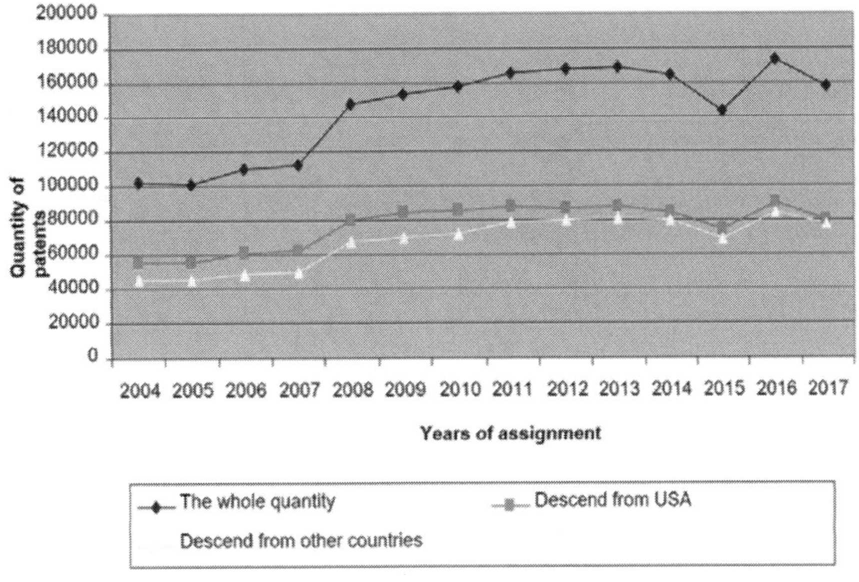

Fig. 1. Dynamics of patents in the world [5]

Some decline in 2015 can be explained by the corresponding recession in the world in 2013, as shown in the graph of NASDAQ index's behavior (fig. 2), which determines the position of high-tech sector companies (all graphs were taken from the site [5]). Analyzing the graphs, we see that most clearly the rise in 2010 is reflected in companies Advanced Micro Devices Inc, Cisco Systems, Inc, Dell Inc, EMC Corporation, Hewlett-Packard Co, Intel Corp, International Business Machines Corp, Lexmark International Inc, Microsoft Corporation, Oracle Corp. Further significant decline is noticeable. Experts who examined the processes of business think that it was due to events of 2014. But, as analysts of business activity point, in connection with the war in Iraq this recession ends in 2021. Then there is the rise. In the patent case, these trends are observed late in approximately 2 years. The author believes this is due to the need of some time from the beginning of investments in innovation till obtaining of scientific and technical results, the long process of obtaining patents.

This trend is also evident in the companies of the information field (fig. 3, 4). Dynamically developing companies are Advanced Micro Devices Inc., Cisco Technologies Inc., Canon Kabushiki Kaisha, Hitachi Ltd., Intel Corp., International Business Machines Corp., Microsoft Corporation. If we analyze the number of patents used, the picture is different. Since 2000, there has been a sharp drop in the number of patents used (fig. 5, 6), which may be related both to changes in economic activity worldwide and the growth of piracy, which is being fight with active measures.

This more clearly manifests in the International Business Machines Corp, Canon Kabushiki Kaisha. In the remaining companies there is steady decline in the use of patents.

Fig. 2. NASDAQ Index's behavior [6]

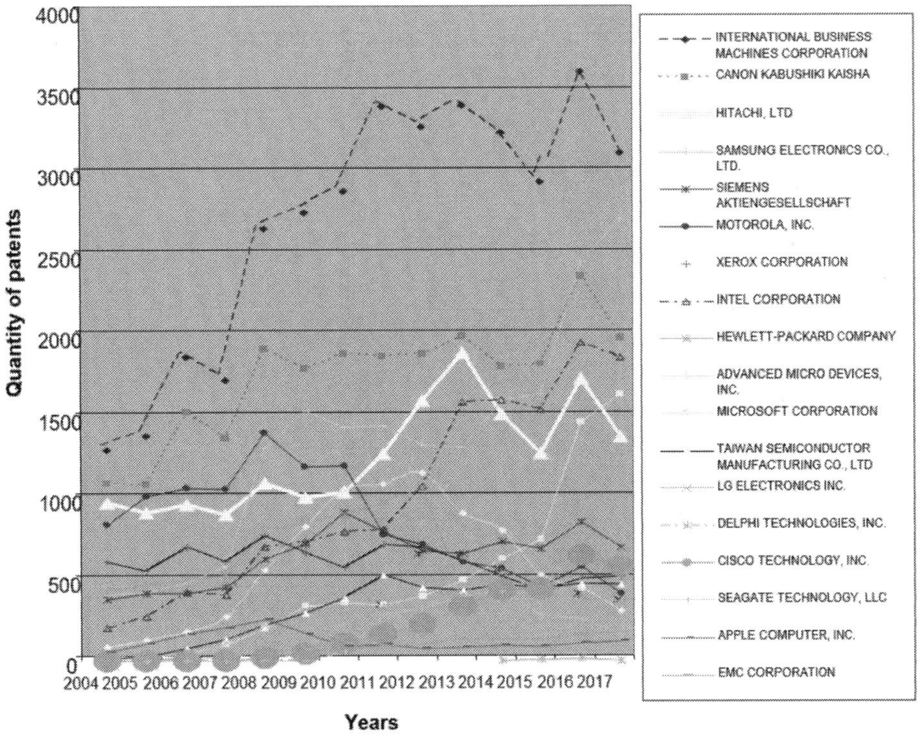

Fig. 3. Number of patents granted to TNCs of information field [7]

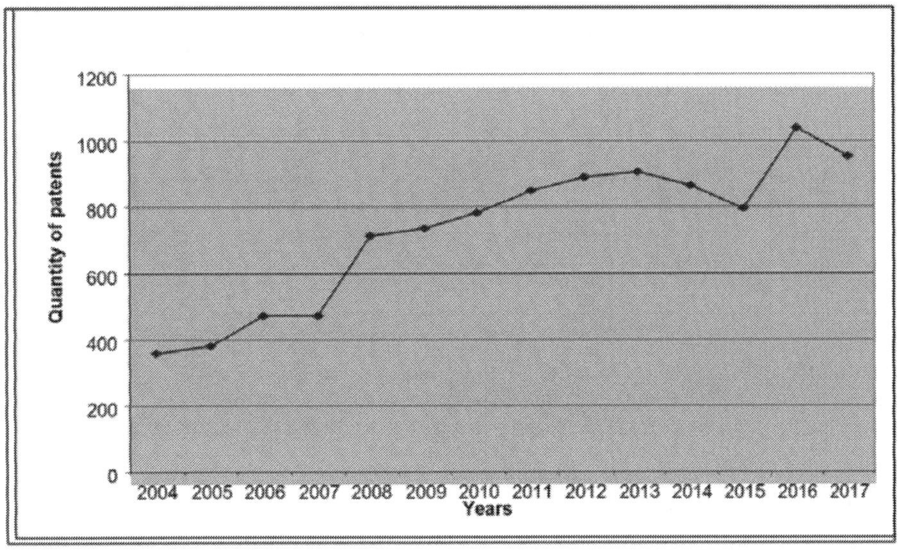

Fig. 4. The average number of patents granted to TNCs of information field [8]

 https://doi.org/10.1051/shsconf/20208905004

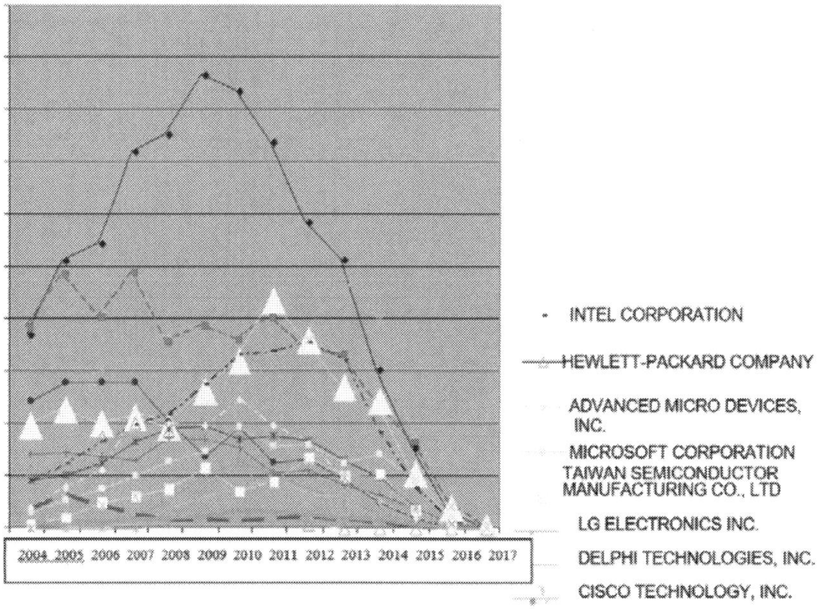

Fig. 5. Number of patents used by TNCs of information field [8]

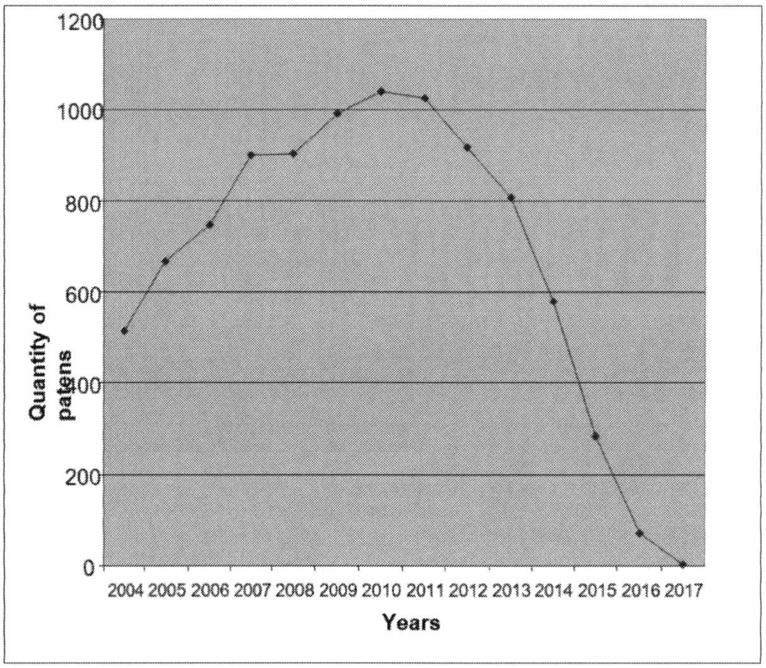

Fig. 6. The average number of patents used by TNCs of information field [8]

In the context of innovative measures there was introduced the concept of information society. Formation of information society is possible only with the consolidation of public and business sector. And that is the basis in implementing the ideas of the project "Electronic Russia", which is the push of consistent informatization of our country's society in all levels of its development.

In international cooperation the perspective direction is coordination of activity in the information field with the countries of Eastern Europe and Central Asia. The priorities of this cooperation are a regional partnership, the introduction of electronic communication at the level of state sector, consulting on information security as in the middle of each country so in cooperation between states, informatization of export, media services and so on.

For Russian companies, the situation is characteristic when it is only the material resources that are evaluated in identifying the innovation project cost – without taking into account the cost of intellectual property rights. Therefore, the economic effect of the innovation projects implementation turns out to be significantly understated. Underestimating the intellectual resources leads to losing the edge over the competitors and affects the company's prestige that is inseparably connected with the competitiveness of the innovations created [7].

The common disadvantage of the existing IPO evaluation methods (the rate of return, the market and the discounted cash flow ones) is the complexity of identifying the input parameters and low validity due to using a large quantity of forecast data, which renders them inapplicable in fulfillment of large-scale projects of knowledge-intensive innovations creation. Hence the methods for evaluating and licensing the IPO are necessary which could be applied to conditions of a manufacturing company in information economy.

In order to efficiently manage the IPO of an innovation creation project, they have to be structured first. During structuring, the list of the IPO is determined, they are classified, the economic efficiency and effect of the IPO implementation at the enterprise are evaluated, and recommendations are given on further commercial use of the IPO in the project.

For finding out the possible further income from using the project's intellectual resources at the stage of its development, the IPO evaluation method has to be selected according to which its further commercialization will be conducted.

The method of identifying the economic efficiency of the IPO is based on calculating the future cost of the IPO at the end of the patent validity period. When using the method, the license payment values are taken as cash flows. As the commercial importance index of the IPO, the cumulative net cash flow (NCF) value was selected.

The net present value (NPV) shows the distribution of cash flows in the course of fulfillment of the information project. Using this index, the project manager monitors the increase or decrease of profits from the project implementation [9]. The project profitability index (PI) serves as the final characteristics of the project in this method.

For the formation of civilized market of information technologies in Russia it is necessary to unite the efforts of developers of information systems. Significant role in this process plays the Association of Computer and Information Technology Enterprises in Russia.

The purpose of the Association is to strengthen efforts to promote competitive products of Russia companies in foreign markets, the merging of scientific and industrial potential of Russia for software creation, which meets international standards, and also conditions for further development in the field of information technologies [10]. The Association operates in the popularization of Russia IT products in the export market, foreign investments, improving the existing legislation, promoting research activities in the field of information technologies.

In the world such functions World Alliance of Information Technologies and Services, which includes the main Associations of 46 countries and covers about 90% of the world market, being, in fact, a monopoly of this market [11]. Through the joint efforts of the Association's members there are implemented many innovative projects in the field of information technologies [12]. One of them was the program "Job Search", established by Association "Softline", which allows free job search with the help of information system designed for mobile phones.

4 Conclusions

The main directions to achieve synergy effects of innovations in the information business are the use of inventions in the form of royalties, globalization technologies' commercialization, balancing of the international ownership system.

It is found that a necessary condition for increasing the profitability of companies of the information field is the consolidation, which is achieved through mergers and acquisitions. Consolidation has several disadvantages. First, companies can not immediately get planned profit. Merger of companies leads to the disappearance of the conditions for effective implementation of competition. These processes lead to market's monopolization by largest companies. An instrument of struggle for the market of information technologies in the world is opened dissemination of a company's information system.

Specific of software products is their biggest insecurity as intellectual property. So, often the solution of this problem is drawing up terms of patenting and licensing of this activity by the state.

References

1. Roderick Hill and Tony Myatt, *The economics Anti* (2010)
2. T. Sakaiya, *The Knowledge-Value Revolution, or A History of the Future* (1991)
3. World Intellectual Property Organization, https://ru.wikipedia.org/
4. Official website World Intellectual Property Organization, https://www.wipo.int/
5. Official web-portal of State Service of Intellectual Property of Russia, http://rosvois.ru
6. Index Nasdag (IXIC) - Investing.com, https://ru.investing.com/ composite –
7. P.N. Zavlin, A.V. Vasiliev, *Evaluation of Efficiency of Innovations, St. Petersburg* (2007)
8. Number of patents used by TNCs of information field, https://yandex.ru/
9. T.I. Krakhmaleva, Radioelectronic and Computer Systems, **2**, 120 (2007)
10. IT Associations in Russia, https://itk.kz/ru/
11. The World Information Technology and Services Alliance – WITSA, http://neweco.ru/international/
12. S.P. Kirilchuk, E.V. Nalivaychenko, Journal of Applied Economic Sciences (Romania), **4(42)**, 662 (2016)

Forecasting Innovative Development of a Company in the Process of Transformation

Olga Shvaika*, and *Marina* Kupriyanova

Ryazan State Radio Engineering University, Gagarin Str., 59/1, 390005 Ryazan, Russia

Abstract. Innovation is traditionally considered to be an effective instrument of development for industrial enterprises, even at the time of crisis. However, as practical observations show, non-systematic and inconsistence innovative changes may lead to poor results. The sustainable innovative growth is gained as a result of in-depth transformation of all the system elements. The purpose of the research is to formulate an approach to evaluating the expected effect of transformation and restructuring an industrial company and to forecast the perspectives of innovative development. The methodological basis for the research refers to the ideas of the fuzzy set theory and the axiomatic methods of analysis. The axiomatic method makes it possible to construct formalized descriptions of a company and to achieve a sufficient level of detachment. Such descriptions, or texts, may be interpreted by managers as concise information for decision-making. The results of the research find a practical application as tools of evaluating the expected effect of innovation and transformation.

1 Introduction

Industrial enterprises meet new challenges in the turbulent environment and face the necessity of introducing innovations in the key structural elements [1-5]. These elements include function distribution, financial and economic policies, corporate culture, and technology. Being an innovative enterprise means having effective policies of organization transformation and restructuring in order to solve such problems as (1) improving the operation and production structure and processes, using effective tools of control; (2) forming efficient organization structures and methods of adaptation to the changing market demands; (3) introducing of new production technologies; and (4) working out appropriate principles of corporate culture [6]. The process of transformation and restructuring may be defined as a system of measures that align the enterprise operations and processes with the market demand and the adopted development program [7].

* Corresponding author: semer-ka@yandex.ru

2 Materials and Methods

The research relies on the methodology of the fuzzy set theory and the axiomatic approach [8-10]. The idea is to formulate a description of an enterprise structure including the interrelated elements and to evaluate the possibility of effective transformation and restructuring process. Transformation and restructuring as one of the instruments of crisis management may take place in different ways. The limitation of the research is that transformation s viewed as a process leading to innovative (and efficient) development of an industrial enterprise. The axiomatic approach makes it possible to differentiate the characteristics of complex mathematical objects and to form membership functions in particular semantic fields. In other words, the approach provides a classification of attributes according to the criteria of their reference to the described structures. As a result, the objects may be analyzed from various perspectives in terms of well-known algebraic, topological and mathematical notions [11].

Let x, y, z,... be elements of the enterprise that is intended for transformation and innovative development. In general, there might exist a set X of such elements, so that x, y, $z \in X$. An enterprise is a complex structure of interrelated elements [7]. To produce a limited nomenclature of products an enterprise needs such inputs as technology, equipment, and qualified labor resources. The simplified model of elements is based on well-know input-output models [12]. The notion of innovative development may be represented as a binary relation of R between the elements x, y, $z \in X$. There may be formulated several properties of this relation. Any element of innovative development is self-equivalent, i.e. xRx, yRy, zRz. Therefore, the reflexive property may be applied to the elements x, y, $z \in X$.

If element x is equivalent to element y, i.e. xRy, and y is equivalent to x, i.e. yRx, then this property of the elements point out at the symmetry of the elements x and y. This makes it possible to switch from one element to another. In other words, there is no difference which of the elements is analyzed first.

If element x is equivalent to element y, i.e. xRy and y is equivalent to z, i.e. yRz, then x is equivalent to z, i.e. xRz. Therefore, the property of transitivity may be applied to the elements of the transformation and restructuring process. According to the fuzzy set theory, the proposed concept of innovative development as a binary relation R is an equivalence relation. The introduced relation R of innovative development makes it possible to split the set X into classes of equivalency, or a co-set of classes $\{x\}_R$. Five co-set classes are further distinguished: $\{x_К\}_R$ - personnel; $\{x_Т\}_R$ - technology; $\{x_О\}_R$ - equipment; $\{x_П\}_R$ - product; $\{x_{С,М}\}_R$ – raw material. The introduced co-set classes have the following properties:

a. any of the classes is nonempty due to the reflexivity R
b. the equality of co-set classes $\{x'\}_R=\{x''\}_R$ is equivalent to x'Rx";
c. two co-set classes either coincide, or are disjoint:

$$x'Rx'' \leftrightarrow \{x'\}_R=\{x''\}_R, \tag{1}$$

$$x'Rx \leftrightarrow \{x'\}_R \cap \{x''\}_R=\emptyset; \tag{2}$$

d. a merge of co-set classes is equal to X:

$$\cap \{x\}_R=X \tag{3}$$

This implies that a multitude of co-set classes represents a split of X, and each co-set class $\{x\}_R$ serves as a block of this split. The opposite statement is true: each split of X corresponds to certain relation of equivalency R, and the co-set classes are the split blocks.

Furthermore, a set with elements being co-set classes with the relation of equivalence R, is a factor multitude X/R of X set by R. The notion X → X/R is such that every $x \in X$ corresponding to a certain co-set class to which this x belongs. This notion is a factor representation X in X/R.

The new sets $\{x_K\}_R$ - personnel; $\{x_T\}_R$ - technology; $\{x_o\}_R$ - equipment; $\{x_П\}_R$ - product; $\{x_{C,M}\}_R$ – raw material are supplementary. Together with the main sets they form a scale of M sets. It means that in pairs they may be subject to any mathematical operations, which result in additional supplementary sets. Definition of a unique element from a scale set is the result of defining a certain number of elements of the scale set, formulating the relations between common elements of these sets, reflecting some subsets of these sets into other ones.

3 Results and Discussion

The results of applying the axiomatic approach to the process of transformation and restructuring an enterprise are forecasts of innovative development for each of the defined elements. Several scales are formulated to evaluate the innovative efficiency. The choice of variants is realized by the criterion of operational complexity in each element. If the state of an element does not correspond to the target, it is necessary to take measures for improving it. The choice criteria may be the following:

a. time needed for introducing changes to a particular element
b. cost of changes, acceptable for the enterprise
c. subjective factors influencing the decision makers, their personal likes and dislikes
d. results of innovation, the depth of qualitative transformation allowing to make a considerable shift in performance.

To give an example, it is possible to apply the method to the Personnel element of an enterprise.

Human factor makes this element the most important and the most complicated one among others. An innovative enterprise has three key categories of personnel: (1) administrative, (2) R&D, (3) main and support staff. The R&D personnel is included in case the enterprise is engaged in the sphere of scientific research and development. The R&D personnel may either generate, or implement innovations. This type of operations demand highly qualified specialists. A concept of innovative human resources (HR) policy is introduced as a binary relation R1 between the elements of the equivalence class $\{x_K\}_R$-personnel, i.e. $x_K, y_K, z_K \in \{x_K\}_R$. Let's consider the particular properties of the introduced relation. Any element of the HR policy is self equivalent, i.e. $x_K R1 x_K$, $y_K R1 y_K$, $z_K R1 z_K$. Therefore, this implies that there is a condition of reflexivity of the elements:

$$x_K, y_K, z_K \in \{x_K\}_R \tag{4}$$

If element x_K is equivalent to element y_K, i.e. $x_K R1 y_K$, and y is equivalent to x_K, i.e. $y_K R1 x_K$, then this attribute of the class of equivalence $\{x_K\}_R$ indicates that fuzzy elements x_K and y_K, have the attribute of symmetry. It makes it possible to switch from one element to another. In other words, the order of analysis is not significant. Furthermore, if element x_K is equivalent to element y_K, i.e. $x_K R1 y_K$ and y_K is equivalent to element z_K, i.e. $y_K R1 z_K$, then it is possible to define that x_K is equivalent to element z_K, i.e. $x_K R1 z_K$.

Therefore, the property of transitivity may be applied to the elements of the class of equivalence $\{x_K\}_R$. Then, according to the fuzzy set theory, the introduced notion of HR policy as a binary relation R1 is an equivalence relation [12]. The introduced relation R1 of innovative HR policy makes it possible to split the class of equivalence $\{x_к\}_R$ - personnel into subclasses of equivalence, as a conjunction of co-set of classes $\{x\}_R$.

Four co-set classes are further distinguished: $\{\{x_{об}\}_{R1}\}$ - education; $\{\{x_{кв}\}_R\}$ - qualification; $\{\{x_{вн}\}_R\}$ – readiness for innovation; $\{\{x_{ск}\}_R\}$ – social compatibility. Depending on the specific features of a particular enterprise, the list of co-set classes may be completed and improved. Thus, according to theses subclasses, there may be differentiated several variants of development for this element (Fig. 1).

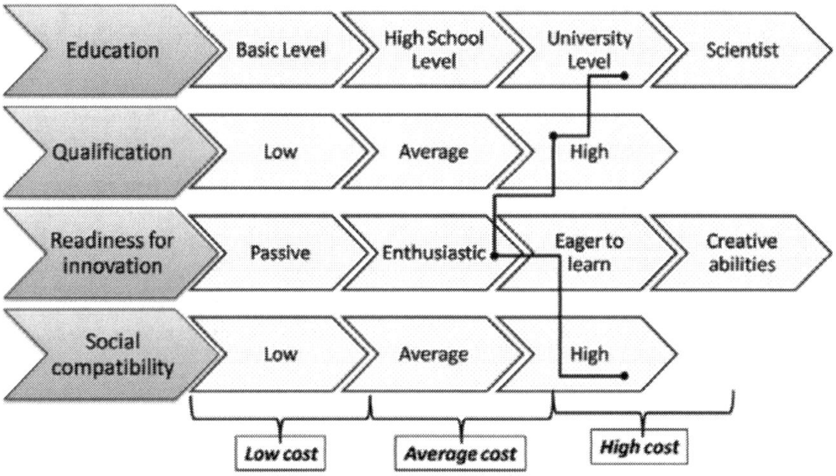

Fig. 1. A scale of evaluating the potential for innovative development, the *Personnel* element.

The introduced subclasses $\{\{x_к\}_{R1}\}$ have the same properties as the classes of equivalence$\{x_к\}_R$.

Thus the key influencing elements of the innovative HR policy may refer to the following fuzzy sets:

a. the level of education
b. qualification
c. readiness for innovation: this criterion is important for the administrative personnel, the tempo of innovative development is to a large extent dependent on the appetite for innovation among decision-makers
d. social compatibility: as practical observations show, team workers are rarely excellent in creative abilities, but they are the drivers of successful and sustainable development

The next step is evaluation of every particular worker according to these criteria and taking into account the functional scope and complexity of the role in the enterprise.

For example, a top-manager is expected to have such attributes as university level of education, a high level of qualification, a high level of social compatibility, a good level of readiness for innovation. For an R&D staff member it is not so vital to have high compatibility with the team. What is expected from them is a sufficient level of creativity and an enthusiastic attitude to innovation.

The main staff members need to have an appropriate level of education, sufficient for their task performance. Average social compatibility and ability to learn are important characteristics. The lower the requirements, the lower the cost for human resources is. At the same time, most experts state that innovation proves to be effective only if the human resources are of high quality. The comparison of every staff member follows a certain pattern (Fig. 2).

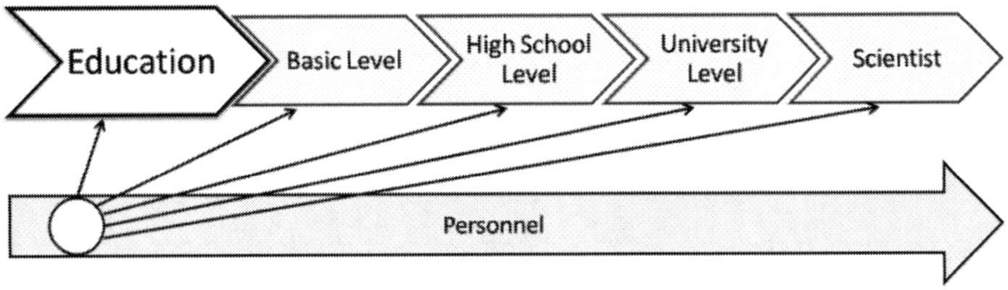

Fig. 2. Variants of comparison.

If a staff member has a high school level of education and if it meets the requirements, it means that the target level of education for this particular job position is equal or higher than the level of high school. In this case the company will develop efficiently. Depending on the requirements, the target level shifts from lower to higher until the minimum sufficient level is chosen.

The higher the level of education, the more results are expected to be gained from the human resources. However, high qualification and work experience may compensate for the lack of proper education. The procedure of appointment for a particular position is different for new personnel and for the already working staff members. If the enterprise chooses a new strategy of development, it does not necessarily mean recruitment of new specialists. There can be introduced innovative measures of human resource management in order to support qualification growth of the already working staff.

Transformation and restructuring the Personnel element of a company is realized via such measures as improvement of requirements to recruitment, organization of educational programs for the specialists who need to acquire new competencies, exchange of successful practices with other enterprises, salary growth and bonuses, support of scientific research and personal development, career management, changes in the administrative system and distribution of rights and obligations, decentralization of authority, etc. Fig. 3 illustrates the equivalence scale according to the criteria of the life cycle.

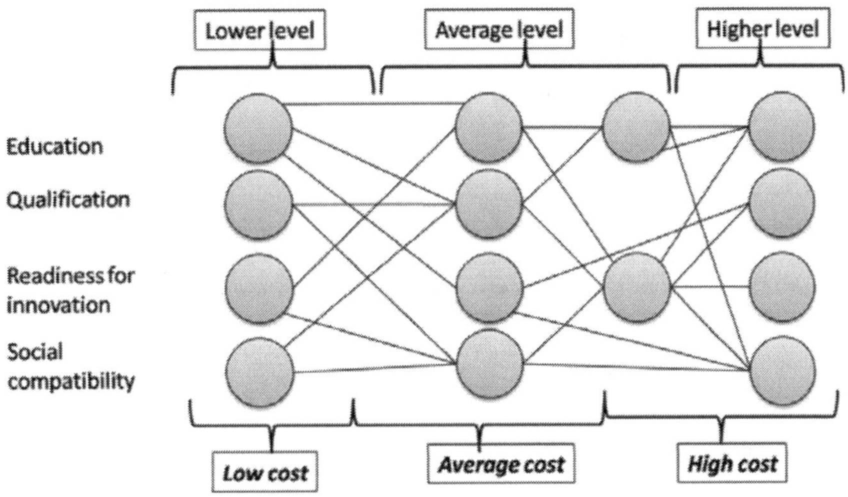

Fig. 3. The equivalence scale for the Personnel element

The other elements of an enterprise may have numerous subsets of properties (Table 1).

Table 1. Subsets of properties for the five key elements of analysis.

$\{x_{\text{к}}\}_R$ - personnel	$\{x_{\text{т}}\}_R$ - **technology**	$\{x_{\text{о}}\}_R$ - **equipment**	$\{x_{\text{п}}\}_R$ - **product**	$\{x_{\text{с,м}}\}_R$ – **raw material**
$\{\{x_{\text{об}}\}_{R1}\}$ education	$\{\{x_{\text{сб}}\}_{R2}\}$ way of implementation	$\{\{x_{\text{сп}}\}_{R3}\}$ way of implementation	$\{\{x_{\text{рн}}\}_{R4}\}$ novelty for th market	$\{\{x_{\text{кч}}\}_{R5}\}$ quality
$\{\{x_{\text{кв}}\}_{R1}\}$ qualification	$\{\{x_{\text{ус}}\}_{R2}\}$ level of complexity	$\{\{x_{\text{нз}}\}_{R3}\}$ depreciation	$\{\{x_{\text{нп}}\}_{R4}\}$ novelty for the enterprise	$\{\{x_{\text{сз}}\}_{R5}\}$ cost of purchase
$\{\{x_{\text{ви}}\}_{R1}\}$ readiness for innovation	$\{\{x_{\text{р}}\}_{R2}\}$ resource intensity	$\{\{x_{\text{ус}}\}_{R3}\}$ level of complexity	$\{\{x_{\text{сс}}\}_{R4}\}$ cost	$\{\{x_{\text{нп}}\}_{R5}\}$ suppliers stability
$\{\{x_{\text{ск}}\}_{R1}\}$ social compatibility	$\{\{x_{\text{э}}\}_{R2}\}$ energy intensity	$\{\{x_{\text{пр}}\}_{R3}\}$ productivity	$\{\{x_{\text{рп}}\}_{R5}\}$ sales rentability	$\{\{x_{\text{рсз}}\}_{R5}\}$ stocks capacity
		$\{\{x_{\text{эк}}\}_{R3}\}$ profitability		

These new defined subsets are supplementary. Together with the five key elements $\{x_{\text{к}}\}_R$ - personnel; $\{x_{\text{т}}\}_R$ - technology; $\{x_{\text{о}}\}_R$ - equipment; $\{x_{\text{п}}\}_R$ - product; $\{x_{\text{с,м}}\}_R$ – raw material the supplementary subsets for the M scale of sets. It means that in pairs they may be subject to any mathematical operations. The operations produce additional supplementary subsets. Definition of a unique element from a scale set is the result of defining a certain number of elements of the scale set, formulating the relations between common elements of these sets, reflecting some subsets of these sets into other ones. At the M scale a particular set of properties is defined as a crispy set of elements belonging to the M set of properties. Let S be the conjunction of subsets of the M set defined by these properties. Then element $s \in S$ is defined in the database $\{x_{\text{к}}\}_R$ - personnel; $\{x_{\text{т}}\}_R$ - technology; $\{x_{\text{о}}\}_R$ - equipment; $\{x_{\text{п}}\}_R$ - product; $\{x_{\text{с,м}}\}_R$ – raw material and X structure S type which is characterized by the M scale of formation and the properties defined by S. Axiom introduction into the type structure is the final stage of defining the core of the axiomatic theory. It is similar to a gen in which the total information of the theory is preserved in a compressed mode. In concept construction axioms play the role play the role of mathematic prototypes of programs. The programs may be realized in the form of database systems that control the database integrity in operations with the data sets (input, alternations, and output). Correspondingly, the programs check the total content of the database including every element at the moment it is put in. The programs check the compliance with the requirements formulated in the axioms. Interpretation of letters referred to the structures of T type is fulfilled in terms of a natural language. This is a linguistic interpretation which defines the names of sets, to which the elements belong. This predetermines either an object, or a fuzzy interpretation. A T type structure corresponds to a lattice with its head comprised of linguistic interpretations of the type structure together with linguistic interpretations of the corresponding projections. The second stage of implementing the method is lattice formation for each element of the enterprise (Table 2).

Table 2. Evaluation criteria for the Personnel element.

Subjective choice	**Time span needed for transformation of the element (here ranged in the order of growth)**	**NPV**	**Innovative result (for the enterprise)**

Table 2. Continued

This part is completed according to subjective views of the decision maker	Short educational courses for specialists	Recruitment of new specialists instead of previous staff members	Experience exchange with other enterprises of the economic sector
	University level of education	Payment for education of a specialist	High education as a requirement for a particular category of specialists
	Science degree	Salary growth for high qualification	Corporate culture improvement

The final step of the method is completing a table with the variants of innovative development in order to improve each of the elements (table 3).

Table 3. Variants of innovation oriented measures for each of the elements.

Personnel	Equipment	Technology	Raw material	Production
Higher requirements to recruitment	Selling part of old equipment	Working out new technology	Searching for new suppliers	Searching for a new product
Corporate culture	Modernization of the equipment	Selling the old technology	Limiting the stocks	Introducing innovation to the production process
Improvement of the administrative structure	Buying new equipment		Leasing part of the warehouse space	Discontinue the production of a product
Change in the methods of HR management	Training the specialists to work with the new equipment			Searching for new distribution channels

The result of the analysis is a list of measures appropriate for each of the elements and a forecast of the expected economic effect. The chosen measures indicate the path of development aligned with the intended strategy of development. Each measure is estimated according to its potential cost and profit.

4 Conclusions

The study discusses urgent problems of forecasting innovative development of an enterprise in the process of transformation and restructuring. Special attention is paid to the innovative development of the key elements of an enterprise. The core of the proposed approach is a comprehensive view of an enterprise as an integral unity of elements (here these are the personnel, technology, equipment, product, and raw material). The choice of the elements depends on the specific characteristics of the enterprise. The analysis includes forecasting of development for each of the elements on the basis of criteria at the scales of innovative development and the scales of their coherence with the life cycle parameters with the help of the axiomatic approach. Similarly, the choice of criteria for the scales is formulated individually for a particular enterprise. The approach has been applied to industrial enterprises in order to work out a program of their efficient transformation.

References

1. A. Coad, N. Grassano, B. H. Hall, P. Moncada-Paternò-Castello, A. Vezzani, Structural Change and Economic Dynamics, **50**, 126 (2019)
2. F. Gault, Res. Policy, **47**, 617 (2018)
3. G. Lu, P. Dai, X. Zhang, CPE, **1**, 284 (2018)
4. X. Song, Y. Ding, Journal of Intelligent & Fuzzy Systems, **37**, 1649 (2019)
5. M. Žižka, V. H. Valentová, N. Pelloneová, E. Štichhauerová, ESI, **5**, 4, 780 (2018)
6. M. D. Aistova, *Restructuring of Enterprises* (Moscow, Alpina Publisher, 2002)
7. O. B. Kazakova, *Opportunities of innovative development* (Ufa, Gilem, 2007)
8. L. R. Burra, P. Poosapati, International Journal of Computer Applications, **11**, 7 (2016)
9. B. K. Wong, V. S. Lai, International Journal of Production Economics, **129**, 157 (2011)
10. O. Kulak, S. Cebi, C. Kahraman, Expert Systems with Applications, **37**, 6705 (2010)
11. F. J. Gouillart, J. N. Kelly, *Transforming the Organization* (1996)
12. A.Y. Bass, *Theoretical fundamentals of a permanent investment process* (2006)

Analysis of the allocative efficiency of social service entities for sustainable development

Ekaterina Ogorodnikova[1,*], *Alexander Mokronosov*[1], *Maria* Kompaniets[2], *Maria* Khokholush[1], and *Maria* Selezneva[1]

[1]Ural State University of Economics, st. March 8, 62/45, 620144 Yekaterinburg, Russia
[2]Slavonic University of the Republic of Moldova, MD-2068 Chişinău, Republic of Moldova

Abstract. The relevance of the study lies in the essential role of the social services sector in matters of sustainable economic development. The purpose of the article is to analyze the allocative efficiency of the activities of the social service entities. The methodology of allocative efficiency makes it possible to identify a model of the organizational and economic mechanism of the social services sphere, which implements the tasks of the most efficient use of limited resources. The research results show higher efficiency of government institutions, higher volume of services provided and lower costs per client. This feature is due to the relative mass of service and low costs for the provision of stationary premises, the use of a tender procurement system.

1 Introduction

At present, methodological approaches to determining the effectiveness of the organizational and economic mechanism of the social services sphere have been formed within the framework of the infrastructure-distribution methodology and are aimed at assessing the provision of the population with appropriate social infrastructure facilities. In the works carried out within the framework of the normative approach, complex indicators can be used that characterize the provision of institutions in the social services sphere, the implementation of certain standards, and budgetary provision. For example, the article by N.I. Yashina [1] provides a set of 24 indicators that allow, according to the author, to assess the final standardized indicator for assessing the effectiveness of budgetary financing of health care. The article by A. N. Ilchenko [2] presents an integral model assessing the development of socio-economic infrastructure, including indicators of the provision of certain benefits.

The second group of methods reflects the assessment of the effectiveness of the organizational and economic mechanism of the social services sector using statistical tools for processing primary data. In the work of N.M. Surnina [3] noted that the integrated monitoring system for infrastructure development is based on statistical tools in the study of analytical indicators of time series. A similar approach is presented in the work of O.Yu. Ulyanova [4], who highlighted such criteria of the model's effectiveness as "the

* Corresponding author: cmb_8@mail.ru

timeliness of the services provision, which preserves the rhythm of production and life and the adaptability of infrastructure institutions to cultural and technological changes in the structure of demand." The group of statistical methods also includes the cluster analysis, which make it possible to differentiate territories depending on the efficiency of the functioning of the organizational and economic mechanism of the social services sphere. Various aspects of the application of this approach are considered in the works of Alter [5], Bartlett [6], Baruch and Ramalho [7].

Survey methods for assessing the effectiveness of the organizational and economic mechanism of the social services sphere are based on the methodology set out in the order of the Ministry of labor of the Russian Federation No. 995n dated 12/08/2014, which presents 52 indicators of an independent assessment of the quality of the relevant organizations activities. Summarizing the results of the application of survey methods, the authors note the overestimation of the positive characteristics of the existing organizational and economic mechanism of the social services sector: the completeness of filling out the questionnaires is satisfactory, the employees of the assessed institutions provide the respondents and interviewers with the necessary assistance, the survey results indicate sufficient satisfaction of the service recipients.

The group of methods, assuming an assessment of the economic efficiency of the existing organizational and economic mechanism of the social services sector, involves the use of investment design methodology, that is, the authors use the principles applicable to the commercial sector, which contradicts the foundations of the formation of the organizational and economic mechanism of the social services sector. For example, the work of A.V. Chisnikov [8] provides a rationale for the development of regional social infrastructure based on comparing budget expenditures and products created on the basis of investment data. In the works of V.O. Evseev and A.V. Materkin [9] there are presented models of the dependence of the volumetric indicators of the social service system on individual factors of production, however, there are no comparative algorithms to identify the most effective form of activity in the specified market.

2 Methods

Despite the high importance of the above approaches, the task is to form the organizational and economic mechanism of the social services sphere within the framework of the allocative efficiency methodology, within which producers will implement the task of the most efficient use of limited resources [10, 11]. This requirement is due to the resource deficit in the functioning of the organizational and economic mechanism of the social services sector. To solve the problem of analysis and the subsequent formation of the organizational and economic mechanism of the social services sphere, a methodology was formed, the algorithm of which is shown in Figure 1.

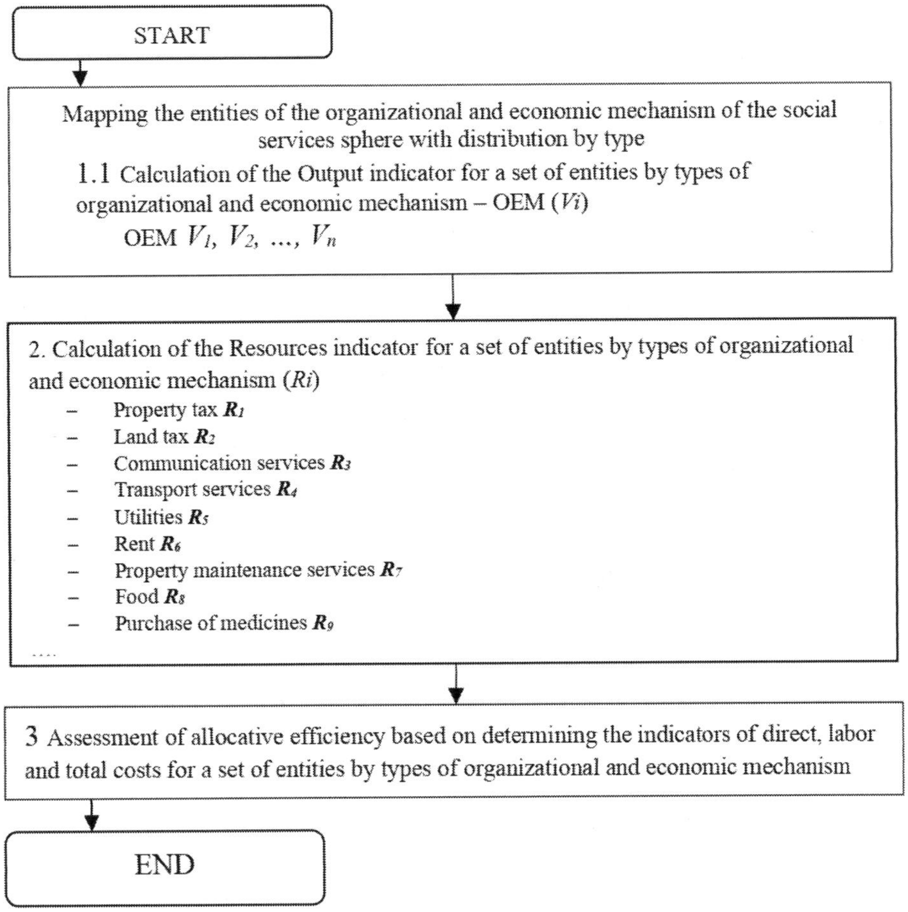

Fig. 1. Algorithm for determining the allocative efficiency of different types of organizational and economic mechanism of the social services sphere

The first stage of the methodology is the mapping of the entities of the social services sphere in the context of industry affiliation and the type of organizational and economic mechanism:

– Infrastructure and distribution OEM;
– Market OEM;
– OEM of activities of socially oriented non-profit organizations;
– Participatory OEM.

The study is carried out for the territory of the Russian Federation as a whole, while the subsequent implementation of the recommendations is carried out based on the aims of designing an organizational and economic mechanism for a specific local area. This approach will allow avoiding distortions of the assessment under the influence of territorial differentiation factors, in particular: the cost of energy resources, access to gas supply, and individual measures of institutional regulation [13]. The sources of data for organizations that are state and municipal property is the Unified portal of the budgetary system of the Russian Federation, to determine the number of services provided by organizations within the market OEM and non-profit organizations, data from the portal of the Unified State Information

System for Social Security are used, as auxiliary sources of information we used analytical SPARK-interfax system and information from the sites of the surveyed organizations.

At the second stage of the methodology implementation, the total resources involved in the activities of entities by types of organizational and economic mechanism are determined. It is advisable to group these resources according to the following sources: financing of the fulfillment of the state task, paid services to the population and donations. As a rule, the structure of attracted resources depends on the form of ownership: organizations that are in state and municipal ownership are financed from the funds of the respective budgets, organizations of private ownership – from payments of the population, non-profit organizations – from donations.

Further, the determination of indicators of individual resources expenditure ($R_{1 \ldots 71}$) is carried out, the structure of expenditures depends on the type of organizational and economic mechanism of the social services sphere.

The results obtained make it possible to proceed to the stage of determining the allocative efficiency of a set of entities by types of organizational and economic mechanism. This approach will make it possible to determine the comparative advantages of various options for the use of resources by the entities of the organizational and economic mechanism of the social services sphere. Understanding the differences in the internal economic mechanism and external institutional regulation of various types of entities, it is possible to subsequently design a hybrid organizational and economic mechanism for the social services sphere, which most effectively uses limited resources.

3 Results

The provision of social services in a stationary form is carried out mainly by state and municipal organizations, at the same time, the private sector is quite developed in the Central, Northwestern and Siberian federal districts of the Russian Federation. The number of participants in this segment did not include charitable organizations, which is explained by the closed nature of information about the distribution of funds, the semi-legal nature of activities in this segment. Municipal-owned market participants include integrated social service centers with inpatient departments for pensioners and disabled people. Non-state participants, on the one hand, specialize in a certain segment of recipients, and, on the other, they dwell on a specific form of their provision: with accommodation, at home, visiting an institution.

A similar situation is observed in the provision of social services in a stationary form for orphans and children with special needs. Despite the prevalence of state institutions, organizations of a private form of ownership are presented in all federal districts, with the exception of the North Caucasus and Far East. The specifics of the activities of entities by types of organizational and economic mechanism, which are in state and municipal ownership, presupposes a comprehensive nature of service, at the same time, social services are not interchangeable for recipients. In this regard, when modeling the activities of market participants that are in state and municipal ownership, activities with and without providing accommodation are also described. Charitable organizations are not represented. The objective reason for this situation is the lack of constant funding of activities in the form of donations.

The research results are shown in Figure 2.

https://doi.org/10.1051/shsconf/20208906002

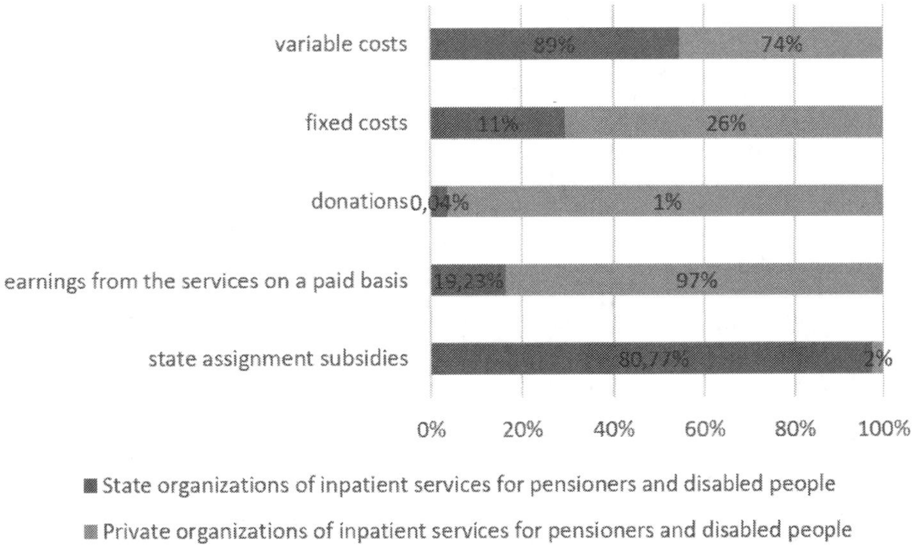

Fig. 2. Parameters of the economic model of organizations providing residential care for the elderly and disabled, %

Characterizing organizations that provide social services with provision of accommodation, one can note a higher concentration of beds in public institutions, on average, this value was 311 units, in private organizations of this type of activity, this value was 40 beds.

Next, let's look at the cost structure, Figure 3.

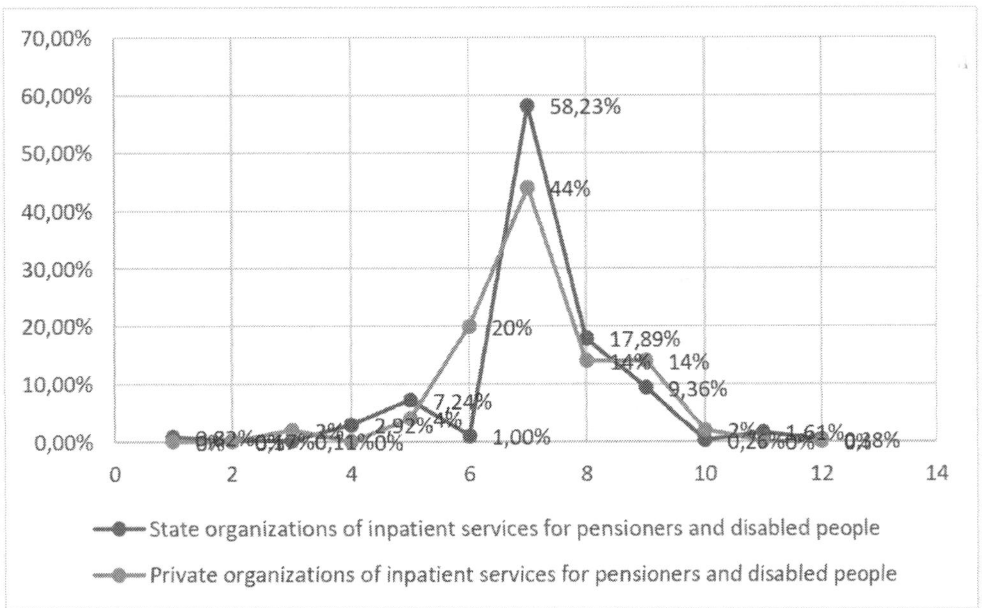

Fig. 3. The structure of expenses of organizations in the social services market, %

Analyzing the results of comparing the cost structure of organizations operating in the field of social services in the form of government agencies and commercial organizations, it can be concluded that the structure of the costs of these organizations is fundamentally similar. Significant differences apply only to the item related to the attraction and maintenance of premises used to accommodate clients. Additionally, it should be noted that organizations of private ownership form an average profit based on performance at the level of 11-23% in relation to expenses depending on the geographical location.

4 Conclusions

The main result of the study is the conclusion about more effective work of organizations with state ownership in the provision of social services with the provision of accommodation. That is, the compulsory stimulation of the development of private organizations in this area of activity leads to ineffective spending of limited resources involved in the social sphere. This situation is primarily due to the recoverable profit in the course of providing socially significant activities, the share of profit, according to our estimates, is 11-23% in relation to the costs incurred. In addition to the formation of profits, private organizations do not have mechanisms to control the expenditure of resources during procurement procedures, the absence of their own stationary premises. As a result of the analysis, it can be concluded that it is necessary to enlarge private social service organizations and form a number of recommendations that limit the use of resources.

References

1. N.I. Yashina, O.V. Emel'yanova, E.S. Malysheva, N.N. Pronchatova-Rubtsova, Finance and Credit, **3** (2018)
2. A. Ilchenko, Xiang Xiao Gang, V. Stepanov, Modern high technologies, Regional application, **4** (2016)
3. N.M. Surnina, A.A. Ilyukhin, S.V. Ilyukhina, Journal of new economy, **5** (2016)
4. O. Yu. Ulyanova, S. O. Yashchenko, E. I. Chebanov, Regional economy: theory and practice, **48** (2011)
5. S.K. Alter, Social entrepreneurship: New models of sustainable social change (2006)
6. J. Bartlett, At your service: navigating the future market in health and social care Demos, London (2009)
7. Y. Baruch, N. Ramalho, Nonprofit and voluntary sector quarterly, **1** (2006)
8. A. V. Chisnikov, *Improving the methodology for assessing the level of development of social infrastructure in the region*: dis (2010)
9. A.V. Materikin, Economics and Law, **2** (2014)
10. T. Bovaird, Public administration, **1** (2006)
11. R.C. Feiock, H.S. Jang, Public Administration Review, **4** (2009)
12. K.P. Kearns, Public administration review, **1** (1994)
13. E.S. Ogorodnikova, A.E. Plakhin, T.V. Kochergina, P.V. Mikhailovsky, T.I.Guseva, M.V. Selezneva, Espacios, **25** (2019)

Corporate volunteering as a form of social innovation

Julia Kuznetsova*

Branch of Kuzbass State Technical University named after T.F. Gorbachev in the city of Novokuznetsk, Ordzhonikidze Str., 8a, 654000 Novokuznetsk, Russia

Abstract. Corporate volunteering (CV) is a new phenomenon in the activities of Russian companies. Based on the use of diachronic analysis, a comparative study of the transformation of the essence and content of corporate volunteering in the works of foreign and domestic scientists and specialists has been carried out. There has been revealed a shift in emphasis from CV as an element of corporate social responsibility, to CV as an independent business process. Due to the novelty of this phenomenon, as well as the complex of characteristics determined on the basis of the use of synchronous analysis, it has been proved that CV can be defined as one of the forms of social innovation.

1 Introduction

The development of the social sphere is an important direction of the internal policy of the state. Since the emergence of the first state formations, the task of finding the most effective approaches and mechanisms for solving social problems of society has remained paramount, especially during periods of exposure to global challenges and threats. In the last decade, social innovation (SI) has been recognized as an important resource for improving the lives of the citizens. However, in our country, a single concept has not yet been developed regarding mechanisms that can provide a level of social security for the population corresponding to the world, which is especially necessary during periods of global social upheavals. Along with this, volunteering is a special sphere of human activity, which makes a significant contribution to increasing the effectiveness of the country's social policy. Volunteering can be individual, group, carried out within the framework of a volunteer organization. The newest form for the Russian companies is corporate volunteering, aimed not only at helping those in need, but also increasing the company's social role in society and its internal stability.

In the context of social changes taking place in recent months, caused by the coronavirus pandemic, the increasing importance of volunteering, while the number of innovative solutions in the social sphere increases, it seems relevant to study the issue of the essence and content of corporate volunteering as a form of social innovation.

* Corresponding author: acanaria2005@yandex.ru

2 Research methods

Based on the use of two methods of analysis - diachronic and synchronous, the study investigates the transformation of the essence and content of corporate volunteering in the works of foreign and domestic scientists and specialists, identifies characteristics of corporate volunteering that allow us to consider its form of social innovation (Figure 1).

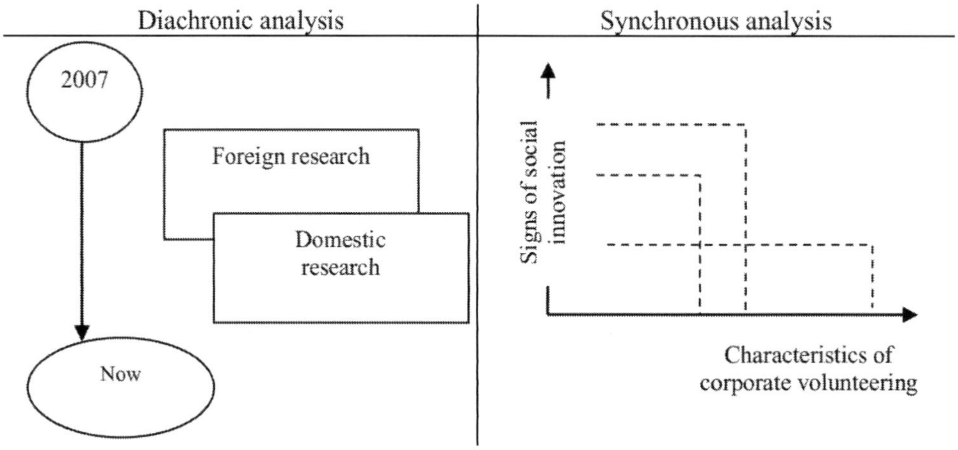

Fig. 1. Research methods of corporate volunteering as a form of social innovation

3 Analysis of corporate volunteering as a form of social innovation

Social innovation is a new idea, method, approach to solving social problems of a society, making significant changes in its life.

The emergence of the term "social innovation" is associated with the research of M. Yunus, winner of the 2006 Nobel Peace Prize. One of the significant works devoted to the study of the methodological foundations for assessing social innovation is the work of Preskill H., Beer T. According to the authors, a social innovation is one that is aimed at solving a social problem not only at the "present moment", but also for the future, and which has the characteristics of efficiency, effectiveness, sustainability [1]. Moreover, the significance of this innovation should grow from year to year not so much for individual individuals as for the whole society. With regard to domestic research, their heyday falls on the early 2000s. Most often, social innovation is understood as the development of new ideas for solving social problems of society, meeting social needs. Social innovation can be expressed as a new product, service, initiative, organizational model, or service delivery approach. Some social innovations are "modest and gradual" [2], others are systemic and fundamental. Some include new services, others - new social enterprises.

Social innovation can take several forms:

1) social innovation in the public sector in the provision of public services (the goal is to increase productivity or save financial resources);

2) social innovation in the non-profit sector (the goal is to improve the quality and increase the volume of public services, or to provide new types of them);

3) social innovation of voluntary and community groups that can be involved in the provision of basic public services.

Social innovation offers practical, budget-friendly solutions for achieving positive social change. In other words, we can talk about two types of innovations: innovations "from above" (initiated by the state) and innovations "from below" (initiated by the population, faced with the difficulty or impossibility of meeting social needs within the framework of existing social mechanisms, as well as by non-profit organizations (NPOs) that have realized the need to solve certain social problems).

Social change can occur at any level, both locally and internationally, as determined by the proven effectiveness of the approach, the potentially large number of beneficiaries, high levels of expectation and sensitivity to social innovation. Most often, an idea arises from below, as an answer to a certain problem, or as a fundamentally different way of realizing the existing social needs of the population. In this context, CV can also be considered as a social innovation, since, on the one hand, it allows a company to achieve a stable position in a competitive market, and, on the other hand, bring benefits to a population in need. In order to determine the full set of characteristics of CV as a social innovation, a study of the transformation of the essence and content of CV in the works of foreign and domestic scientists and specialists was carried out on the basis of diachronous analysis.

The concept of "corporate volunteering" is relatively new in the studies of domestic authors. According to the United Nations Volunteers (UNV) program, almost 30% of the total number of people in the world who volunteer on a permanent basis (and this is 109 million people) are from the CV sector. The emergence of the concept is associated with such a phenomenon in the activities of companies as the provision of gratuitous assistance by employees to those in need in the territory of their presence. Most often, CV is perceived as a stage in the development of corporate social responsibility (CSR) and the quality level of CV is directly dependent on the level of CSR development. Today, when the external reputation of the company is of paramount importance in the competitive struggle, CV allows to strengthen the corporate culture, increase the loyalty of employees, develop their social responsibility, strengthen the team spirit, which positively affects the quality and results of the entire company.

2007 is highlighted in the content of the research as the starting point of the analysis because the first work on the topic under study was posted in the Scientific Electronic Library eLIBRARU.ru exactly that year. D.V. Chernov considered the change in the structure of social responsibility in the context of the transformation of the model of society [3]. The appeal to this topic is due to the growing number of private and public initiatives focused on charity during that period, which became significant for the entire developing system of social work. According to D.V. Chernov, “the strategy of modern social work, regardless of the model, is aimed at enhancing a person's position in the process of accepting help by increasing his independence, the ability to control his daily life and his problems. Under certain conditions, corporate social responsibility may well become one of the mechanisms for solving social problems within the framework of the state-social model” [3, p. 160]. At that time, CV was not yet an independent element of the company's business processes. Until 2012, domestic researchers had published only 5 works on CV, the most notable of which is the work of N.Yu. Bondarenko, who clearly positions CV as a tool for the effective development of the company, including at the international level [4].

The period of 2012-2014 in domestic studies of CV should be characterized as a search for CSR consistency [5], psychological aspects of the behavior of corporate volunteers [6]. D.P. Frolov, A.A. Shulimova consider social responsibility from the point of view of “the institutional form of the forced adaptation of corporations to the growing demands of civil society and regulators (defensive reaction), marketing technology to strengthen the brand image (response), the way of expanding norms and practices on counterparties and partners in global supply chains (backlash) ”[5, p. 124]. At that time, the authors recommended that domestic entrepreneurs act in three directions: 1) analyze and take into account in their activities the experience of transnational corporations and enterprises with foreign investments in the field of reputation management and brand management; 2) adapt innovative practices in the field of business social responsibility.

In 2015-2016, there was a shift in emphasis from CV as an integral part of CSR to its role as a new instrument for the company's strategic development [7]. In addition, the interest of researchers in the foreign experience of CV has significantly increased. It seems that it is 2016 that can be called the year of active interest in CV on the part of domestic authors. In 2017, the authors turned to the mechanisms for organizing and implementing CV [8], works began to appear defining CV as an innovation in certain areas of the company [9]. Interest in foreign experience was manifested in the study of the main features of CV in the European Union (EU). It was determined that the coordination of CV in the EU countries is maintained by non-profit organizations, but the role of the companies themselves is increasing. This is due to the fact that CV is included as an independent element in the range of possible areas of business social responsibility.

In 2018, a lot of studies appeared that define CV as a tool for promoting a company, a landmark for modern business, as a socio-cultural phenomenon, a tool for communication with the external and internal environment. Approaches for the activation of CV are considered [10]. In 2019-2020, there is a clearly expressed tendency for the development of CV models and the formation of trends in its development [11]. So, S.A. Galchenko determined that "the experience of leading companies shows that corporate volunteering is a driver of social change, both for the organization itself and its internal environment, and for the external society" [10, p. 121]. One of the drivers is Pro Bono - an outstaffing site or a new direction of the market "Learning 2 Business", the main idea of which is active interaction of representatives of educational organizations (young scientists, teachers, students) with entrepreneurs (small businesses).

Despite the fact that a definite date for the appearance of CV in Russia has not yet been designated, according to a group of analysts with the support of the National Council for Corporate Volunteering, "the financial crisis of 2008 acted as a catalyst for development" [12, p. 13], when many companies reduced their CSR costs. Undoubtedly, an important milestone in the formation of CV in Russia is the International Forum "Corporate Volunteering: Business and Society", whose mission is "to identify and replicate best practices and innovative practices of socially responsible business conduct" [13]. A notable tendency is that a large number of various works on CV appears on the Internet sites of various expert and analytical agencies, recruitment portals (for example, the RosBusinessConsulting portal, HH.ru).

The history of CV abroad is about 50 years old, and the pioneers are large US corporations [12, p. 12]. The study of foreign experience shows that two approaches to CV have emerged in the world: global (development of signature projects or signature partnerships by transnational corporations, replicated in the countries of presence) and local (development by regional divisions of their own volunteer programs) [14]. The many times greater experience in organizing CV in foreign companies, as well as the presence of a significant backlog of research work (for example, the portal of the world-famous publishing house Springer contains more than 5 thousand works on researched and similar topics), allows us to talk about the existing systemic understanding of CV.

Due to the fact that the starting point of the study of domestic works on CV was adopted as 2007, in order to obtain comparable results of the study of the essence and content of corporate volunteering, the analysis of the works of foreign authors was carried out starting from the same year. In 2007-2009, many foreign researchers are united in the opinion that not only large, but also medium-sized enterprises, being under constant scrutiny of the media, have come to the conclusion that it is necessary to use CV as an instrument to increase their social attractiveness in society [15]. This became relevant, since the activities of the leadership in the media more often began to be viewed not from the point of view of a certain process, but from the position of a certain problem. In this context, studies by G. Mutz, M. Fischer, D. Schöffmann, presented to the public in 2008, made it possible to look at CV as a method of creating sustainable corporate value [16]. In 2009-2012, the content of CV was revealed from the position of its capabilities in ensuring the legitimacy of organizations [17]. In addition, CV is seen as an innovative way to adequately respond to the challenges that arise in civil society. Until 2015, one of the most pronounced aspects of the study of CV had been the national aspect. J. Bellamy's 2015 study reveals a clear pattern: the more religious an employee is, the more likely he is to volunteer [16]. Since 2016, the following areas of research on CV in the works of foreign authors have been distinguished: change in the content of CV in connection with interaction with non-profit organizations [19], CV as a personnel management tool [20] and an element that interconnects all business processes.

Within the framework of the use of synchronous analysis, it was determined that CV can act as one of the forms of social innovation (SI) (Table 1).

Table 1. Assessment of corporate volunteering as a social innovation.

SI signs	Compliance degree *	Characteristic of CV as SI
A fundamentally new idea, approach, method, technology	CC	Volunteering by itself is not a new phenomenon. The novelty is revealed in the application to the activities of the Russian companies
Orientation towards achieving a social goal	BC	CV is focused on helping those in need in the external environment and ensuring the stability and cohesion of employees in the internal one
Striving for cross-sector collaboration	BC	The sectors are business and society
Functionality of an innovative solution	CC	CV is gradually becoming a separate business process. However, the solutions themselves for interacting with the external environment differ in the level of functionality
SI contributes to solving socially significant problems	CC	CV is aimed at solving social problems at the local level, often targeted
SI causes significant social changes in society	CC	CV causes multi-level changes at the local level. Significant changes are achieved in the internal environment of the company
SI causes the revision of existing standards of activity	HC	CV does not significantly affect the transformation of established norms in society
SI defines a new order of interaction with society	BC	Those in need get help from a new source – company employees
SI change the form of activity	BC	CV make changes to the implementation of CSR

* the following scale is used for evaluation: BC – high degree of compliance, CC – medium, HC - low

4 Conclusion

The results of the diachronic analysis obtained by the author suggest the presence of a pronounced evolution in the study of the essence and content of corporate volunteering as a form of social innovation. This is reflected in the transition of CV from one of the functions of corporate social responsibility, to CV as an independent business process. Moreover, this trend can be traced in the works of both foreign and domestic authors. The main distinguishing feature of foreign studies that allows thinking about CV as a social innovation is the emphasis on the innovativeness of the volunteer activities of company employees. Taking into account this aspect of foreign studies, with the help of synchronous analysis, the analysis of CV as a form of social innovation based on a set of SI features has been carried out. It has been determined that according to the correspondence of features with a high and medium degree of compliance, CV is a form of SI by 89%.

References

1. H. Preskill, T. Beer, *Evaluating social innovation* (2012)
2. Social innovation: official website of the Center for Social Innovations Nesta, https://www.nesta.org.uk/
3. D.V. Chernov, Social sphere and everyday life of a Siberian city (XX - early XXI centuries), 160 (2007)
4. N.Yu. Bondarenko, Motivation and Wages, **4**, 258 (2009)
5. D.P. Frolov, A. A. Shulimova, Journal of Institutional Research, **1**, 124 (2013)
6. A. B. Gosteva, E. V. Bogomolova, Man. Community. Management, **3(105)**, 264 (2014)
7. D.V. Skulches, V. V. Burlyaeva, E. V. Ledneva, Bulletin of the Institute of Friendship of the Peoples of the Caucasus. Theory of Economics and National Economy Management, **2**, 89 (2016)
8. Z. A. Troska, Plekhanov scientific Bulletin, **2(12)**, 220 (2017)
9. T.V. Skotskaya, V.V. Zotov, Social and humanitarian problems of education and professional self-realization, Social engineer-2017, 124 (2007)
10. S.A. Galchenko, Questions of Economics and law, **122**, 121 (2018)
11. S.A. Galchenko, Politics, Economics and innovation, **1(24)**, 19 (2019)
12. R.G. Zakiev, I. Yu. Shvets, M.A. Belogurova, *Corporate volunteering in Russia. From practices to the standards. Manager's guide* (2018)
13. Ninth Moscow international forum "Corporate volunteering: business and society", https://corpvolunteers.ru/
14. A. V. Kuznetsov, International processes, **4**, 103 (2015)
15. W. C. Zimmerli, K. Richter, M. Holzinger, *Corporate Ethics and Corporate Governance* (2007)
16. A. Habisch, R. Schmidpeter, M. Neureiter, *Handbuch Corporate Citizenship. Corporate Social Responsibility für Manager* (2008)
17. C. K. Schäfer, *Corporate Volunteering und professionelles Freiwilligen-Management* (2009)
18. J. Bellamy, *Religion and Volunteering* (2015)
19. L. Hustinx, *Changing Nature of Formal Service Program Volunteering* (2016)
20. S. Dreesbach-Bundy, B. Scheck, *CSR und Corporate Volunteering. Mitarbeiterengagement für gesellschaftliche* (2018)

SHS Web of Conferences **89**, 06004 (2020) https://doi.org/10.1051/shsconf/20208906004
Conf-Corp 2020

Game mechanics application in the optimization and automation framework of personnel management processes

Igor Kotsyuba[1,*], *Konstantin* Loginov[1], and *Aleksey* Shikov[2]

[1]ITMO University, Kronverksky Avenue, 49, 197101 Saint Petersburg, Russia
[2]Russian Academy of National Economy and Public Administration under the President of the Russian Federation (North-West Institute of Management - a branch of RANEPA), Sredny prospect V.O., 57/43, 199178 Saint Petersburg, Russia

Abstract. The article examines the processes of applying game mechanics to optimize and automate the management and development of company personnel. This allows one to free up resources for solving tasks that require high qualifications, enhances the quality of work performed by reducing the factor of human errors in monotonous actions, and also ensures the engagement and motivation of employees. It is concluded that solving problems of optimization and automation of personnel management processes is an integral part of the competitive companies' activities. Employees' motivation and engagement in these processes increase their efficiency and loyalty to the company. The use of game mechanics in the implementation of corporate HR management processes increases the motivation and engagement of employees, and, as a result, greatly enhances their efficiency.

1 Introduction

An indisputable the modern reality fact of enterprises' existence and competitiveness is the presence of highly qualified motivated and involved personnel. For this reason, it is fundamental to have well-built, highly organized business processes for management and personnel development. Like any business processes that are scalable for a large number of people, HR processes involve the prevailing amount of routine work and have a number of requirements and limitations. Optimization and automation of these processes allows freeing up resources for solving tasks that require high qualifications, improves the quality of work performed by reducing the factor of human errors in monotonous actions, and also involves and motivates personnel.

Motivation is viewed as a driver that encourages efforts to achieve goals and is a significant factor in the effective activities' implementation.[1] Moreover, motivation directly affects the time and volume of these very efforts that a person is willing to spend on solving problems [2]. On the other hand, motivation can be considered as a result of activities,

* Corresponding author: igor.kotciuba@gmail.com

which is why it is so important to carefully consider all the processes in which employees are involved.

Recently, researchers in the field of training implementation, including corporate training, have been paying more and more attention to the introduction of gamification as a tool to increase the engagement and motivation of students in training [3, 4, 5, 6, 7]. It is believed that the training implementation using gamified services and platforms has the potential to increase the training efficiency, based on the following aspects: providing students with the opportunity to overcome difficulties and challenges; the ability to aggregate knowledge obtained in various ways and from different sources, as well as use it to make prudent decisions; forecast results based on decisions made; encourage students for their activities; provide interactive feedback; stimulate gradual systematic training at a calm comfortable pace [8, 9, 10, 11].

2 Materials and methods

In general, gamification is understood as "the use of video game elements in non-gaming systems to improve user experience and user engagement" [12]. The game mechanics are that kind of game elements. Game mechanics mean certain scenario elements that describe how to change the user's game state. In other words, game mechanics are a set of actions that a user can perform, as well as a set of rules and restrictions.

The following main areas of gamification application can be distinguished within the framework of the personnel management processes automation:

1. At the stage of candidates' selection.
2. As part of the adaptation process.
3. In order to support the personnel development process.

The candidates' selection process is one of the most important in the company, determining the staffing of the organization. The main task of a hiring specialist is to find the most competent candidates for a specific position, as well as their engagement in the company.

Let there be N number of vacancies and M employees. Each i-th employee can perform each j-th task with a certain efficiency coefficient u_{ij},, the value of which can be increased by using one or another set of game mechanics. Then the task of selecting candidates can be formalized as follows:

$$\text{f(x)} = \sum_{i=1}^{m} \sum_{y=1}^{n} u_{ij} x_{ij} \ \rightarrow max \tag{1}$$

where u_{ij} is the coefficient of utility of the i-th worker for the j-th job role;

$$D = \{x_{ij} \in \mathbb{R}^{m*n} \mid \sum_{i=1}^{m} x_{ij} \ \leq \ a_j, \ \forall j = \overline{1,n}, \quad \sum_{y=1}^{n} x_{ij} \ \leq 1 \ \forall i = \overline{1,m}\} \tag{2}$$

Each employee can take no more than 1 role in the team.

$$x_{ij} = \begin{cases} 1, if \ i - th \ \text{со}\text{т}employee \ is \ in \ the \ j - th \ role, \\ 0, in \ case \ he \ is \ not. \end{cases} \tag{3}$$

One of the key existing issues in personnel management is the staff turnover problem. Relatively high employee turnover rates negatively affect jobseekers' loyalty to the company, which leads to a decrease in the likelihood of hiring highly qualified professionals. This problem is most acutely perceived by innovative companies and enterprises, for which the time spent on the adaptation process of new employees should be reduced to a minimum, and the adaptation process itself should be the most effective. Automation of the personnel adaptation process through the development and implementation of an information system for personnel management and development can significantly reduce the time spent by experienced employees on getting up to speed.

The task of implementing the employees' adaptation can be formulated in terms of the optimization problem (4):

$$f(x) = \sum_{i=1}^{m} eff_i \cdot x_i \to max$$
$$D = \{ x \in R^m | x_i \in [0,1] \forall i = \overline{1,m}, \quad (4)$$
$$\sum_{i=1}^{m} t(edu)_i \cdot x_i \le T(edu),$$
$$\sum_{i=1}^{m} t(adapt)_i \cdot x_i \le T(adapt) \}$$

where eff is the training efficiency of the i-th employee, t(edu) is the training time of the i-th employee, t(adapt) is the adaptation time of the i-th employee, T (edu) is the total time allocated for training, T (adapt) - the general fund of time allocated for adaptation. X(i) - a plurality of trained / adaptable employees

Along with the above-mentioned tasks of personnel management, the issue of managing employees' development and training is highlighted. Ensuring the development and continuous maintenance of proficiency level in the competencies necessary for the production process for innovative enterprises is one of the most important conditions for competitiveness and market leadership.

One of the main approaches to organizing training is to focus on the employee competency matrix. The competence matrix is understood as a list of competencies that are valuable for the company and help in the implementation of its tasks. It covers various knowledge, skills and abilities and is organized in the form of competency profiles - sets of competencies and knowledge levels of these competencies necessary for the successful completion of specific tasks. The competence matrix is shown in table 1.

Table 1. Matrix of competencies

	c_1	c_2	...	c_p
k_1				
k_2				
...				
k_r				

where c_1, c_2, ..., c_p – a set of competencies corresponding to a specific profile;

k_1, k_2, ..., k_r - level of competencies' possession.

Implementation of the concept of the current matrix of employee competencies and the target one allows one to select the most suitable candidates for a vacancy, build training paths for the purpose of vertical or horizontal career growth of employees, track its progress, and form a base of interchangeable personnel. The calculation of deviations of the current competency matrix from the target can be calculated as:

$$\begin{cases} k_{1goal} - k_{1cur} \le \delta_1, \\ k_{2goal} - k_{2cur} \le \delta_{2,} \\ \dots \\ k_{pgoal} - k_{pcur} \le \delta_p \end{cases} \quad (5)$$

where k_{pgoal} is the target value of the proficiency level in the p-th competence, k_{pcur} is the current value of the level of proficiency in the p-th competence, δ_p is the permissible deviation of the current level of proficiency in the p-th competence from the target. Thus, the vector will act as a criterion for achieving the target competency profile:

$$\begin{pmatrix} \delta_1 \\ \delta_{2,} \\ \dots \\ \delta_p \end{pmatrix} \quad (6)$$

However, the process of training implementation and increasing the competence of personnel must meet a number of requirements: minimizing the time spent on training; minimization of the working hours of the expert teacher; ensuring the maximum efficiency of the training provided. Then this problem can be represented as a multiobjective optimization problem:

$$\begin{gathered} f_1(x) = \sum_{i=1}^{m} eff_i \cdot x_i \to max \\ f_2(x) = \sum_{i=1}^{m} t(exp)_i \cdot x_i \to min \\ f_1(x) > f_2(x) \\ D = \{ x \in R^m | x_i \in [0,1] \forall\, i = \overline{1,m} \\ \sum_{i=1}^{m} t(edu)_i \cdot x_i \leq T(edu) \} \\ x_i = \begin{cases} 1, & if\ i - th\ if\ applied\ game\ mechanics \\ & 0, otherwise \end{cases}, \end{gathered} \quad (7)$$

where eff is the effectiveness of training using the i-th game mechanics, t(edu) is the training time using the i-th game mechanics, t(exp) is the time of the expert's work using the i-th game mechanics. X(i) is the plurality of all game mechanics.

During the operation of the corporate training system, it becomes crucial to assess the effectiveness of the developed system. There are various approaches, that take into account different metrics and indicators, used to assess the effectiveness [13-15]. The most common model for assessing the effectiveness of the corporate training courses implementation is the Kirpatrick's four-level model [16].

The task of conducting corporate training is to ensure its maximum efficiency, while minimizing training costs and time. Then, this kind of problem can be formulated in the form of a multi-criteria optimization problem:

$$\begin{gathered} f_1(x) = \sum_{i=1}^{m} firstLev_i \cdot x_i \to max \\ f_2(x) = \sum_{i=1}^{m} secondLev_i \cdot x_i \to max \\ f_3(x) = \textstyle\sum_{i=1}^{m} thirdLev_i \cdot x_i \to max \\ f_4(x) = \sum_{i=1}^{m} fourthLev_i \cdot x_i \to max \\ D = \{ x \in R^m | x_i \in [0,1] \forall\, i = \overline{1,m} \end{gathered} \quad (8)$$

$$\sum_{i=1}^{m} t(edu)_i \cdot x_i \leq T(edu)$$
$$\sum_{i=1}^{m} cost_i \cdot x_i \leq COST \},$$

where x_i is the i-th game mechanics from the set X, t(edu) is the training time using the i-th game mechanics; T(edu) - general time fund allocated for training; COST - general cost fund, $firstLev$ - $fourthLev$ - achievement indicators of the corresponding four levels according to Kirpatrick. The result of solving this optimization problem is the vector of game mechanics used, which makes it possible to maximize the efficiency of corporate training, taking into account the existing restrictions on the cost of training time.

3 Results and discussion

As part of a study to assess the effectiveness of the gamification introduction in the process of personnel corporate training, the authors of this article set up an experiment consisting in passing the developed electronic training course "Agile Methodology", presented in several modifications: with the use of a different set of game mechanics and without them. This course was intended for employees of an innovative software development company. The experiment involved 32 people aged 22 to 34 years. When applying for this course, all participants were randomly divided into three control groups for three types of e-learning courses, depending on the type and number of game mechanics used.

The e-learning course was implemented in three versions:

1. Without using game mechanics (course type 1);
2. Using the mechanics of reward / punishment, the mechanics of competition, the mechanics of recall (course type 2);
3. Using the mechanics of reward / punishment, mechanics of competition, mechanics of reminder and mechanics of prompts (course type 3);

The training course was divided into modules. Two types of deadlines were used as time constraints for the delivery of the final tasks in the module: soft and hard. The soft deadline assumed a decrease in points for completed tasks by 5% of the points scored for each day of delay, thus implementing the punishment mechanics. The tough deadline meant that there was no way to evaluate assignments if it was violated.

Data analysis obtained on the results of passing the e-learning course showed that the values of the average time for completing course tasks in the case of using game mechanics turned out to be 25% - 29% less (depending on the type of course) the time for completing tasks in the absence of gamification elements (Fig.1). This fact is explained by the implementation of the impact of the game mechanics of reward / punishment and competition, which stimulate the students' desire to get the maximum number of points, while observing the existing deadlines.

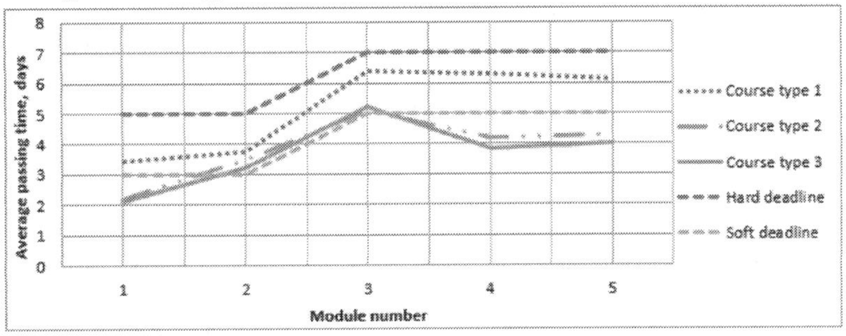

Fig. 1. Dependence completing time of modules tasks on the course type

We can also conclude that the use of game mechanics of reward / punishment and competition, which gives the possibility of obtaining more points in case of meeting the deadlines for completing the course tasks, contributed to earlier completion of tasks and adherence to deadlines, which ensured almost halve the number of strict deadlines violations.

The effectiveness of the application of the hint game mechanics is also confirmed in the course of the data analysis. The number of correct answers on the second attempt in the case of using the game mechanic of the hint turned out to be 11% higher than in the case without it, and on the third - by 16%. This fact proves the effectiveness of the application of the hint game mechanics in the framework of the electronic training courses implementation.

Also, the retention rate was chosen as the basic criterion for assessing the effectiveness of the application of the developed solutions. This metric reflects information about what percentage of the initially registered users interacted with the system on day N. The classic calculation of the metric is carried out as follows:

$$ret_N = \frac{k_N}{p_N} * 100\% ,$$

where ret_N – day retention rate N;

k_N - the number of users logged in per day **N;**

p_N - the number of users who started using the system N days ago.

In the case of taking courses with the use of gamification elements, on average, the value of the N-day retention rate is higher than in the case of passing the course without introducing game mechanics (Fig. 2). The value of the standard deviation of the retention rate for the period excluding the first and the last two days was 38% of the average value of this metric in the case of completing the training course without the use of game mechanics, and in the case of their application - 10-13%.

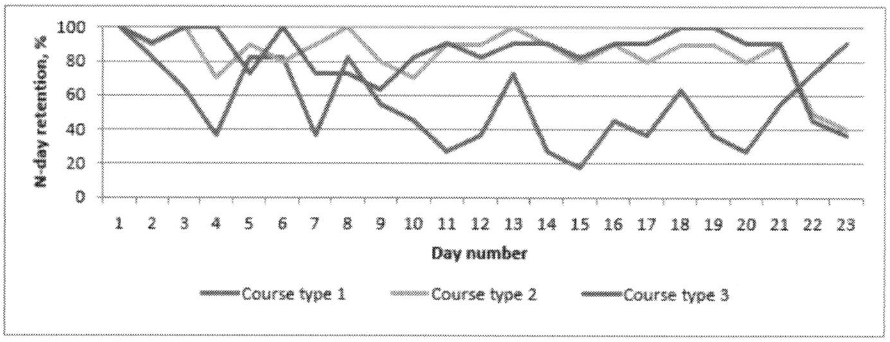

Fig 3. Graph of changes in the N-day retention rate for three types of courses

This fact truthfully indicates that the use of the game mechanics of rewards / punishments contributes to the constant systematic work on the passage of the training course.

4 Conclusions

Evaluating the effectiveness according to Kirpatrick's model also showed that all four levels of the model were successfully achieved. The fact that the value of the average upload time of software releases was reduced by 7%, the rate of the number of urgent fixes (hotfixes) made after the release of releases decreased by 17% indicates the achievement of the fourth level of the model, namely, an improvement in business performance as a result of training.

Thus, it can be concluded that solving the problems of optimization and automation of personnel management processes are an integral part of the competitive companies' activities. Motivation and engagement of employees in these processes increase their efficiency, staff loyalty to the company, and create a positive brand image. The use of game mechanics within the framework of the implementation of corporate HR management processes allows to increase the motivation and engagement of employees, and, as a consequence, increase their efficiency.

References

1. W. Harlen, In J. Gardner (Ed.), *Assessment and Learning*, 61 (2006)
2. J.D. Bransford, A.L. Brown, R.R. Cocking, *How people learn* (2004).
3. S. Tobias, J.D. Fletcher (Eds.), *Computer games and instruction* (2011)
4. M.S.Y. Jong, Curriculum Journal, 26 (2), 249 (2015)
5. M. Kebritchi and A. Hirum, Computers & Education, 51 (4), 1729 (2008)
6. S. Erhel and E. Jamet, Computers & Education, 67, 156 (2013)
7. J.L. Plass, B.D. Homer and C.K. Kinzer, Educational Psychologist, 50 (4) , 258 (2015)
8. C. Bouras, V. Igglesis, V. Kapoulas, I. Misedakis, O. Dziabenko, A. Koubek, Journal of Intelligent Games and Simulation, 3 (2), 70 (2004)
9. M.I. Cicchino, Interdisciplinary Journal of Problem-based Learning, 9 (2) (2015).
10. J. McCall, The History Teacher, 46 (1), 9 (2012)
11. B. Kim, H. Park and Y. Baek, Computers & Education, 52 (4), 800 (2009)
12. S. Deterding, D. Dixon, R. Khaled, L. Nacke, *From game design elements to gamefulness: Defining gamification 15th international academic MindTrek conference* (2011)
13. J.J. Phillips, *Handbook of Training Evaluation and Measurement Methods* (1997)
14. A. Adelsberg, E. Trolley, *Running Training Like a Business The Forum Corporation* (1999)
15. T. Waddington, B. Aaron, R. Sheldrick, *Guerilla Evaluation: Adapting to the Terrain and Situaion. Instructional Desing in the Real World* (2004)
16. D. Kirkpatrick, Training and Development, 50, 54 (1996)

The role of the concept of corporate social responsibility in promoting goods and services in the economically developed countries

Marina V. Andriianova[1] and *Valeriia V.* Kruchinina[2]

[1]Moscow State Institute of International Relations (MGIMO University), Pr. Vernadskogo, 76, Moscow, Russian Federation
[2]ISS – Intelligent Security Systems (ISS-Soft Ltd), Skolkovo innovation center, Bolshoy boulevard 42, bld 1, 143026 Moscow, Russian Federation

Abstract. As companies lost their reputation amid ethical scandals, socio-ethical marketing as well as social responsibility concept, became vital components and a fundamental difference of successful companies. This shift towards socially responsible marketing is indicative of shifting the focus from production and goods/services to the consumer and his philosophy of life, as well as to social ethics. However, with the changing environment especially in digitization encompassing all spheres of society, companies also have to improve their marketing tools including marketing communications as a key factor in the formation of the socially oriented relationships between manufacturer and consumer.

1 Introduction

Conventionally, marketing communications are considered as the relationship formed by a firm with an external environment. As marketing tools can be also considered the product itself, its packaging and labeling; advertising and PR; trade marketing and direct marketing; branding; online promotion; loyalty campaigns; sponsorship and philanthropy; personal sales and after-sales service; various kinds of events (exhibitions, seminars, conferences). Ultimately, the purpose of all marketing communications is to convince the consumer (potential or actual) that the product or service provides the fullest possible satisfaction of its needs, which should give an impulse to acquire the proposed product or service.

In recent decades, major changes in the economy and society have taken place; these changes have a significant impact on all spheres of activity of any organization. They are also particularly important to ensure effective interaction with the target audience. Marketing communications are an organization's outpost in its relationship with the consumer. Often actual customer appeal of the goods forming the relationship to the producer and the product or service offered is less important than the values embedded in the concept of the social responsibility of the manufacturing company. The effectiveness of the organization's market strategy, its competitiveness, and, ultimately, its long-term viability is directly dependent on the understanding of changes in the external environment and the ability to create both the concept of social responsibility and communication with stakeholder groups.

2 Methods

In classical economic theory, market availability is a systemic factor that determines the existence of an organization. [1] From the point of view of producers of goods and services, the markets of economically developed countries are the most attractive. The high standard of living peculiar to the economically developed countries, shows, on the one hand, a high level of capacity to pay and, on the other, a high level of consumption, which is becoming the norm and a social value. As a result, a sustained high level of demand for a wide range of goods and services is created (fig. 1).

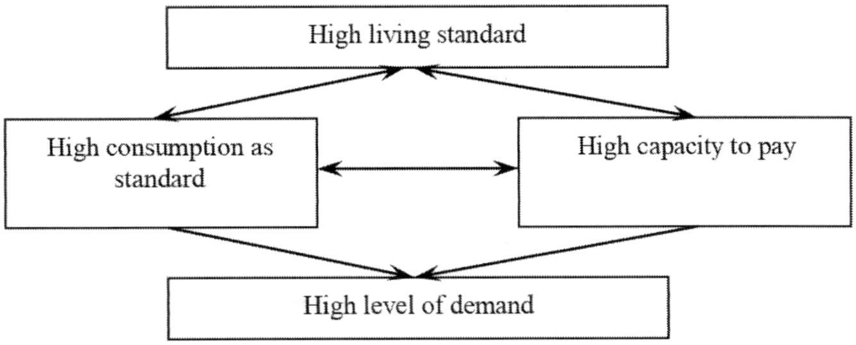

Fig. 1. The link between high living standards and high demand.

It is evident that a high standard of living is a multi-factor concept, it would be extremely narrow and one-sided to assess it only by the income level of the population. One suitable way of considering living standards is through an indicator such as the Human Development Index for which an entire set of indicators is taken into consideration.

Table 1. Top 20 countries by human development level. [2]

Rank	Country	Human Development Index / HDI (value)	Inequality / Inequality adjusted HDI	Education / Life expectancy of schooling (years)	Gender Development Index / GDI (value)	Gross national income (GNI) per capita ($)	Mobility and Communication / Internet users, total (% of population)
1	Norway	0.954	0.889	18.1	0.990	68059	96.5
2	Switzerland	0.946	0.881	16.2	0.963	59375	89.7
3	Ireland	0.942	0.872	18.8	0.975	55660	84.5
4	Germany	0.939	0.861	17.1	0.968	46946	89.7
5	Hong Kong	0.939	n.a.	16.5	n.a.	60221	n.a.
6	Australia	0.398	0.862	22.1	0.975	44097	86.5
7	Iceland	0.938	0.885	19.2	0.966	47566	99.0
8	Sweden	0.937	0.874	18.8	0.982	47955	92.1
9	Singapure	0.935	0.810	16.3	0.988	83793	88.2
10	Netherlands	0.933	0.870	18.0	0.967	50013	94.7
11	Denmark	0.930	0.873	19.1	0.980	48836	97.6
12	Finland	0.925	0.876	19.3	0.990	41779	88.9

Continuation of Table 1. Top 20 countries by human development level.

13	Canada	0.922	0.841	16.1	0.989	43602	91.0
14	New Zeland	0.921	0.836	18.8	0.963	35108	90.8
15	United Kingdom	0.920	0.845	17.4	0.967	39507	94.9
16	United States	0.920	0.797	16.3	0.991	56140	87.3
17	Belgium	0.919	0.849	19.7	0.972	43821	88.7
18	Liechtenstein	0.917	n.a.	14.7	n.a.	99732	98.1
19	Japan	0.915	0.882	15.2	0.976	40799	84.6
20	Austria	0.914	0.843	16.3	0.963	46231	87.7

The data presented in the table clearly shows that, in addition to the high spending capacity in economically developed countries due to the income level of the population, the nature, and structure of demand is influenced by many other factors related to individual, group, and societal attitudes that are difficult to analyze, assess, forecast, and manage by economic entities. Adding the religious aspect of the population, it becomes apparent that the developed economies are currently experiencing a strong tendency for needs to be fragmented and demand at a combined high level to be tremendous detailed about the target audience. Account must be taken of the fact that in economically developed countries have historically been both the most stringent requirements for goods and services producers so as the tradition and practice of social responsibility of business. For this reason, businesses are expected to meet new social conditions, and will not limit itself by a purely applied production function. [3]

The structure and composition of the target audience are constantly changing; the target audience ceases to be stable with permanent characteristics. In such circumstances, the situational approach is particularly important, so the producer needs to know and understand the specifics of his target audience at any given time. [4]

In economically developed countries, most consumers have equal access to the same set of goods and services. [5] Personification is possible only through the formation of a unique combination for the individual. The stronger the consumer's desire for individuality, the more fragmented the needs, the wider the range of goods, and services available to meet them, the less likely the exact repetition of a consumer's personal choice, the higher the consciousness of one's personality. Thus, a vicious circle emerges, leading to the increasing fragmentation of needs. The phenomenon of identity through the pool of needs has led to new challenges for manufacturers of goods and services, and problems in social life. The wait-and-see attitude or a lie-up position in such a situation results to be costly; a paradoxical phenomenon in a modern society that aspires to humanistic values, tolerance, and multiculturalism is the phenomenon of cancel culture and rejection of cultural appropriation. Some examples are given below:

- Reputational scandals in both companies were linked to the statements of their directors, which were humbling in regards to people with excess weight. The Abercrombie&Fitch refusal to manufacture and sell XL- and XXL- sized clothing instigated an active protest with the slogan "Fitch out Abercrombie (please)", which has seriously damaged the reputation of the company and significantly worsen its financial performance. The Victoria's Secret brand refused to cooperate with its director a few days after his interview, brought public interest. The plus-size collection was quickly developed and the first-ever Victoria's Secret plus-size model Ali Tate Cutler was invited.
- Flora vs Starbucks. In a Flora margarine campaign in South Africa, Unilever banked on conservative values and failed, because of not taking into consideration that the target audience has already undergone a significant transformation. The dudgeon caused by ambiguous advertising around the world forced the company to invest a lot of effort to

settle the situation in order to keep the customers. Starbucks, by contrast, acquires additional customers through an effective information campaign so everyone knows that Starbucks, according to the chairman emeritus in Starbucks Howard Schultz, is trying to cover “all existing varieties”. Howard Schultz said, “If you feel, respectfully, that you can get a higher return than the 38% you got last year, it's a free country. You can sell your shares in Starbucks and buy shares in another company”.

Such examples demonstrate that the market rules for “the game” are irretrievably changing. Levitt’s Total product concept (TPC) is a well-known concept that describes a product as a set of mandatory and additional attributes that influence consumer perception and choice. It should be admitted that those attributes that reflect the social position of the manufacturer in the economically developed countries are already firmly in the mandatory category. This fact is beyond all question; however, it is interesting to consider the causes and consequences of such transformation.

The peculiar must-have of a modern product or service is its value component. This may be compared, for example, with the existence of a pool of policy rules and prohibitions in religious systems, such as Islam or Judaism. This is a well-structured, strict system, understood and accepted by all members of the community. It is a system of rules that interferes with all spheres of life and all activities committed by a person. A person with certain religious beliefs cannot afford to commit or acquire something that contravenes the established pattern of behavior. Therefore, the modern consumer considers a product not simply a means of satisfying basic needs but as a carrier of values and beliefs. The purchase of a product or service is possible only if the value systems of the manufacturer and the consumer coincide; in the worst-case scenario, the value system of the producer should at least not contradict the system of consumer beliefs. Such consumer behavior is common to consumers in economically developed countries for various reasons but from an economic point of view, this is possible because of market saturation and lack of monopolists. [6] In fact, the consumer always has a choice – you can simply purchase a product or service with similar consumer properties of another manufacturer with a value system that is considered acceptable to the consumer. Manufacturers used to create new needs by addressing consumers like the characters of Sorokin’s “Ice trilogy”, who to awaken their sisters and brothers with sleeping hearts were beaten on their chest with an ice hammer and cried out, “Speak straight from your heart!” Now the insistent demand “Speak straight from your heart” is addressed already from consumers to manufacturers, and the ice hammer will go down on anyone who refuses to respond. It is no longer enough to offer a good product or service to the market; the producer is often obliged to justify the right of the good or service existing from the perspective of public interest.

3 Results and subsequent discussion

As a result, it is the value of a product or service that should represent an individual projection of the social and personal interest of the consumer. [7] In fact, the range of goods and services favored by the modern consumer forms his (consumer’s) social avatar. This consumer avatar could tell people who are the consumers, what are their values and beliefs, and their role in society. The modern human tells the world what he is in the public sense through the goods and services he purchases. [8] Hence, the concept of social responsibility of the manufacturer is so important today. The brand in today’s world is firmly and closely linked to those values and preferences underlying the concept of corporate social responsibility. This relationship has long been known, however, now it has acquired a qualitatively different content, the changes have affected both the concept of corporate social responsibility and marketing communications.

Previously, the social concept of business was based on the principles of a responsible and conscientious approach to production and meeting the requirements of the target audience. Actually, this is not enough – the manufacturer should demonstrate his stance on the hottest social topics. Any deviation in this situation would be interpreted as unfavorable to the producer.

Marketing communications used to serve as a means to inform consumers about the manufacturer's values, it just was not so important – there were the consumer properties of goods and services in the first place. [9] The value-added information function has now become a priority for marketing communications. It should be noted that active digitization processes particularly affecting economically developed countries are further affecting marketing activities. First of all, it is necessary to be aware of such features of modern digital marketing communications as instant distribution of marketing materials on the Internet, the impossibility of their total elimination, and their accessibility to a wide range of users. It turns out that marketing materials made according to the specific characteristics of the target audience become available to multiple user groups for which they were not purposed before; however, the information diffusion process is extremely rapid, almost instantaneously, and if necessary, the materials cannot simply be withdrawn and may become a stain on the reputation of the company forever. This means that the importance of both the concept of corporate social responsibility and marketing communications is increasing in the modern world, but the corresponding risks are also increasing. [10]

By analogy to strategic management, modern companies should change their approach to social responsibility management, so a prescriptive approach should be abandoned and a diverse, flexible, adaptive system of social responsibility should be created. Ideally, this flexibility should come out through marketing communications in the form of special signals, which would prevent the situation when customers raise questions. For example, questions about the name of the Fazer chocolate "Geisha" and the presence of a drawing of a man in a fez on the packaging of the Valio Greek yogurt caught the manufacturers unawares, because they have never thought that such small details shall deserve attention, and it was not clear to consumers how manufacturers feel about such hot social topics as an objectification of women and discrimination on the basis of nationality and religion.

The variability of the social responsibility concept should not be revealed in evasiveness, indulgence, and hypocrisy in dealing with consumers; the manufacturer shall be ready to state his position on hot social topics. This approach should be implemented through a well-designed system of marketing communications that gain additional value as the most important link between the manufacturer and the consumer. New demands on the social responsibility concept create an additional load on marketing communications functionality. The varying nature of these functions should be mentioned. For example, signal/emergency function – in a peaceful atmosphere, marketing communications should fully transmit to the consumer the necessary weak signals about the manufacturer's value system, taking into account specific features of communication (such as the working language, the presence of slang, etc.); if the situation worsens, the marketing communication system should identify immediately the source and essence of the tension and transmit the necessary information to the manufacturer. The balancing/crisis management function – in a peaceful environment it means the function of adapting the manufacturer's concept of corporate social responsibility to the needs of all stakeholders (i.e. identification of attitudes, stakeholders' opinions, and identification of appropriate weak signals clarifying the concept of social responsibility); in a crisis situation, this function should ensure the immediate mobilization of the marketing communications system as well as to represent the company's position, which should at least not to impair its market position. The need for the qualitative implementation of these functions explains the increasing importance of the components of the marketing communication system that ensures direct and operational communication with the stakeholders, especially it is referred to the social media marketing and community management.

4 Conclusions

The analysis of development trends and the transformation of the interaction between the social responsibility concept and the company's marketing communications system testify that their role in shaping consumer image and consumer relations that have a direct impact on purchasing decisions has increased noticeably. This is linked to the general improvement of well-being and quality of life of the population in economically developed countries as well as to the processes directly related to the current market conditions (for example, to the digitization process). The manufacturers can no longer confine themselves to a declaration of commitment to the basic humanistic values; they should become full participants in contemporary social processes and state their position and attitude to the hottest social topics. Goods and services in economically developed countries are no longer a means to meet the immediate needs, so they have turned into the consumer's expression of his public position. Since the process of redesigning the concept of social responsibility and the marketing communications systems already been launched but in the initial stage, the technical implementation of these changes is not yet evident. Rather, at this point in time, there is a broad area of concern that requires appropriate solutions.

References

1. N. Capon, R. Capon, J. Hulbert, *Managing Marketing in the 21st Century* (2009)
2. Human Development Report Office (2019), hdr.undp.org
3. K. Jerath, M. Brown, How the marketing technology landscape will transform in the new year (2019), adweek.com
4. A. Aribag, T. Otter, D. Zantedeschi, G. Allenby, T. Bentley, D. Curry, M. Dotson, T. Henderson, E. Honka, R. Kohli, K. Jedidi, S. Seiler, X. S.Wang, *Customer Needs and Solutions*, 1, 82 (2018)
5. J. Kagan, *Digital Marketing: Strategy & Tactics* (2018)
6. M.V. Andriianova, Innovations and investment, 11, 103 (2019)
7. K. Jerath, P. Fader, B. G. S. Hardie, European Journal of Operational Research, 1, 340 (2016)
8. D. Lerman, R.J. Morais, D. Luna, *The Language of Branding: Theory, Strategies, and Tactics* (2017)
9. K. Wilcox, J. Laran, S. Sen, *Does Imitation Benefit the Imitated Brand? The Effects of Target Ambiguity and Processing Mindset on Judgment* (2018)
10. V. Swaminathan, A. Sorescu, J. E.M. Steenkamp, T. Clayton, G. O'Guinn, Bernd Schmitt, Journal of Marketing, 2, 24 (2020)

The Analysis of Mining Companies' Corporate Sustainability Assessment Methods (International and Russian Approaches)

Oksana Marinina[1,*], *Tatiana* Ponomarenko[1], *Marina* Nevskaya[1], and *Punsalmaa* Ochirbat[2]

[1]Saint Petersburg Mining University, Saint Petersburg, Russia
[2]Mongolian State University of Sciences and technology, Mongolia

Abstract. The study presents the classification and analysis of corporate sustainability assessment methods. We identified the strengths and weaknesses of the methodological approach to determining the resulting indicators and the possibility of applying these methods for administrative objectives. The results of the study prove that modern international and Russian methods for assessing corporate sustainability can be classified according to the goal, the choice of key indicators, the way of collecting information and industry specifics

1 Introduction

By various estimates, the world's practice introduces about 150-200 ratings and indices of corporate sustainability, which differ both in terms of their thematic scope and in terms of assessment criteria and indicator composition.

An analysis of scientific publications and official materials of organizations that provide methodological guidance for corporate sustainability (CS) and related areas (for example, corporate social responsibility) showed that the methods differ in [1]:

– the number, composition and weights of assessment indicators;

– the degree of aggregating indicators;

– the methods of calculating the resulting value;

– the choice of information sources for calculating indicators.

The main goal of the work is to analyze corporate sustainability assessment methods for the purpose of their further application in mining companies. By means of content analysis, the study identifies the strengths and weaknesses of the methodological approach to determining the resulting indicators and the possibility of applying the methods for managing mining companies' CS.

The main objectives of the study are:

– To identify the research area by collecting data on the most common methodologies used by industrial companies or rating agencies.

– To evaluate the selected methodologies according to their goals, objectives, the sequence of collecting information and disclosure of the approach to assessment, a chech-

* Corresponding author: Marinina_OA@pers.spmi.ru

out system, verification and the possibility of including the specifics of mining companies in the resulting indicators.

2 Methodology

The methodology of the research is based on the principles of an integrated approach to the assessment and analysis of Russian and international methods for assessing CS. It represents the sequence of collecting, processing and verifying the results (Fig. 1).

Data Collection	**Data Analysis**	**Credibility of Research**
A set of Russian and international CS methods	✓ Analysis of information on the application of the methodologies; ✓ Selection of the most significant methodologies for assessment and analysis; ✓ Qualitative analysis of their strengths and weaknesses	The oppositional method: ✓ regulatory documents, standards in CS; ✓ reports of Russian and foreign agencies (Bloomberg, RSPP, etc.); ✓ scientific articles on the acceptance of modern methods and tools of analysis and assessment in the field of CS

Fig. 1. Research Design

For analysis purposes the following methodologies were selected:

– International methods for assessing CS and social performance
– Methods for assessing SC and social performance with the industry specifics
– Methods and approaches for assessing social performance
– Russian rating methods for assessing CS and social performance
– Individual researchers' methods of assessing companies' performance.

3 Result and discussion

By CS assessment methods companies improve their CS results choosing effective ways to achieve them. This requires not only the transparency of the assessment methodology, convenience of comparing and analyzing information and simplified integration of CS indicators into the process of making investment decisions, but also linking the ratings to global requirements, taking into account the industry specifics, identifying CS factors in the formation of value chains and supply chains. Companies use CS ratings to make decisions and internal assessments, to identify trends and to support stakeholder communications.

It should be noted that, despite the controversial nature of the concept of corporate sustainability, a number of documents have been developed that regulate the CS requirements and serve as guidelines. The official methods are universal; therefore, they are applied to all companies without focusing on the industry sector. But methods developed in scientific studies are not officially recognized and are not applied, although they often consider industry specifications and show progress in the scientific understanding of CS [1].

Table 1. Classification of CS methods, ratings, indices [2-9]

Kinds of Methodologies	Methods
Russian methodologies (ratings) for assessing CS and social performance	Basic indicators of RSPP (Methodology of the Russian Union of Industrialists and Entrepreneurs) RSPP indices (Interfax-Era methodology) CCI Social Reporting Standard (Chamber of Commerce and Industry) Fundamental efficiency ratings (compiled by the environmental and energy rating agency Interfax-ERA) Methodology for assessing energy efficiency class from E to A ++ (RERA) (RUSSIAN ENERGY RATING AGENCY) Russian Regional Network for Integrated Reporting (RRS) Employer Attractiveness Rating NRA (National Rating Agency)
International methodologies for assessing CS and social performance	Dow Jones Sustainability Index (методология RobecoSam). FTSE4GOOD Global 100 Index methodology. Global Reporting Initiative (GRI) methodology. Bloomberg ESG Index methology. DEA (Data Envelopment Analysis) KLD (Kinder, Lydenberg, and Domini) Econometric Impact Index (economic benefit index) proposed by Smith O`Brien. Carbon Disclosure Project (CDP) (carbon reporting and performance leaders: Carbon Disclosure Leadership Index, CDLI; Carbon Performance Leadership Index, CPLI)
Methodologies for assessing CS and social performance with industry specifics	Oil and Gas Industry Guidance on Voluntary Sustainability Reporting. IPIECA/API (in the field of oil, gas production) Rating agency Tomorrow's Value Rating (TVR) (oil, gas) Equitable Origin has released the EO100 Standard (in the field of oil, gas) Rating of openness of oil and gas companies in the field of environmental responsibility (organized by KREON Group) Rating of the openness of mining and metallurgical companies in Russia in the field of environmental responsibility
Methodologies and approaches for assessing social performance:	SIA (Social Impact Assessment), SRA (Social Return Assessment), SCBA (Social Costs-Benefit Analysis), SVA (Stakeholder Value Added) Total Impact Measurement and Management (PWC - developer, approach to assessing the impact of business on the economy, society and the environment, as well as the tax contribution of the company) Coalition for Environmentally Responsible Economies (CERES) Institute for Social and Ethical Responsibility (ISEA)
Social Reporting Standards	AA1000 KCO Institute of Social and Ethical Accountability SA8000 Social Accountability International Social Responsibility Guide ISO 26000
Methodologies of Russian researchers for assessing performance (more than 100)	Skliar E., Zverkovich I. [10]; Penc I. [11], Saprikina O. [12]; D. A. Endovitsky, Borzakov [13], Ponomarenko T.V., Marinina O.A. [14] etc.

CS assessment methods can be classified according to the following criteria [15-19]:
– the effectiveness of CS and ratings of companies in terms of sustainable development;
– target orientation: environmental, social or complex indicators of CS assessment;
– the key indicators or the main index;
– the method of collecting information: statistical methods and surveys;
– the inclusion of industry specification.

Studies on the acceptance of modern methods and tools for CS analysis and assessment, as well as the assessment of statistical data on the quality of ratings and methodologies, are reflected in the study Sustain Ability [20]. CS ratings of RobecoSAM, MSCI, CDP, Sustainalytics are considered the most useful, convenient and of the highest quality according to a survey of North American (35%), European (39%), Asian (12%) and other groups (13%) of experts in 2018 (Fig . 2.). Despite a significant increase in the number of ratings over the past decade, they remain the leaders. Many companies prefer their assessment and control tools [20].

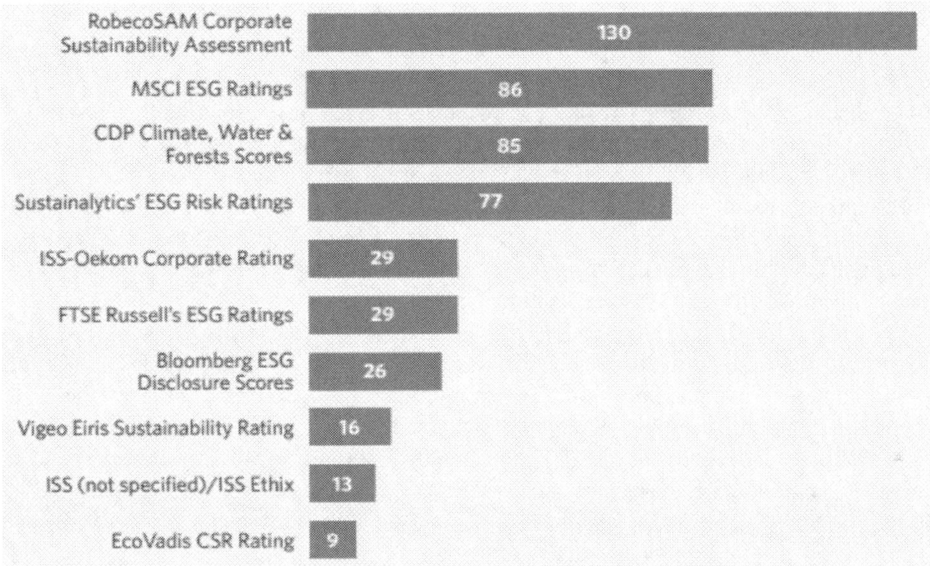

Fig. 2. Rating of CS methodologies by the number of companies using them [20]

The review and analysis of the most well-known methods for assessing CSR and CS showed that, despite their significant number and variety, the methods can be grouped according to the following characteristics.

1. Research area: the area of companies' activities the methods are aimed at. They involve not only methods which are focused on assessing environmental sustainability and the social component of SD in the CSR system and on giving a comprehensive (socio-ecological) assessment of CSR, but also integrated methods for assessing CSD and analyzing the environmental, social and economic components.

2. Goals: monitoring the situation in the research areas; assessing the social and environmental consequences of project implementation or the activities of companies; constructing company ratings (comparative assessment).

3. Evaluation criteria and indicators and ways of their construction:
– quantitative indicators (for example emissions when assessing environmental impact);
– cost indicators of companies' performance;
– quantitative indicators of companies' areas of activity;
– qualitative indicators for assessing social and environmental effects;

– aggregate indicators (indices);
– point-rating indicators.
4. Information sources for assessing CS:
– secondary sources (open source);
– primary sources (special surveys of companies);
– primary sources (special surveys of local communities).
5. Express assessment and detailed assessment of CS. Express assessment forms a general idea of the CSR by the company's past activities and non-financial reports and serves as an information base for further analysis, identification of potential risks and making justified decisions. The list of indicators for express assessment reflects the social, economic and environmental significance of the company.

4 Conclusions

CS assessment methods vary in the number and composition of indicators, the degree of aggregation of indicators and the ways of calculating the resulting value, weights of indicators, and the degree of the company's progress in CS. Official methodologies are universal and can be applied to all companies without considering their specifics. The methods of individual researchers are rarely used in practice and are not officially recognized, although many of them consider industry specifications.

The characteristic feature of CS assessment methods is that the scheme for choosing criteria, weights and rules for the calculation is not always unknown; the consistency and reliability of the procedures is not confirmed by the disclosure of the approach to collecting relevant information and of the verification system.

Most of the industry methodologies are aimed at solving a narrow problem. For mining companies, these are the methods of environmental or social licensing practices. The rating of environmental responsibility of Russian mining companies is based on the criteria formulated in the environmental requirements for the sector, the sectoral strategy of the World Bank for Development (EBRD), the IRMA Responsible Mining Standard, the GRI Sustainability Reporting Guidelines, etc.

Acknowledgements

This research was funded by RFBR and MCESSM, grant number № **19-510-44013\19.**

Author Contributions

Conceptualization, methodology, writing—original draft preparation - O. Marinina; validation, resources, visualization – T. Ponomarenko; formal analysis, investigation – M. Nevskaya, writing—review and editing – P. Ochirbat. All authors have read and agreed to the published version of the manuscript

References

1. Complex of indices of corporate social responsibility and reporting (2015) https://media.rspp.ru
2. Rate the Raters 2020: Investor Survey and Interview Results (2020) https://sustainability.com
3. C. Wong, A. Brackley, E. Petroy. Expert Views on ESG Ratings February (2019)

4. ESG factors in investment (2019) https://www.pwc.ru
5. DJSI: Corporate Sustainability Assessment (2012) https://www.spglobal.com
6. Corporate Sustainability Assessment & Corporate Public Reporting Annelies Poolman (2020) https://www.spglobal.com
7. RepRisk provides ESG intelligence for Dow Jones Sustainability Indices (2014) https://www.reprisk.com
8. Leadership GRI SRS (2013) https://www.globalreporting.org
9. International Integrated Reporting Council ('IIRC') (2013) https://integratedreporting.org
10. E. N. Sklyar, Management in Russia and abroad, **2** (2008)
11. I.Sh. Pence, Nedra Communications LTD, **1** (2008)
12. O.A. Saprykina, Transport business in Russia, **5** (2012)
13. D. A. Endovitsky, L. M. Nikitina, D.V. Borzakov, Economic analysis: theory and practice, **8** (2014)
14. T.V. Ponomarenko, O.A. Marinina, Journal of Applied Economic Sciences, **12(6)** (2017)
15. A.O. Nedosekin, E.I. Rejshahrit, A.N. Kozlovskiy, Journal of Mining Institute, **237** (2019)
16. A. Tsvetkova, E. Katysheva, SGEM, **19**, 573 (2019)
17. V.S. Litvinenko, P.S. Tsvetkov, K.V. Molodtsov, Eurasian Mining (2020)
18. A. Cherepovitsyn, A. Ilinova, Journal of Ecological Engineering, **17**, 2 (2016)
19. N. Pashkevich, T. Tarabarinova, Vide, Tehnologija, Resursi - Environment, Technology, Resources, **1** (2017)
20. Rate the Raters 2019 Expert Views on ESG Ratings (2019) https://sustainability.com

Implementation of CSR in the organizations'strategies: experience of Russia and Belarus, directions of methods and tools development of CSR in the conditions of digitalization of economies

Galina Yaheva[1,*], *Andrey* Plakhin[2], *Yulia* Vailunova[1], and *Maria* Selezneva[2]

[1]Vitebsk State Technological University, 210035 Vitebsk, Republic of Belarus
[2]Ural State University of Economics, 620144 Ekaterinburg, Russia

Abstract. The purpose of the article is to analyze the experience of introducing CSR in the Union State of Russia and Belarus and to develop directions for the development of CSR methods and tools in the context of the digitalization of the economies of Russia and Belarus.The article systematizes methods for assessing CSR at the macro and micro levels: compliance with international CSR standards; involvement in the UN Global Compact; indices for assessing the effectiveness of companies in the field of CSR; ratings and indices for assessing the effectiveness of CSR, developed by public and state organizations.The analysis of the experience of CSR implementation in the Union State of Russia and the Republic of Belarus is carried out. Identified barriers to the development of CSR in Russia and the Republic of Belarus. The directions of development of CSR methods and tools in the context of digitalization of the economies of Russia and Belarus are studied. Scientific significance lies in the development of CSR methods and tools in the context of digitalization of economies. The study used modern scientific literature on the digital ecosystem and CSR, statistical and analytical materials found on thematic Internet sites and in print media. The practical significance of the results lies in the possibility of their use by state and regional authorities, business in CSR strategies, as an integral part of the strategy for sustainable development of the economy, for making decisions aimed at increasing the stability and competitiveness of the economies of the Union State.

1 Introduction

The establishment of the institution of corporate social responsibility (CSR) took place in the 70s of the last century, when leading companies in the West, trying to protect their reputation from anti-corporate-minded social groups, began to declare their responsibility to society. Currently CSR helps to improve the company's image in the eyes of consumers, partners, and society as a whole. Thus, according to the McKinsey report, 78% of the

* Corresponding author: gala-ya@list.ru

experts they interviewed, say that CSR activities improve the company's image. According to a CONE study, 53% of shoppers say a company's CSR is an important factor when choosing a product. In a Deloitte study, 38% of CEOs claim that CSR increases employee loyalty [1].

Thus, CSR contributes to sustainable development, including the health and well-being of society; takes into account the interests of various stakeholders; complies with applicable law and is consistent with international norms of behavior; integrated into the activities of the entire organization and applied in its relationships.

An organization's overarching goal of social responsibility should be seen as contributing to sustainable development, the goal of which is to achieve the sustainability of society as a whole. It is with the implementation of the CSR concept that the goals of sustainable development can be achieved in the new realities.

The peculiarity of the external environment is the active development of information and computer technologies and their implementation in various fields of activity. The COVID-19 epidemic, the difficult economic situation (falling GDP in all countries) are reflected in the CSR projects of organizations, including Russia and the Republic of Belarus. This circumstance creates an opportunity to use digitalization tools to stimulate CSR strategies, which determines the relevance of the topic of this study.

The purpose of the article is to analyze the experience of introducing CSR in the Union State of Russia and Belarus and to develop directions for the development of CSR methods and tools in the context of the digitalization of the economies of Russia and Belarus.

Based on the goal, the following tasks have been identified:

- Systematization of methods for assessing CSR at the macro and micro levels;
- Analysis of CSR implementation experience in Russia and Belarus and identification of barriers to CSR development;
- Study of directions for the development of CSR methods and tools in the context of digitalization of the economies of Russia and Belarus.

2 Materials and Methods

The theoretical and methodological basis of the study is the scientific works of foreign researchers that had a significant impact on the formation of the foundations of social management: F. Taylor (school of scientific management), the ideas of M. Weber, A. Fayol (school of administrative management), E. Mayo (school of human relations), the behavioral school (A. Maslow, K. Aldarfer, F. Herzberg, D. McGregor), the theory of loyalty and fair treatment of an employee by R. Husman and J. Hetfield, consideration of CSR as an element of corporate culture (ideas of E. Shane), the theory of investment in human capital by G. Becker and T. Schultz. The ideological origins of the modern concept of CSR in the theory of G. Bowen, M. Friedman (the theory of corporate egoism), A. Kerrol (the pyramid of CSR levels), E. Freeman (the theory of stakeholders) are also considered.

Many scientific works are devoted to the study of the necessary degree of social responsibility of business, where the problems of responsible behavior of enterprises have been studied for a long time. Thus, the works of A. Burleigh, G. Bowen, A. Caroll, J. McGuire, G. Means, M. Friedman, E. Freeman and others played an important role in the development of the concept of CSR. In the scientific literature of researchers from the Union State of Russia and Belarus, the issues of corporate social responsibility are covered to a lesser extent, however, there are works of such researchers as Yu. Blagov, I. Belyaeva, M. A. Eskander, G. Tulchinsky, E. Wittenberg, A. Kostin, S. Peregudov, Y. Petrunin, V. Purlik, I. Semenenko. A special role is also played by the research of international organizations in the development of new approaches to CSR and the formation of standards for social reporting of business, including the UN, UNESKO, the World Bank and others.

The study used modern scientific domestic and foreign literature on the digital ecosystem and CSR, statistical and analytical materials found on thematic Internet sites and in print media. The main methods of study are empirical and theoretical: observation, comparison, graphical, statistical and econometric analysis.

3 Results and Discussion

3.1 CSR assessment at macro and micro levels

CSR of business is a concept that reflects the voluntary contribution of business to the development of society in the social, economic, environmental spheres, aimed at improving the standard of living of society and protecting the environment. The concept of stakeholders forms the basis of CSR, since the main goal of any business is to satisfy the interests of a wide range of interested groups. According to the studies [2-6], the main priorities of CSR organizations in Russia and Belarus are usually stated as follows:

- Production of high-quality and competitive products;
- Compliance with standards and laws;
- Payment of taxes;
- Introduction of mechanisms against environmental pollution;
- Implementation of measures to take care of the employees of the organization.

Thus, the CSR strategy becomes part of the sustainable development strategy of the entire economic system. It is quite obvious that in order for the system to develop steadily, it is necessary to ensure a balance of interests of different interested groupsor stakeholders. If it is a macroeconomic system, then not only the state, but also international organizations, business, public associations, the population and other stakeholders should be interested and act in its sustainable development. If this is a microeconomic system, then, consequently, it is important to take into account the interests of not only company shareholders, but also investors, representatives of state authorities, personnel, suppliers and consumers of products, local communities and non-profit organizations (NPOs).

The analysis and systematization of methods and tools for assessing CSR available in science and practice made it possible to identify the following methods for measuring involvement in CSR.

1. Method – Compliance with recognized international CSR standards (table 1), which include the following [7]:

- SA8000 (Social Accountability 8000) - a standard for assessing the social aspects of management systems. In Russia, only a few enterprises have been certified for compliance with SA800 (those who are part of the supply chain of European companies that require such a certificate);
- AA1000SES. The leading format for non-financial reporting on CSR is the AA1000 Stakeholders Engagement Standard, created by the Institute of Social and Ethical Accountability. The Russian version of the standard has been prepared by the Institute for Urban Economics Foundation, taking into account broad world practice. At present, the AA1000SES standard in Russia has been successfully implemented by more than 30 leading companies;

Table 1. Interrelation of international, Russian and Belarusian CSR standards.

<table>
<tr><th>International Standards</th><th>Russian Standard</th><th>Belarusian Standard</th></tr>
<tr><td>ISO 26000</td><td rowspan="2">IC CSR-08260008000</td><td rowspan="4">Most of the social reports are compiled according to international standards AA1000 and GRI, in short form they contain two sections: "internal corporate responsibility" and "social investments"</td></tr>
<tr><td>SA8000</td></tr>
<tr><td>GRI</td><td>Basic indicators of the Russian Union of Industrialists and Entrepreneurs</td></tr>
<tr><td>AA1000SES</td><td>None</td></tr>
<tr><td>None</td><td>Social Reporting Standard of the Chamber of Commerce and Industry</td><td>None</td></tr>
</table>

Source: compiled by the authors based on [1, 3, 7].

- GRI – the Global Reporting Initiative standard is based on the concept of sustainable business development, in which companies need to balance economic, environmental and social activities;

- ISO 26000 "Guidelines for social responsibility" - a standard in the development of which experts from Russia also took part, provides a guide on the principles, main themes and problems underlying social responsibility, ways of integrating socially responsible behavior into strategies, practices and organization processes.

In the Republic of Belarus and Russia, CSR reporting is not mandatory (only about 16% of companies keep reporting [1], therefore it is difficult to carry out a statistical analysis of the share of enterprises and organizations certified according to CSR standards).

2. Method – Involvement in the UN Global Compact, which was founded in 2000 and is one of the largest CSR initiatives in the world. Today the UN Global Compact unites more than 10 thousand companies and about 3 thousand non-profit organizations from more than 160 countries of the world [8].

Since 2006, the Republic of Belarus has started the processes of joining the UN Global Compact by those companies that seek to improve their public image, and by now more than 70 Belarusian organizations have joined the UN Global Compact [8].

The Russian network of the UN Global Compact was created in 2008 and currently includes 74 organizations [9]. At the same time, it is quite obvious that the assessment of the degree of organizations' involvement in CSR can be assessed by the share of organizations that are members of this Global Compact. Although these statistics do not provide an assessment of the effectiveness and efficiency of organizations in the field of CSR.

3. Method – Indexes for assessing the effectiveness of companies in the field of CSR. Currently, scientists offer various approaches to calculating certain indices in the field of CSR [10-13]. However, none of the considered indices aggregated the information in such a way that it was possible to determine the level of sustainable development not only of individual companies, but also of the country in which they operate. In addition, the calculation of the indices themselves is rather difficult due to the complexity and lack of mandatory statistical reporting.

4. Method – Ratings and indices for assessing the effectiveness of CSR, developed by public and state organizations.

The following ratings are used in the Republic of Belarus:

- A socially responsible brand. This indicator was developed within the framework of the USAID project in 2005 and is used in the "Brand of the Year" competition [2];

- The Goodness Index. Developed by the Belarusian Foundation of Goodness, which is the official representative of the UN Global Compact in Belarus since 2017 [14]. As part of the calculation of this index, business is evaluated in the following areas:

- Labor practices and human rights;
- Good business practices;
- Consumer welfare and community participation;
- Environment.

The Goodness Index consists of a total of 134 indicators characterizing the four areas of corporate sustainability mentioned above. The result of the Index for each of the directions can be in the range from 0 to 3 points, and the overall indicator of the Index is the sum of indicators in the directions and can be in the range from 0 to 12 [14].

In Russia, to assess the activities of organizations in the field of CSR, are used indices on sustainable development of the Russian Union of Industrialists and Entrepreneurs, such as the Responsibility and Openness Index – the information disclosure index (to what extent the volume and quality of information in public corporate reporting of the largest Russian companies reflect the impact of companies on the social and the natural environment), and the Sustainable Development Vector Index (2015), which allows to assess what the environmental "price" of production is and how much social benefits are provided [3]. The Sustainable Development Vector Index consists of indicators such as wages and social support of staff, staff turnover, injury rate, staff training, investments in communities, energy consumption and energy efficiency, impact on water sources, air emissions, including greenhouse gas emissions, waste generation and handling [3].

The analysis of methods and tools for assessing CSR available in science and practice allowed us to conclude that there is no generally accepted method for assessing the involvement of organizations in CSR and assessing the effectiveness of organizations in the field of CSR. Therefore, to analyze the level of CSR of business in the Republic of Belarus and Russia, the results of studies with methods and tools comparable in the countries of the Union State will be used.

3.2 Analysis of the experience of CSR implementation in the Union State of Russia and Belarus and identification of barriers to CSR development.

The methodological basis of the study was the statement on the differentiation of CSR strategy at the macro and micro levels.

The quantitative assessment of CSR at the macro level is proposed by the consulting company "CIVITTA" [1], and includes 3 stages:

1. Determination of key indicators and degree of significance: ISO 14001 (0.25), ISO 45001 (0.25), SA 8000 (0.5), Global Reporting Initiative (1), Global Compact (1), Carbon Disclosure Project (1);
2. Collecting information on the number of organizations certified and supporting initiatives, separately by country;
3. Calculation of the index of each country according to the formula 1 [1].

$$CSR\ index\ of\ country\ A = \sum_{i=1}^{6} \frac{\frac{number\ of\ companies\ in\ indicator\ i\ of\ country\ A}{total\ number\ of\ companies\ in\ indicator\ i}}{\frac{GDP\ at\ purchasing\ power\ parity\ in\ country\ A}{Worldwide\ GDP\ at\ purchasing\ power\ parity}} \quad (1)$$

Where: GDP – Gross domestic product

The results of the assessment carried out by the consulting company CIVITTA are presented in Figure 1.

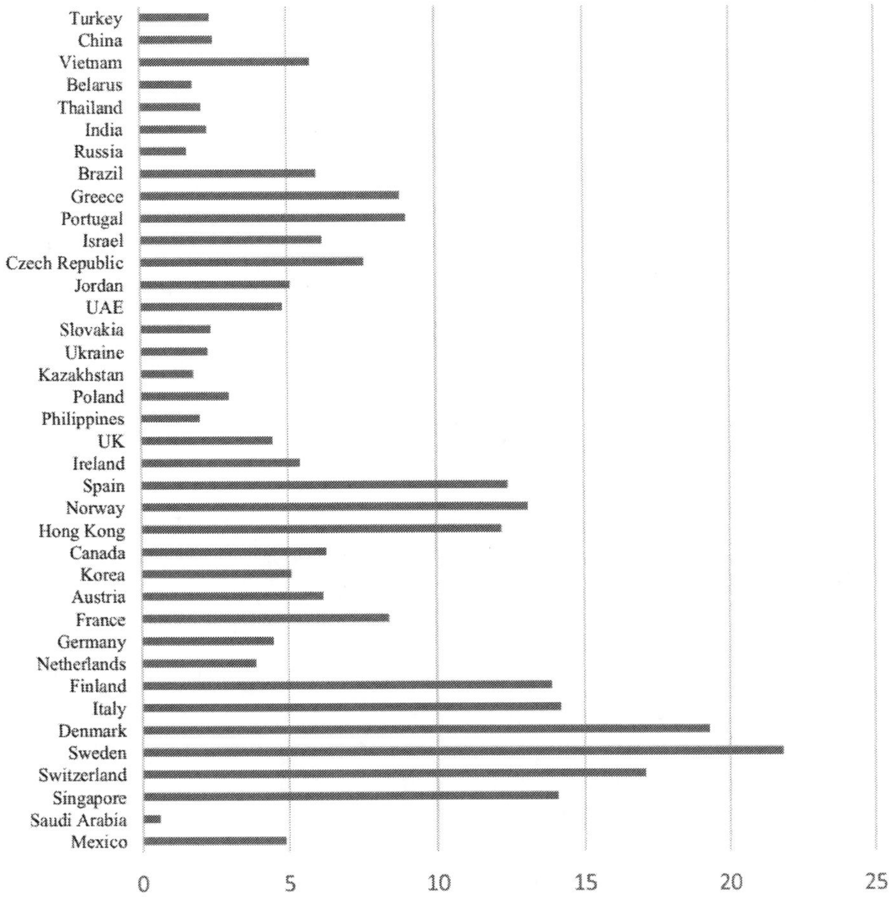

Fig. 1. CSR development index [1].

As can be seen in the graph, in general, the development of social responsibility of business in Russia and the Republic of Belarus is rather slow and lags significantly behind the developed best practices of Western countries. CSR leaders are Sweden, Finland, Great Britain. The Republic of Belarus has an estimate of 1.8, Russia - 1.6. The deviation from the national average CSR level is 14.9.

To study the factors influencing CSR, an analysis of the relationship between GDP and the CSR index was carried out. The results are shown in Figure 2.

The study has shown that there is a direct relationship between the CSR development level index and GDP per capita. Therefore, it is possible to formulate a hypothesis that the development of CSR is required for GDP growth.

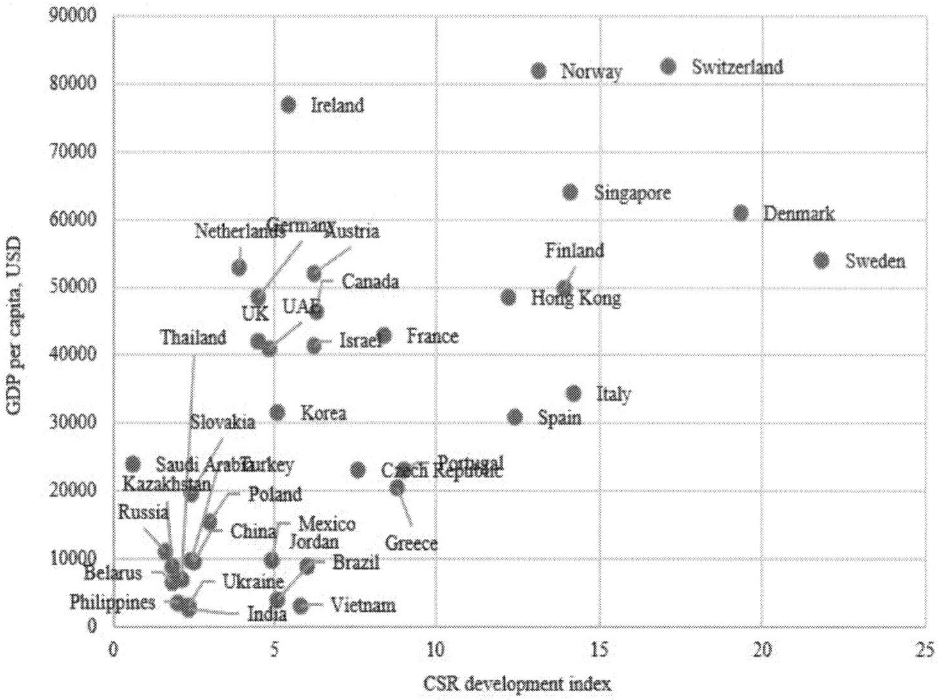

Fig. 2. Dependence of GDP and CSR index [1].

Further, at the micro level, the assessment of CSR efficiency was carried out on the basis of the Goodness Index criterion (Republic of Belarus) According to the results of the 2019 assessment, the highest Goodness Index in the Republic of Belarus was assigned to the following companies:

- large business (more than 500 employees are employed) - assessment of 15 Belarusian enterprises, of which IBA Group 11.8; JSC ASB "Belarusbank" 11.6; Priorbank OJSC 11.2; OJSC “BMZ - managing company of BMK holding” 10.8; OJSC "Belinvestbank" 10.0 and others); the average score of the Good Index for the large business category is 9.5 out of 12 maximum possible (advanced level according to the assessment methodology);

- medium-sized business (100 to 500 employees are employed) - 12 companies were assessed, including the ProLiv group of companies 11.2, SoftTeko 9.9, Qulix Systems CJSC 9.5, Intetiks Bel JLLC 8.8, Limited Liability Company HERE BY MEDIA "8.1 and others; the average score of the Goodness Index for the medium-sized business category is 8.1 (advanced level according to the assessment method) out of 12 maximum possible;

- small business (up to 100 employees are employed) - 14 companies were assessed, of which OOO SAP CIS 9.9, Closed Joint Stock Company HERMITA 9.3, OOO KREITEX 8.0, OOO CSBT SATIO 7.0, OOO BelVneshReklama 7.0; the average score of the Good Index for the small business category is 7.2 out of 12 maximum possible (active level according to the assessment method) [14].

It can be concluded that the assessment of organizations in the Republic of Belarus (41 organizations) using the Goodness Index - assessment of corporate sustainability and business responsibility is quite high.

Another method for assessing the effectiveness of CSR at the micro level is used in Russia based on the Sustainable Development Vector Index. This index is based on the analysis of quantitative information presented in dynamics for a number of indicators that reflect the economic, social and environmental dimensions of sustainable development and CSR (Figure 3).

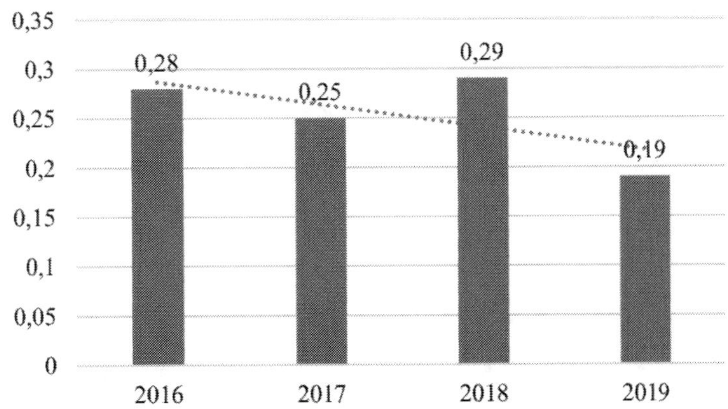

Fig. 3. Average value of the Sustainable Development Vector index in Russia [3].

In accordance with the figure 3, one can observe a negative fact of a decrease in the Sustainable Development Vector Index for the period 2016-2019, by 0.09.

The leading companies in the Sustainable Development Vector Index – 2019 were 30 companies that showed a positive balance of performance over 3 years (individual index above 0): ALROSA, Gazprom, EVRAZ, EuroChem, Inter RAO, LUKOIL, Metalloinvest, MMK, MTS, NLMK, NOVATEK, Norilsk Nickel, OMK, Polyus, Russian Railways, Rosatom, Rosneft, Rosseti, Rostelecom, RUSAL, Sakhalin Energy, Sberbank, Severstal, SIBUR, AFK Sistema, SUEK, Tatneft, TMK, PhosAgro, FGC UES [3].

An important aspect of the analysis from a practical point of view is the analysis of the areas of CSR activities of companies. The results of the study by the consulting company CIVITTA showed that the priority area of CSR in Belarus is to support vulnerable categories of citizens (68 companies), science and education (46 companies), culture (45 companies) and sports (43 companies) [1].

The main objects of activity of Russian organizations in the field of social responsibility are employees of the organization and their families, attention is paid to the problems of ecology, culture and sports, events that are held within the framework of corporate volunteering.

Based on the analysis of the institutional environment, we can formulate the following problems of implementing CSR strategies in Russia and Belarus [15]:

1. Imperfection of legislation. At present, the legal norms and rules in the field of CSR are not reflected in either the Civil Code or the Tax Code, and there is no legislation on charity in Russia or Belarus.

2. Absence in the current legislation of the countries of the Union State of a clear system for granting tax benefits. Benefits apply only to some types of activities, but their mechanism is not clearly spelled out, which leads to the rejection of charity or the use of "shadow" schemes. Until recently, there was a double taxation of philanthropy.

3. Lack of targeted government policy in the area of corporate social responsibility. Lack of public and media attention to CSR.

4. Weak, uncoordinated and unsystematic support by state structures for the activities of companies in the field of CSR.

5. Lack of understanding by the top management of communication companies of CSR with the possibility of improving market positioning and the role of a positive image.

6. Underestimation by management of the economic benefits of CSR implementation in practice.

7. The conviction of management that CSR diverts resources from solving important tasks of modernizing the economy.

8. Low level of intersectoral partnership (state-society-business).

9. Lack of mechanisms to educate stakeholders about the benefits of CSR.

At the same time, the external environment is dynamically developing and creates prerequisites for the development of CSR in the countries of the Union State. An analysis of the external environment based on a review of various studies revealed the following trends in the development of the world economy and society, which influence the implementation of a CSR strategy:

1. Development of ICT and digitalization of society, leading to increasing transparency and information accessibility for various stakeholders;
2. Development of network forms of relationships between stakeholders;
3. "Democratization" of knowledge thanks to the Internet.

A significant factor influencing the development of the economy and society as a whole is the activation and penetration of ICT into all spheres of activity and all business processes. The digitalization of the economy provides for the transformation of business as a result of the introduction of digital technologies. In fact, digitalization leads to the development of new CSR methods and tools that can be used in CSR business strategies at the macro and micro levels.

3.3 Directions for the development of CSR methods and tools in the context of digitalization of the economies of Russia and Belarus

Based on an analysis of the experience of introducing CSR in Russia and Belarus, as well as barriers and prerequisites, a number of methods can be proposed at the micro and regional levels aimed at enhancing the process of introducing CSR into business activities in the context of digitalization.

1st direction - Development and distribution of digital communications between government and business for the development of CSR. This direction can be carried out using such tools as:

- Conducting webinars and events for the development of intersectoral social partnership with the participation of a wide range of stakeholders: government, business and society;
- Joint programs with local authorities for additional training and professional retraining of the released labor force - online universities, open universities.

2nd direction - Introduction of digital tools into public-private partnership (PPP) mechanisms [16], since PPP is one of the most common forms of interaction between business and the state in the framework of CSR. It is proposed to select participants on the basis of an electronic auction for the participation of private business in projects of such a partnership, which will speed up the procedure and make it more transparent.

3rddirection - Development of interaction between society and business based on digital charity. Charity is especially relevant in the context of the coronavirus pandemic, when the support of medicine and patients is so necessary. The algorithm for creating a digital charity business model can include the following steps:

- Creation of a unified information platform;
- Opening of a charitable foundation;
- Business organization (social support) - creation or purchase of a digital platform, offering users a service, a product (consultations, dispensing drugs, a lottery, a card, etc.);
- Choice of payment system;
- Creating an account on a social network;
- Launching a widget, messages;
- Advertising placement;
- Provision of services, transfer of funds; transfer of benefits, etc .;

- Evaluation of results.

4th direction - Organization of volunteer work in the regions, forms of work with youth, interaction with non-profit and religious organizations on the basis of digital platforms for the implementation of CSR principles in the public life of the region. During the period of economic downturn and economic crisis, business initiatives move to the territorial and local levels. New approaches are needed to launch mechanisms for the work of regional partnerships (business - regional and local authorities, civil society organizations) and public-private partnerships in the social sphere and medicine. For example, the creation of such digital platforms - "Anti-Coronavirus", "Healthy Children", etc. All of the above will save jobs, create new and develop existing socially-oriented business models, increase production and sales, protect low-income people, and in general comply with CSR principles.

4 Conclusions

Modern Russia and Belarus are characterized by the gradual development of the sphere of corporate governance, an integral part of which is the social responsibility of business to external and internal stakeholders. Its main trends are: the leading role of the corporate sector in the development of CSR; transition from free reporting forms to reports oriented to international standards; enhancing the role of sustainable development as the basis for CSR.

The downturn in economies as a result of the COVID-19 pandemic is forcing businesses to look for new approaches and methods of "survival". At the same time, the external environment provides new opportunities for development - these are digitalization mechanisms. Use of new directions of CSR strategies, such as the development and dissemination of digital communications between government and business for the development of CSR; introduction of digital tools into public-private partnership mechanisms; development of a tool for interaction between society and business based on digital charity; the development in the regions of volunteer work, forms of work with youth, interaction with non-profit and religious organizations in the implementation of CSR principles based on digital platforms, makes it possible to predict the growth of the efficiency of organizations and economies of the Union State as a whole.

The practical significance of the results of the study lies in the possibility of their use by state and regional authorities, as well as by business in CSR strategies, as an integral part of the strategy for sustainable economic development, including for making decisions aimed at increasing the stability and competitiveness of the economies of the Union State.

References

1. CIVITTA BY conducted a study on the development of CSR companies in Belarus (2019), https://civitta.by
2. Professional competition Brand of the Year in the Republic of Belarus (2019), https://brendgoda.by
3. Indices for Sustainable Development of the Russian Union of Industrialists and Entrepreneurs - 2019: "Responsibility and Openness", "Vector of Sustainable Development", https://media.rspp.ru
4. S. B. Dokholian, *Corporate social responsibility of business in Russia and the United State,*. Bulletin of the S. Yu. Witte Moscow University. Series 1: Economics and Management, **8** (2014)

5. M.A. Chernousova, F.R. Eshugova, N.A. Asanova, Naukovedenie Internet magazine, **8** (2) (2016), http://naukovedenie.ru
6. E. A. Sorokina, *Trends of corporate social responsibility of business in Russia.* Bulletin of the Chelyabinsk State University Economic sciences, **9** (431) (2019)
7. L. S. Kiseleva, CSR standardization: Russian and international realities in the field of sustainable development, https://ria-stk.ru
8. National network of the UN Global Compact in Belarus, http://globalcompact.by
9. National network of the UN Global Compact in Russia, http://www.globalcompact.ru
10. A.A. Okorochkova, *Management problems: theory and practice. Measurable Macroeconomics: New Indices for Sustainable Development and Corporate Social Responsibility*. Public administration. Electronic bulletin **53** (2015)
11. A.A. Gullo, N.G. Filimonova *Analysis of methods for assessing the effectiveness of corporate social responsibility of enterprises,* Socio-economic and humanitarian journal of Krasnoyarsk State Agrarian University, **1** (2018)
12. A.P. Zhoidik, *Development of methods for assessing corporate social responsibility of Russian companies: Dis. Cand. Econ. Sc.*(2015)
13. D.V. Bobrov, *Development of methodological support of corporate social policy: Dis. Cand. Econ. Sc.* (2015)
14. Index of Goodness in Belarus, https://indexdobra.by
15. Corporate social responsibility in Russia, https://spravochnick.ru
16. G.A. Yasheva *Cluster concept of increasing the competitiveness of enterprises in the context of network cooperation and public-private partnership* (2010)

Corporate social responsibility in Russia

Elena Zavyalova*, and *Elena* Ostrovskaya

Moscow State Institute for International Relations (MGIMO University), Prospect Vernadskogo, 76, 119454 Moscow, Russia

Abstract. Nowadays companies try to strike a balance between maximizing their profit and being socially responsible. On the one hand, a lot of scientists and economists admit that the primary goal for any company has been and still is to maximize profits of shareholders. On the other hand, some other experts believe that a company nowadays should not concentrate only on pursing profits. Some other objectives are connected with interests of stakeholders, ethical and moral behavior of a company and others. Based on the analysis of theoretical approaches, the paper offers to unify different groups and their interests and goals. The authors during the research come to the conclusion that the development of CSR in Russia is directly connected with the country's peculiarities and at the same time all European best practices are taken into consideration and relied upon by Russian companies in terms of CSR.

1 Introduction

The idea that companies should be socially responsible for their activities appeared in the 20 century, but such ideas have become widespread and have been introduced into companies' everyday practice since the beginning of the 21 century. Large multinational companies admit that lives of millions of people all over the world depend on these companies' activities. Moreover, these companies have an impact on the economic and political climate of the countries where they operate.

Every time when there are debates about social responsibility of business, a question arises concerning the role and place of business in real social and economic conditions of a certain country. We may suppose that the level of involvement of any business in social problems mostly depend on the political and economic system in a certain country, and this system imposes typical norms of ethical behavior, especially in business. So, socialist background proves that the development of CSR is similar in former Soviet countries (Kazakhstan) and BRICS countries, such as China, i.e. CSR is developing fast in large corporations, which were founded during the Soviet period, and is developing at a much smaller pace in small and medium-sized business.

The authors prove that in Russia CSR is gaining ground. The majority of Russian scientists and experts in this field admit this fact. CSR programs are mainly introduced by industrial giants in accordance with the world practice, but have a number of peculiarities, connected with Soviet background of these companies.

*Corresponding author : e.zavyalova@inno.mgimo.ru

2 Materials and Methods

The presented analysis is based on the concept of the stakeholder theory. Using the methods of inductive and deductive approach, general-purpose approach, namely the comparative analysis: synthesis, generalization, classification, systematic and institutional approaches, as well as statistical analysis we will analyze both theoretical basis and practical implementations in Russia.

This theory implies that any business should not concentrate only on maximizing profits for its shareholders, but it should also serve interests of different stakeholders. So the idea runs as follows: aims of a corporation are connected with interests of other groups apart from shareholders.

3 Results and Discussions

In May 2018, a new executive order on national goals and strategic objectives through to 2024 was announced, and the primary goal is socially oriented growth. It is obvious that the huge volume of social entitlements from the budget is required, and the government cannot cope with this task alone, that is why there is a primary objective to attract business to help the authorities with social issues. Especially it is worth doing in such spheres that may enhance competitiveness of the business: education, healthcare, all types of internal CSR.

The modern situation has its roots deep in the Russian history. Social ideas have always been very popular. In the nineteenth century, for instance, representatives of the big business supported artists, musicians or founded hospitals or orphanages. They were known as "*maecenas*" and the Russian society still enjoys some of their deeds. The most famous are the Russian railways built by the Morozovs and Tretyakov Art Gallery. After the revolution in 1917 the whole policy was socially oriented, though it was more like the state requirement and not a voluntary company initiative. But now in modern Russia the attitude of the different social layers to corporate social responsibility is controversial.

In general, the term CSR cannot be applicable to those companies, which existed during the USSR period, since CSR is the occurrence of the market economy; moreover, it is the voluntary decision of a company that is pursuing its market goals. There was not such a term as CSR in the USSR.

However, the USSR really provided its citizens with social guarantees and in many cases people's welfare depended largely on the company where they used to work. There was such a term as "concern for demands of workers". Soviet enterprises were responsible for providing their workers with accommodation, food, healthcare, organizing their leisure time and rest. The enterprises owned houses where their workers used to live, hospitals, kids' summer camps, rest houses, cultural centers, etc. Such "benefits" were available both to workers of the enterprise, and to city dwellers of nearby territories.

According to the statistics, by 1980 32 million people had lived in apartments that were owned by enterprises, 30 million people had used healthcare, and 1.5 million kids had spent their vacations in summer camps that belonged to enterprises.

We should start analyzing the issue with the statement that there is no unified approach, in the Russian scientific or in the business society. The most common approach appeared in the 90-s. That time can be characterized as an initial capital accumulation period. It means that the main task of the business is seen only as getting maximum profit. According to this approach corporate social responsibility (CSR) is associated with the following pattern. An efficient company develops successfully constantly increasing the volume of its activity. It leads to the creation of new working places, which in turn leads to the increase in the social welfare. So, this approach to corporate social responsibility is similar to the classical concept of the "*corporate egoism*". This point of view is still popular, though it is quite different from the modern perception of CSR in the world leading economies, namely the EU countries or the USA.

However, today in the modern Russian society there is a clear shift to the "*corporate altruism*" theory. According to the theory companies are expected to finance socially important programs voluntary limiting their potential profits. The Russian society gradually starts to understand that corporate social responsibility is not just charity or vise a versa profit optimization (Aßländer&Curbach, 2017). Corporate social responsibility can be characterized as a mutually beneficial development strategy of the company and its surroundings. Any company that voluntary invests in the social sphere becomes a "good partner" that leads to the profit increase in the long run. If this attitude to CSR prevails, the Russian approach will be quite similar to the definition given by the European Commission "*the responsibility of enterprises for their impacts on society*". (European Commission)

This difficulty in finding the consensus in the definition in Russian reality can be mainly explained by the fact that the state, business and the Russian society have extremely different expectations of the socially oriented programs.

We shall start our analysis with the *Russian business*. Relatively young Russian business is accustomed to working in rather tough conditions. It means that business has to accumulate all resources to survive, not to mention the increase in volume and quality of production. It took more than 20 years for the Russian business to realize the necessity and "profitability" of CSR. The numerous surveys done by different public organizations [17, 13], show that businessmen see their CSR more as a responsibility first of all for their shareholders and employees. They agree to meet the requirements of the local community or to finance some socially or ecologically oriented programs on the state level only if there are some extra resources left. Another specific feature is that corporate social programs are more feasible when they are conducted by strategically oriented companies, mostly having tight connections with the state.

It is also important to mention that the country has been pursuing the liberal economic policy for more than twenty years. The classical understanding of the liberal policy logic implements that social responsibility of any company can be measured by tax payments and creation of new working places.

All this led to two major ways of understanding CSR by Russian business. CSR is divided in external and internal. The internal one is more popular with the modern Russian businessmen especially from the small and medium sized segment. It includes all forms of staff programs: from creation and maintenance of the social infrastructure to creation of the good family like psychological climate at the working place.

External CSR programs usually focus on the creation of a company's positive image. It is even more important when the company decides to enter foreign markets. Another frequent direction of the external CSR of Russian companies is the minimization of the tension with the municipal authorities and the local communities.

To make the picture of the Russian business mentality complete, we need to see the difference between small and big business attitude. If in Europe and other developed countries the small and medium-sized business is an active and significant player in the national economy, which participates in different economic activities, including CSR programs, then in Russia SME (small and medium-sized enterprises) account for only 12 per cent of the GDP. It means that even if they become interested in the development of the CSR programs and achieve some results it is not that visible compared to the big business activities.

So, it can be stated that CSR is the sphere for the Russian big business, and the situation is not likely to change in the near future. There are some reasons for this. First of all, only big business has additional financial resources that can be spent on socially oriented programs. Secondly, the big business is more involved in the state socially oriented projects. And the last reason is that the big business is much more concerned with the maintenance of the social (and political) stability and is ready to pay for it.

The Russian society has quite the opposite position.

By the waning years of the Soviet period, people were used to the fact that social guarantees provided by the government were an integral part of their life. So when the situation in the country changed dramatically and the market economy replaced the Soviet planned economy, and as a result huge social inequality and making everything commercial (that used to be provided free of charge and was guaranteed to any person), made people think that they were cheated and insulted.

The most popular point of view is the following. During the privatization period (in the 90-s) the business got extra huge assets, so now they are responsible for the poor population which still feels like being robbed or deprived due to the liberal reforms. Social policy of a company, which became not a responsibility, but a voluntary decision of the board of the company, is treated by the majority of people as money laundering.

In some cases, it can be seen even as a kind of public pressure on the company, especially if a company's history goes back to the soviet past. The owners of such companies have to pay more attention to the social programs, especially external ones.

There is another controversial point of view in the Russian society. It is quite frequent when the active philanthropic activity of a company, even when it has a clear socially oriented goal, is seen by ordinary people as a way of laundering money. It will take a lot of time and effort to change the attitude. Such efforts should ensure, first of all, higher transparency of CSR programs to increase mutual trust.

The tendency is even more evident in small cities, where the majority of the population works for one company (in Russia such cities are called "*monotowns*").

In Russia, there are nowadays at least 319 monotowns. These towns are built around one or several enterprises that are linked by a line of production. All the infrastructure of this town is aimed at providing continuous operation of such a city-forming enterprise in the monotown. During the privatization process the enterprise was not supposed to take care of city life. Still there was an open question connected with the future of people living in such a city. So the enterprise was forced to take care of its workers, infrastructure and ecology of the city. Such cities became real leaders of CSR in Russia. Successful development of towns where these enterprises operate is directly linked with the success of the enterprise. That is why there are even depressed regions (around 149). These monotowns and their city-forming enterprises need state support in order to preserve and improve social and economic welfare of citizens.

An excellent example of such a monotown in Russia is Magnitogorsk, which was founded around joint-stock company Magnitogorsk Iron and Steel Works (MISW or MMK). MMK is an example of a fast-growing business today; its shares have been traded on the London Stock Exchange since 2007.

As the management of MMK states CSR is the most important principle of its work. CSR of the company includes creating effective and safe working places, professional staff development, support of cultural programs and sport events, environmental protection, multi-level system of healthcare, helping to develop the regions where it operates. Building hospitals, schools, roads and houses, maintaining at a certain level cultural centers, rest houses and kids' summer camps, creating a number of events aimed at improving lives of citizens of Magnitogorsk is the achievement of MMK.

The efficiency of CSR depends not only on companies' and the society perception. One of the most important factors is *the state* attitude to the issue. In the ideal model, the role of the state should be limited to the creation of the necessary framework and encouragement of business to launch and support social projects (Aßländer&Curbach, 2017). Unfortunately, in the real Russian everyday life the government interference in the corporate social activity looks more like an ultimatum. It undermines the main principle of CSR - its voluntary nature. The state doesn't only push the business to CSR (which is good, since it gives quite strong impulse to business to launch some socially important programs), but more often than that the government also points out the main spheres for CSR (which is bad as the companies might have some other priorities). As a result state initiated CSR projects have limited efficiency and look more like a show off. Although some experts believe that, CSR should not always be voluntary decisions of corporations, but the initiatives in CSR introduced by the state will be more effective rather than voluntary activity of corporations (Brejnholt et al. 2019). For instance, such a scheme is used in China, where the government persuades companies to take part in state programs to eradicate poverty, improve educate, etc. (Hofman, Moon, & Wu, 2017).

On the other hand, both business and society agree on the point that the social development is the main concern of the state and the government is responsible for the public welfare. The state is the only power that can strategically estimate the volume and the directions of the necessary funding for the solution to the most important social concerns. That is why the center of the discussion about the role of the state in CSR is gradually shifting to the question of finding the right balance of interest and most appropriate forms of mutual cooperation of all parties concerned.

The question of the right balance of interest is one of the most acute. To make CSR programs more efficient business, the state and the society should understand their potential profits, costs and obligations. Additionally, it is important to remember that in real life it is impossible to divide the interests and costs of the parties.

The typical logical chain looks like the following. The main benefits for the business is the higher quality of the personnel. Internal social programs prevent the qualified workers from leaving the company, attract new experts and give new impulse for workers to increase the productivity without salary increase. It leads to the better image of the company, which makes it more interesting to potential investors. Simultaneously, external CSR programs contribute to the social stability at least on the local level. When the number of such companies reaches the certain level it leads to the higher social stability in general, and eventually results in higher investors' confidence in the country.

It is obvious that the scheme is a kind of an "ideal model". In real life there are there are a lot of spheres where the necessity and the profitability of CSR is not that evident. Traditionally, military complex is excluded from CSR logic. There are industries such as internet companies, scientific centers and other that do not produce any negative effect on the ecological or social environment. So, they do not feel obliged to pay additional costs on CSR.

In practice, we may say that now *Russian own model of CSR* is being formed, which is in line with the western understanding of CSR, however, this model has its clear-cut national features. Therefore, we will concentrate on these aspects.

Speaking about the directions of CSR in Russia, we can outline the following: environment issues, sustainable development, territories' social development, labour protection and human resources development.

It is interesting enough to compare briefly volumes and the scale of Russian and western CSR programs. An average figure for companies from Fortune 500 equals to $50 million. Top five leaders spend on these goals more than $1 billion a year. In Russia, as there is no standard procedure for mandatory publishing financial statements of CSR, it is difficult enough to assess the volume of investments in these programs. The data of the annual survey "Leaders of corporate charity" (a joint annual project initiated by Donors Forum, Vedomosti newspaper and PwC) can be considered as the indirect measure of the scale of CSR-market. Fifty-two Russian and international companies operating in Russia took part in this survey in 2017. Only forty-six companies that participated in the survey revealed the information about their expenses on social support and charity. They spent more than 43.8 billion rubles on social support and charity (in 2015 fifty-four companies spent 19.9 billion rubles on the same needs). The largest volume of social investments for all years surveyed, i.e. 7 billion rubles, was recorded by PJSC MMC Norilsk Nickel [3].

Virtually, we may indicate four typical participants of CSR in Russia:

- Big companies from former Soviet monotowns;
- large strategically-oriented companies that work and have close links with the government;
- small and medium-sized companies;
- Philanthropists. They are not treated as a real participant of CSR, but at the same time, it is the most widespread form. Patronage and sponsorship help business become involved into socially oriented projects. In majority of cases, philanthropists are motivated more by their own interests rather than business interests.

The main difference between the Russian model of CSR and practices of western countries is that there is no definite goal and system in Russia. Since 2015 (the year when the United Nations General Assembly set Sustainable Development Goals), the majority of multinational companies in the world as well as Russian ones have started to adjust their CSR programs to these goals. However, taking into account the data given by RSPP, on the Russian market three quarters of the companies do not link their main goal with sustainable development goals, most organizations believe that they already make a contribution to the development of society, economy and environmental protection within the activity. For instance, only 35% of companies mention Sustainable Development Goals in their development strategies. Nevertheless, it is worth mentioning that both Russian companies and the Government develop in line with the indicated strategies. Therefore, if we compare the contents of the abovementioned national priority projects and the main goals of CSR programs, we may notice a clear correlation between them and such goals as Goal 3, Goal 8, Goal 11, Goals 13-15. According to the research, the obtained data confirm insufficient understanding by representatives of business of interrelation of commercial success and sustainable development, those opportunities that are available on the way of the movement to sustainable development goals when they are built in the business strategy. We hope that even in the absence of the well-defined aim by business to pursue goals of CSR programs and adhere to Sustainable Development Goals, Russia will be able to move forward to sustainable economic growth.

One more specific feature of the Russian model is the system of establishing relationship between business and the state with respect to CSR programs. Moreover, this model is not typical of western countries. On the one hand, the government urges business to launch socially oriented programs. In some cases, we may even use the term "pressure" that is put on business by authorities. As a result, CSR is not considered to be voluntary, but to some extent an obligatory condition of operating on certain markets and in certain regions. On the other hand, there is no support from the state. For instance, there are tax exemptions only for those companies, which do charity work only as a part of definite activity. What is more, companies are allowed to do some charity work using only net profit. Today there is an opportunity for them to reduce corporate tax from 18% to 13.5% on the regional level and get some VAT incentives. That is why there is a small advantage of charity over traditional CSR programs. The authors believe that establishing firm relationship between business and authorities with respect to this issue would lead to a considerable increase in efficiency of CSR programs both for business and for the whole country.

4 Conclusions

In conclusion, we may say that taking care of interests of different partners, the management of a company efficiently fulfils its primary task connected with maximizing profits of shareholders. Such a strategy, in authors' opinion, makes closer two positions. Shareholders should not worry that their money may be controlled inefficiently until stakeholders' policy hampers such a strategy. Moreover, stakeholders' policy is beneficial to a company. First of all, the management of a company aims at building mutually beneficial relations with all its partners, who may explicitly or implicitly influence the thriving of a company. Then, such management creates favorable working conditions for workers who show willingness to work and climb the career ladder, paying attention to quality control of goods and services. As a result, the management in question brings more benefits than the management, which is interested only in maximizing profits.

Disregarding the concerns resulting from the current economic situation or left from the reform of the past, Russian big business is ready to realize full-scale CSR programs. Russian big companies already conduct numerous CSR programs on the internal and external levels. As the relations between the state and business are becoming more and more transparent and predictable, the attitude to corporate social responsibility will shift further from seeing it as an additional "social tax" to mutually beneficial well-balanced work to achieve stable social welfare.

Nowadays more and more businessmen and the society in Russia realize that in order to achieve sustainable economic growth it is necessary to have tight link and mutual contribution of business and the state. It is obvious that the private business activities, investment and innovation are major drivers of productivity, including economic growth and job creation; they create jobs, innovate, generate tax revenues. However, at the same time, the success of business depends on the health of the environment on local, national, and global levels, and helping the government to solve these problems such companies increase their competitiveness and profit.

References

1. M. S. Aßländer, J. Curbach, Business & Society, **56(4)**, 617 (2017)
2. J. Bakan, *The corporation: the pathological pursuit of Profit and power*. London: (2005)
3. Corporate Social responsibility, industry specific supplement to the daily business newspaper RBC, **159(2883)** (2018)
4. T. Donaldson, L. Preston, Academy of Management Review, **18(1),** 65(1995)
5. Ecological report. "MMK. With care for the future", http://mmk.ru
6. G. Eweje, *Introduction: Trends in Corporate Social Responsibility and Sustainability in emerging economies*. Corporate social responsibility and sustainability: emerging trends in developing economies. Emerald Group Publishing Limited, **8,** 3 (2014)
7. E. Feoktistova, G. Kopylova, M. Ozeryanskaya, M. Moskvina, N. Hofmann, D. Purtova, *Compiled by authors, based on Russian Business and Sustainable Development Goals. Corporate practices collection* (2018)
8. R. Freeman, J. Harrison, A. Wick, B. Parmar S. de Colle, *Stakeholder theory: the state of the art*, Cambridge: Cambridge University Press (2010)
9. M. Friedman, *Capitalism and Freedom. Fortieth Anniversary*, Chicago: University of Chicago Press (2009)
10. J. P. Gond, N. Kang, J. Moon, Economy and Society, **40(4)**, 640 (2011)

11. C.M. Hoevring, S.E. Andersen A.E. Nielsen, Journal of Business Ethics,**152(3)**, 627 (2018)
12. P. S. Hofman, J. Moon, B. Wu, Corporate Social Business and Society, **56(5)**, 651 (2017)
13. Materials on the project «Creation of conditions for development corporate social responsibility business in the Pskov region», http://csdpr.ru/
14. J. Jahn, R. Brühl, Journal of Business Ethics, **153 (1)**, 41(2018)
15. J. Jung, H. Cho, Journal of Business and Retail Management Research, **11 (1)**, 7 (2016)
16. D. Matten, J. Moon, Academy of Management Review, **33(2)**, 404 (2008)
17. Report based on the survey of 2015 of managers working in the field of CRS in Central Europe, https://www2.deloitte.com
18. F. Perrini, A. Russo, A. Tencati, C. Vurro, *Deconstructing the relationship between corporate social and financial performance.*(2011)
19. A. G Scherer, G. Palazzo, Journal of Management Studies, **48(4)**, 899 (2011)

Development of innovation policy: case study of north and West African countries

Leandre Yedehou Dossou*, and *Itri* Hanaa

Peter the Great St. Petersburg State Polytechnic University, Polytechnicheskaya, 29, 195251 St.Petersburg, Russia

Abstract. Since the work on Joseph Schumpeter's entrepreneurial theory in the process of creative destruction, many works have been devoted to innovation. However, this previous work often approached innovation from an organizational point of view with a technical orientation. But today innovation covers all sectors of economic activity or not. In addition, we can see that this research on innovation has been more active and implemented in developed countries. On the other hand, some work; have been done in developing country, which shows a low level of innovation development. Therefore, to fill the gap in African literature, the purpose of this article is timely to develop a model of innovation policy for low-income countries in general, and for West and North African countries in particular. To achieve this goal, it is necessary to make the diagnosis at the level of African countries in terms of innovation, in order to solve problems related to the challenges of innovation on this continent. Finally, relevant recommendations ranging from the meaning of an effective implementation of the proposed innovation policy through a mechanism and a verification algorithm for a positive economic and social impact.

1 Introduction

Innovation policy is progressing in African countries. It is seen as a lasting solution to the many economic and social challenges Africa is facing and an effective policy for promoting innovation needs to be put in place. Innovation is essential for companies if they are to survive in the current local, regional and global market environment of the 21stcentury. On this, the scientific studies of the authors such as: (Elder J.; Fagerberg J.; Coenen, C. Deblonde, M Medvecky, F. Phillips, W. Martin, B. Isaksen, A.) [1-8] have studied innovation in developing countries in its aspect of the general indicator on economic performance and competitiveness. (DOSSOU Y. and Khvatova Yu), [9] in our

* Corresponding author: segnor2@yahoo.fr

previous work have addressed innovation in West Africa in its aspect of entrepreneurial activities and related difficulties. In this context, why is it so difficult to develop, finance and implement real innovation policies in West Africa? This research aims to study innovation in its political or institutional aspect. Based on the inventory of the existing level of innovation, we make relevant recommendations for a more reasonable innovation policy and adapted to low-income countries in general and to West and North Africa in particular.

2 Methodology of the research

The main objective of this article is to provide new concepts or new knowledge in the specific area of innovation policy making, based on knowledge theory. Then, using the documented results, we highlight the different key factors in the development of innovation policy to have a reasonable conception of a policy for the development of innovation activities. The policy design suggested in this article is intended to be a toolbox to support effective implementation, and therefore the management of business innovation activities. From the appropriation and the collection of data and information, the different methods were used, namely (document analysis method, structured interview methods, observation method in the evolution of companies). The data and information collected, the methods of evaluation, comparison, and statistical analysis were used. Finally, to solve the problem related to development, financing and implementation of the policy innovation, methods of synthesis analysis, abstraction, algorithm, modeling, induction, and deduction have been applied.

3 Innovation

Science, like innovation, is always historic. But in capitalism, science is seen as a reservoir of knowledge that feeds innovation. The evolution of the size, the influence of the economic and financial power of the company allowed the integration of innovation in science and therefore in immediate production. In the different literary works of authors from African, European, Asian, American and other horizons, we face multitudes of definitions, of which the common point is based on something new (creation, thought, ideas, methods, processes, products, management, etc.) For J. Schumpeter, innovation can be seen as a new thought or a new vision of a known process, the successful application of a new creation, a job or a discovery in the economy and other fields of human activity [10]. In N.D. Kondratyev's works, Innovation, is at the heart of wave theory. Thus, following a new idea well-thought-out and implemented, an innovation is replaced by another, more recent, which ensures the constant satisfaction of the social system with development [11]. As for F. Barnu, innovation is the economically successful implementation on the market of a new idea for this market. Innovation is an essential element for the creation of economic value, considered as an imperative by companies; it nonetheless remains a complex phenomenon that is difficult to control [12]. Finally, according to the Oslo Manual of the OECD, Innovation is the implementation (implementation) of a new or significantly improved product (good or service) or a (production) process, a new marketing method or a new organizational method in business practices, workplace organization or external relations [13, 14].

By synthesizing, we see that if Kondratyev's work led to the conclusion that innovation is the basis of the development of economic cycles, then Schumpeter defined it as the birth of new goods and the opening of new markets, thanks to the combination of resources (productive, scientific, technical and financial) by one or many companies. We can also understand that from the appearance of innovation, is born the call for social resistance which constitutes an obstacle to something new, not understood, ranging tendencies to erase old habits. On the other hand, the Oslo Manual, dating from 2005, is more explicit on the definition of innovation by presenting a new product or service, a new process, new marketing method or a new organizational method, relations of partnerships by a company.

3 Results obtained

3.1 West Africa innovation policy

If according to the report entitled "Perspectives of African innovation" in 2010 reveals that during the intergovernmental meeting of African countries on science, technology and indicators of innovation in Maputo, a decision was taken to adopt also well the Oslo Manual for use in measuring innovation and R&D in Africa [15, 16]. This then shows the absence of a specific reference document for innovation adapted to realities even if we live today in a world where Globalization is a reality. Thus, speaking of the innovation policy in West Africa, involves talking about the different periods and different economic policies implemented since the 1960s. The various analysis perspectives of the economic policy and development strategies implemented can be summed up as following [17]: (i) from 1960 to 1970: State-oriented economic policies as main development actors; (ii) 1980: begining of economic policies focused on the withdrawal of the state; (iii) the 1990s: structural adjustments to policies aimed at the withdrawal of the state; (iv) the 2000s: emergence of international initiatives and the New Partnership for Africa's Development (NEPAD). From these various facts, it is necessary to say that these different economic policy strategies which are linked to innovation activities have not been able to reduce poverty and participate in real economic growth in West Africa. A comparative analysis of the fifteen countries in this region (Benin, Burkina Faso, Ivory Coast, Mali, Niger, Senegal, Togo, Nigeria, Ghana, Liberia, Sierra Leone, Guinea, Guinea Bussau, Gambia, Cape Verde) provides empirical substance with theoretical arguments. It is also during these different periods, that can be noticed the failure of imported innovation policies dictated by regional and international actors such as the International Monetary Fund, the World Bank, the United Nations Programs and the New Partnership for the Development of Africa [18].

After these trial and error decades, innovation policy is finally taking off in Africa. The African Union has recently developed the "Strategy for Science, Technology and Innovation in Africa 2024 (STISA-2024)", which adds to the efforts of each member country to master Science policies, Technology, and Innovation (STI) [19]. Despite this platform or program presence to identify the common challenges of implementing Science, Technology and Innovation policies, the current level of innovation activities in West Africa remains low. Interesting results on barriers to innovation in West Africa are contained in the Global Innovation Index (GII) developed by the World Intellectual Property Organization (WIPO) [20, 21]. The Global Innovation Index in 2019 was calculated for 129 countries in the world, it included 80 indicators that assess the institutional environment for innovation, namely political conditions, education level, infrastructure development, etc. In 2019, the highest GII was 67.24 and the lowest was 14.49. For West African countries, Senegal had 26.83; Ivory Coast 25.55; Ghana 25.27; Mali 24.03; Nigeria 23.93; Burkina Faso 23.30; Benin 20.42; Guinea 19.50; Togo 18.50, Niger 18.13; [22]. It can easily be said that West African countries have very low Global Innovation Index values - therefore a low level of innovation development. By Comparing the 2019 results to the previous year, the innovation policy of six countries is progressing positively (positive score): this is Ivory Coast (+5.55); Burkina Faso (+4.40); Nigeria

(+1.53); Ghana (+0.77); Mali (+0.73); Senegal (+0.33); while the other countries fell with negative score.

3.2 North Africa innovation policy

The North Africa countries seem to have accumulated a significant backlog over the years in innovation, while the lack of knowledge acquisition, absorption and use seems to be among the main problems that limit human development [23]. However, since the 1990s, the North African region has been given special consideration, especially by the European Union, with a view to strengthening cooperation and partnership relations, with the aim of reducing conflicts in the region and creating an atmosphere of stability and prosperity, ensuring that the countries in the area catch up in the medium and long term. In this regard, many initiatives have been undertaken and various association and agreements have been signed within the framework of the Barcelona process. This integration process and its prolongation still represent an opportunity and a challenge for the countries of the south shore of the Mediterranean, as these countries now find themselves in direct competition with their northern neighbors, which necessitates reconsidering their comparative advantages and an industrial change focused on the production of higher added value goods.

The aim of Tunisian and Moroccan innovation policies in the last two decades has been to emphasize new enterprise development or new business creation. Also, to create a formal mechanism for coordinating an innovation policy which the Higher Council for Scientific Research and Technology (HCSRT) that have intended to produce some coordination of industry-oriented initiatives. Morocco and Tunisia show a common turn towards innovation in their science and technology policies. Most measures are aimed at linking research, science, and universities to the manufacturing sector. This does not apply specifically to these two countries or to MENA; (MENA countries consist of Algeria, Bahrain, Egypt, Iran, Iraq, Israel, Jordan, Kuwait, Lebanon, Libya, Morocco, Oman, Qatar, Saudi Arabia, Syria, Tunisia, United Arab Emirates and Yemen.). But, as we have already said, MENA countries entered this orientation of innovation policy much later than other countries and focused on a specific part without a common national innovation strategy. Often, it is the techno parks initiative that was the first visible initiative. Tunisia was a forerunner in the region with the El Ghazala Technopark in Tunis, mainly focused on new information and communication technologies. Morocco opened its technology Park in Casablanca, Egypt opened its smart village near Cairo, and in Lebanon, Saint Joseph's University created Berytech. They are notable for their focus on new information technologies, focus on small startups, and link an educational institution (University or engineering school) with businesses [24]. Various innovative strategies that have also been implemented can be summarized as follows [25]: (a) Technology transfer departments at universities and engineering schools; (b) Engineering Networks; (c) Promotion of intermediate technical centers; (d) Business associations related to innovation and technological development. Over the past five years, both countries, as well as all countries in the Mediterranean, have developed many policies and measures, the main goal of which is to establish links between enterprises and public research centers and universities. Thus, from the policy point of view, innovation is largely related to the development of technical and economic networks.

3.3 Development of innovation activities

Developing innovative strategies involve considering various aspects of public innovation management. Thus, the implementation of innovation strategies must consider the general strategic directions of economic development at a national or regional level. In this context, an innovation strategy can be considered as one of the most important factors for ensuring the efficient development of enterprises in accordance with the overall economic development strategy. This shows that innovation is only an important link in economic and social development policy. As such, the following elements can be considered, considering:

– the available resources (human resources, natural resources, financial resources);
– the extent of government coverage of science, technology and areas of innovation;
– the nature of technology transfer (acquisition of machinery, equipment and software, implementation of research and development, partnership, and specific training abroad, etc.);
– the nature of the generation of ideas (incubation centers for ideas and projects, startup development centers…);
– technology networks, such as tech companies laboratories, and engineering organizations;
– the conduct work that are in demand for research of technological and economic information
– the nature of support for innovation processes.

Then, organization is one of the pillars of functions for the basic management of innovation activities. This implies that established or new businesses must acquire not only knowledge of the organization but also of managing innovation - as an asset for their development and survival in the market. Finally, the main priority of the State is to put in place legal and administrative regulations to help companies, whether public or private, to achieve their specific objectives which fall within the framework defined by the national or regional objective.

3.4 Development of an innovation policy model

The innovation policy model suggested in this article is valid for West and North African countries, but also all low-income countries. It is composed of a diagnostic, the commitment of the State, goal, tasks to be carried out (see fig.1).

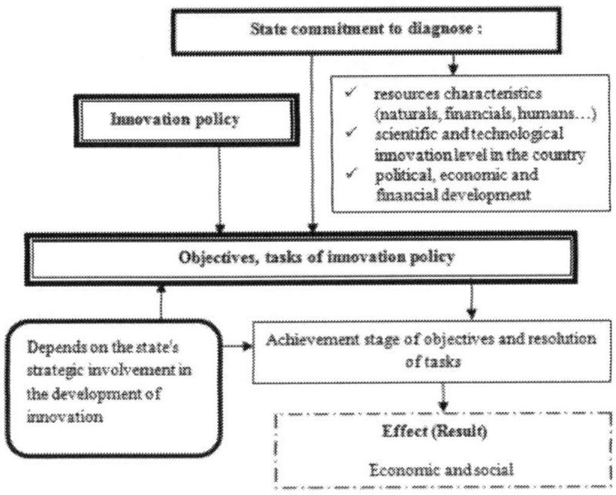

Fig. 1. Scheme for development and implementation of an innovation policy.

Our suggestion for an "innovation policy" was based on a few key points of strategic and tactical approaches and is accompanied by a commitment from the State. The idea of suggesting this policy lies in the history and the level of economic and industrial development of African countries. Thus, the objective is to make the base of the commitment of the State, a solid lever to stimulate economic growth thanks to an innovation option towards the manufacturing and energy sector in order to improve the economic and social conditions of the population. For so doing, in a reasonable way, we must translate the content of our suggestion, which contains: Objectives and missions of innovation policy. The objectives set in the current context are as following: - selectivity in the manufacturing sector to increase production, particularly on the domestic and regional market; - use the State's commitment to the production of electric energy by the method of liberalization of the sector and to make it a security and legal priority which would find positive echoes among private sector actors.

3.5. Suggestion of a verification algorithm

In the context of protecting companies and the State in the same way, in order to minimize the risks involved in the implementation of innovative activities, it is appropriate for the success of our innovation policy suggestion, to set up a verification institutional and administrative algorithm consisting of four mandatory steps (see Fig. 2)

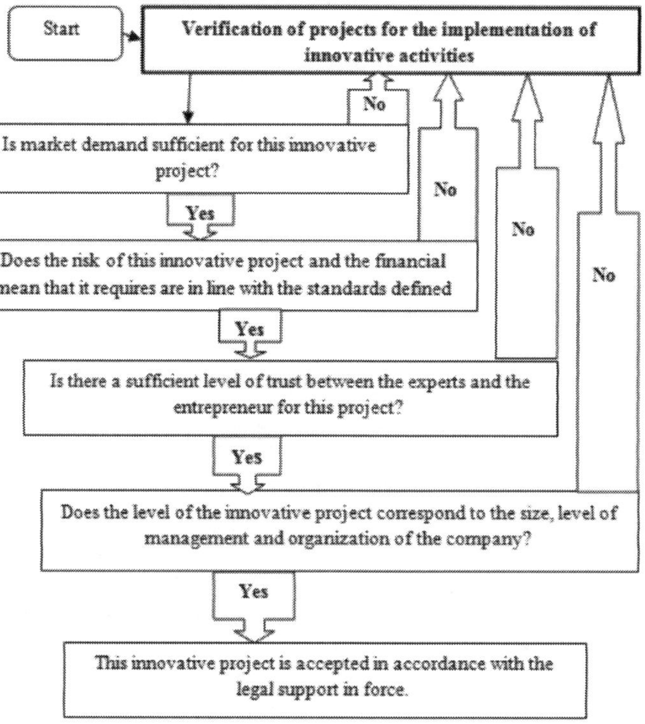

Fig. 2. Verification of innovation projects by algorithm before their implementation.

3.6. Development of an Implementation Mechanism

It seems reasonable that the mechanism suggested for the implementation of an innovation policy in the case of our study is possible thanks to the commitment of the State, to take the legal, financial risks of its priority projects, and also those of semi-public or private companies in the context of institutional and administrative verification. This mechanism is presented as following for companies and for the State:

– **For companies (a)** orientation through the approach of projects through startups, research centers, results of university research; **(b)** a creative approach based on the employees of the company without external appeal. (Thus, many studies have established a lack of correlations between the emergence of scientific knowledge, its materialization and its commercialization. Consequently, from this point of view, all companies do not need to carry out several complex fields of activity innovative entrepreneurial, from R&D to marketing).

– **For the State (i)** implement a partnership approach with foreign companies in specific areas requiring the transfer of technology through the purchase of licenses, equipment and software, etc. within a legal and legal framework ; **(ii)** Establishment of a common innovation office framework composed of experts (State experts, experts from the private sector, social protection experts) for institutional and administrative verification of all priority projects by the algorithm before their implementation.

Conclusion

The success of the innovation policy in the case of this article depends on important factors such as, the availability of resources (financials, materials, humans,...), research and development centers, the more or less stable market, creativity, know-how, etc. Also, this success depends on the organization of innovative activities within companies, management support, teamwork, optimal use of qualified human resources, delegation of authority and responsibility for the monitoring and evaluation. On sight, these factors are accessible if the legal commitment of the State exists because companies in low-income countries face difficulties in taking enough risk despite their ability and possibility to implement activities of innovation.

Innovation is not always synonymous with the creation of a technology of breaking up or introducing radically new products. This can lead to a rise in the range and to the creation of more added value for existing products. However, the absorptive capacity depends inter alia on learning mechanisms, Human Resource qualifications and managerial quality. However, our results show that in North African countries, whether on innovation incentives or productivity levels, this reveals underutilization and inefficient allocation of human capital in North African countries.

In order to boost innovation a special attention should be paid to certain aspects of the national innovation system. In the current context, faced with competitive threats from a number of emerging countries, notably Asian countries, offering unbeatable production costs in traditional sectors, North African countries are now being forced to target a wider range of technologies requiring significant absorptive capacity and a highly skilled workforce, to strengthen their non-price competitiveness in traditional sectors and diversify their productive activities into other sectors with higher added value.

References

1. J. Elder, J. Fagerbergm, Innovation policy: what, why, and how. Oxf. Rev. Eco. Pol. **33,** 2 (2017)
2. K. Bronson, Journal of Responsible Innovation, **2(1),** 62 (2015)
3. C. Coenen, Broadening discourse on responsible research and innovation (RRI), N. E., **10(1),** 1 (2016)
4. M. Deblonde, Journal of Responsible Innovation, **2(1),** 20 (2015)
5. S. de Saille, F. Medvecky, Eco. & Soc. **45(1),** 1 (2016)
6. W. Phillips, H. Lee, A. Ghobadian, . N. O'Regan, P. James, Social Innovation and Social Entrepreneurship: A Systematic Review, Grp. & Org. Mgmt, **40(3),** 428 (2015)
7. B. R. Martin, Sc. P. Pol., **43(3),** 432 (2016)
8. A., Isaksen, K. Onsager, Eu and Reg. Std., **17(3),** 227 (2010)
9. Y.L. Dossou, T.Yu. Khvatova, SPbU. J. Eco., **12 (4)** 144 (2019)
10. M Iizuka, P. Mawoko, F. Gault, Innovation for Development in Southern & Eastern Africa: Tech. and Innov.Pol. (2015)
11. J. Schumpeter, "Capitalism, socialism and democracy: term creative destruction, 119 (1974)
12. N. D. Kondratiev, Yu.V. Yakovets, L.I. Abalkin, *Great economic cycles and prospective theory of selected works*, 766 (2002)
13. F. Barnu,The true nature of innovation, Ed. Tee & Doc, (2010)
14. A. Groff, Innov. a answer to the challenges of sustainable dev.? Lib. in Tlse, F. (2011)
15. The Oslo Manual of the OECD: Guidelines for Collecting and Interpreting Innovation Data, 3 (2005)
16. UA–NEPAD AU, New Partnership for Africa's Dvpt. **77** (2010)
17. Y. L. Dossou, Yu. Khvatova, Analysis of innov. activity of companies in dev, countries. Ex. West African countries, SPBU, **13 (1) 79** (2020)
18. S. Landry, Innovation in dvpt strategies in Africa. P. Kart., **296** (2015)
19. A. Diop, Af. and Dev. **41,** 159 (2016)
20. M. Iizuka, P. Mawoko, F. Gault: Innov. for dev. in East and South Africa, 1 (2015)
21. Global Innovation Index (2019)
22. Data Global Innovation Index 11th ed, 222 (2018)
23. ANIMA Invest. Netwk.: The Med. between growth and rev. FDI and Part, 21 (2010)
24. Global Competitiveness Report (2017-2018)
25. J. N. Assad, B. Saleh, M. Piron and R. Arvanitis (2008)

Financial and economic aspects of corporate social responsibility

V.V. Ilyashenko*

Ural State University of Economics, 620144 Ekaterinburg, Russia

Abstract. The article shows the importance of corporate social responsibility (CSR) in ensuring sustainable development of the country. The types of CSR and its features in various states are considered. The author describes the economy of the Russian Federation and its impact on the system of corporate social responsibility in the country. The high profitability of resource-extractive industries and their use of the country's national wealth defines their special role in CSR not only towards their employees through wages and the allocation of social benefits from profit, but also to the society. It is shown that the established country's political system significantly influences the possibility of implementing a system of corporate social responsibility. The author characterizes the significant regulatory and stimulating role of the state in social development through taxation and the structure of government spending. When assessing the financial conditions of CSR, the author analyses the impact of capital outflow on its development. Corporate social responsibility also includes the responsibility of organizations to the environment. The author provides a rating assessment of Russian oil and gas, mining and metallurgical companies openness in terms of environmental responsibility.

1 Introduction

Corporate social responsibility (CSR) is an integral part of the modern economy and a prerequisite for sustainable development. It has its own characteristics in different countries, on which the combination of private and public interests and the degree of the state's regulatory role in this process depend. The implementation of CSR largely depends on the financial and economic conditions existing in a given state, and it is fully applicable to the Russian Federation. When characterizing corporate social responsibility, it is necessary to take into account the profitability of organizations, distribution of profits, taxation of the country, the structure and dynamics of government spending, capital movement, costs of implementing environmental programs and other economic indicators.

*Corresponding author: k-45409@planet-a.ru

2 Materials and methods

When researching corporate social responsibility, we used scientific articles devoted to this topic; as well as statistical materials of Rosstat and the Bank of Russia; statutory documents. In writing the article we applied comparison methods, groupings of indicators, expert estimates, index method, etc.

3 Results and discussion

The development of corporate property (CSR) is an objective process that reflects the process of concentration and centralization of capital. At the same time, the important role of corporations in the industries that determine the country's economic potential also involves an appropriate level of corporate social responsibility in connection with the high level of profitability of many of these organizations. The objective economic situation in modern Russia demonstrates that enterprises in the raw material sector of the economy play a significant role. So, according to Rosstat, in 2019 the profitability of goods sold, products (work, services) in mining operations was 29.6%, including oil and natural gas extraction - 31.9%, in metal ores extraction of - 66.3%. According to the RBC rating, the total revenue of the 500 largest companies in Russia in 2018 amounted to 85.5 trillion. rub. and was 18.2% more than in 2017, amounted to 82.3% of the GDP of the Russian Federation. At the same time, 52.9% of the total profit of these companies was received by oil and gas sector organizations - 4.9 trillion. rubles, 13.4% - by mining and metallurgical enterprises, or 1.2 trillion. rub. [1]. It is obvious that the problem of increasing social responsibility is relevant specifically for these industries not only because of the high level of profitability, but also because their activities are associated with the use of national wealth belonging to the entire population.

Corporate social responsibility is understood as the responsibility of an organization for the impact of its activities on society and the environment, which contributes to sustainable development, health and well-being of society; takes into account the interests of interested parties and the state; corresponds to the applicable legislation and is in line with international norms of behavior [2, 3]. If we formulate the definition of corporate social responsibility more briefly, then in the scientific literature this concept is interpreted as the voluntary participation of business in the society development [4]. In accordance with the International Standard ISO 26000 "Guidance on social responsibility" "an integral characteristic of social responsibility is the desire of an organization to include social and environmental factors in its [processes] decision-making and to be accountable for the impact of its decisions and activities on society and the environment." The International Standard ISO 26000 emphasizes the close relationship between social responsibility and sustainable development, which involves meeting the needs of society without compromising the ability of future generations to meet their own needs. At the same time, there are three components of sustainable development - economic, social and environmental, which are interdependent [5]. This does not mean that a high level of corporate social responsibility necessarily ensures sustainable development. The economic development of various countries is cyclical. As society develops, crises become global in nature. This is also noted in the International Standard ISO 26000 "Guidance on social responsibility". Economic and financial crises have the greatest impact on the most vulnerable segments of the population and, therefore, necessitate strengthening the social responsibility of companies. It should not be forgotten that crises may occur in the world, caused not by economic reasons, but by unexpected events, which is observed in modern conditions in connection with the coronavirus pandemic. The spheres of tourism, trade, and air transportation found themselves in the most difficult conditions. According to our calculations, based on forecasts of the International Monetary Fund, most of the 12 leading countries in the world, nominal GDP in 2020 will decrease compared to the previous year in Brazil - by 28.4%, India - by 12.4%, in Italy - by 12.4% and Russia - by 11.5%. And only in China, where the coronavirus epidemic began, the IMF predicts an increase in nominal GDP from $ 14.17 trillion in 2019 to the expected $ 15.22 trillion, i.e. by 7.4%. The Ministry of Economic Development evaluates the economic development of the Russian Federation more optimistically, believing that real GDP in 2020 will decrease by 3.9%, the Central Bank of the Russian Federation tentatively estimates a decline in real GDP this year by 4-5% with an expected inflation rate of 3.8% Obviously, in crisis conditions, the determining regulatory role of the state in economic policy in general and in social policy in particular is all the more important.

When analyzing the peculiarities of corporate social responsibility in different countries the author distinguish 2 main types of CSR. Their specificity lies in the degree of independence of organizations in determining the boundaries of their responsibility to society: 1) business itself determines the level of its responsibility to society; 2) state institutions have a decisive influence on business in determining social requirements. In this regard, there are "open" and "hidden" approaches [6,7].

An open approach is largely prevalent in the United States, where organizations independently determine the boundaries of their social responsibility, which is associated with a certain established entrepreneurial culture and the voluntary use of the principles of corporate social responsibility. Government incentives for CSR in the United States consist of the provision of income tax incentives that are used for charitable purposes [8].

The state plays a significant rolein European countries, it determines regulation of the organizations'activities in the CSR area by laws, norms and standards. In particular, we are talking about compulsory health insurance, pensions, environmental protection; consolidating the principles of social responsibility in the Integrated Product Policy, ISO 26000 Social Responsibility Standards, Environmental Audit and Management Scheme. In Japan, the social responsibility of companies is of decisive importance with the important role of the state. The Chinese approach to corporate social responsibility in modern conditions presupposes a focus on the development of social services, increasing taxes and increasing the number of jobs with their qualitative renewal and ensuring the need to benefit society. The Islamic approach to CSR is based on religious moral and ethical norms. It is obvious that the CSR concepts of European countries, Japan and China are most suitable for Russia as a European-Asian country.

The Guidelines of the Global Reporting Initiative (GRI) [8] continue to be the most widespread voluntary corporate reporting standard in the CSRarea. The development of the corporate social responsibility system is evidenced by a significant increase in the publication of non-financial reporting by various companies (social and environmental reports, reports on sustainable development, etc.). So, if in 2011 about 30,000 reports from more than 7,500 organizations were submitted to the International Register of Non-Financial Reports (Corporate Register), at the end of 2018 this figure was 98,000 reports received from more than 16,300 companies [4].

The possibility of implementing a system of corporate social responsibility largely depends on the political system that has developed in a given country. So, Naumov A.S. notes that the constitutions of European states stipulate that all types of the country's wealth, regardless of what property they are in, serve the common interests. As for our country, in his opinion, in Russia "reforms continue to be carried out in conditions of preservation of the existing oligarchic political system" and are aimed not at changing this system, but at its modernization. The financial, monetary, foreign economic and social policies of the state and monetary authorities in Russia in modern conditions are such that

 https://doi.org/10.1051/shsconf/20208907002

arguments can be made to soften this point of view, but it cannot be completely refuted. In these political and economic conditions, corporate social responsibility in the Russian Federation is objectively implemented to a lesser extent than is required for the development of society [9]. Nevertheless, there are certain successes in this area in Russia. According to the results of the annual research Reputation Institute Global CR RepTrak, in 2018 The LEGO Group, Natura, Microsoft, Google, The Walt Disney Company, BOSCH, Havaianas, Intel, Lavazza and IKEA were recognized as world leaders in CSR. At the same time, two Russian organizations, Gazprom and Sberbank, are among the 500 largest companies in the world [2]. The rating of the Russian Union of Industrialists and Entrepreneurs, compiled in 2018 by the Russian Union of Industrialists and Entrepreneurs, identified such companies as ALROSA, Aeroflot, Gazprom, EuroChem, Zarubezhneft, KAMAZ, MTS, as leaders in the indices of corporate stability, responsibility and openness, NOVATEK, NLMK, Norilsk Nickel, InterRAO, LUKOIL, Metalloinvest, Rosatom, Rosneft, RUSAL, Russian Railways, Russian Networks, Rostelecom, RusHydro, SIBUR, AFK Sistema, Severstal, SUEK, Sberbank, Sakhalin Energy, Tatneft, Transneft, Uralkali [2, p. 432]. Some authors, analyzing the growth rates of sales, net profit and social investments, note that the largest share of social expenses is observed in the metallurgical industry and the electric power industry. At the same time, 85% of companies plan to allocate funds for charitable programs, which are an integral part of corporate social spending. More than 80% of organizations either did not change or increased the share of social investments in sales, which positively characterizes CSR practice. At the same time, the authors did not establish a direct relationship between changes in financial indicators and social spending [10].

Based on the definition of CSR, it includes two sides: 1) social responsibility of the corporation to its own employees; 2) social responsibility of the corporation to society. As for the corporation's social responsibility to its own employees, it is determined, firstly, by the level of wages received by the employees of this enterprise and the social benefits that are available to them from the profit of the enterprise; secondly, the fulfillment of social obligations to society. If we analyze in this respect CSR of resource-extracting enterprises, whose social responsibility in Russia should be higher due to their high profitability and is largely necessary for the reasons noted above, then the following forms of its manifestation can be noted. First of all, CSR is realized through a higher level of wages in these industries. So, according to Rosstat, in 2019 the average monthly nominal accrued wages in the Russian Federation amounted to 47,867 rubles. At the same time, in the extraction of minerals it was significantly higher - 89344 rubles, including in the extraction of crude oil and gas - 135,364 rubles, in the extraction of metal ores - 78,520 rubles. As for the public benefit of CSR, it really manifests itself primarily in the charitable activities of these organizations. According to the publications, one of the leaders in the field of public benefit of CSR is the company "Rosneft", which, according to the authors, conducts a unique permanent work in the area of corporate social responsibility. This activity is carried out in several main areas: the development of social infrastructure, support for health care and the development of medicine, financing of educational programs, support for professional and amateur sports, implementation of environmental programs, sponsorship projects in the field of culture [11, 12]. Some authors characterize the relationship of corporate social responsibility with innovation, with corporate ethical standards, and also note its necessity in non-profit organizations [13, 14, 15].

However, it should be borne in mind that the scale of CSR development in Russia largely depends on the regulatory influence of the state, due to the mentality and characteristics of economic thinking that developed in the 1990s. The main means of increasing corporate social responsibility of organizations in the field of financial policy is taxation, which allows to expand the opportunities for organizations to influence the well-being of society and the environment through an increase in state budget revenues.

In this regard, the decision of the Government of the Russian Federation to increase the mineral extraction tax (MET) for producers of fertilizers, ferrous and non-ferrous metals,adopted on September 16, 2020, should be positively evaluated. This decision, firstly, largely restores justice in the receipt by the state of natural rent not only from the oil industry, but also from other resource industries, and secondly, it expands the financial possibilities for solving social problems of society. It should be added that an increase in prices for these economic resources and the development of cost inflation in the country may become a negative effect of these tax measures [16, 17]. In our opinion, the financial possibilities of social responsibility of Russian business in the development of society in the 2000s would have been higher if the progressive scale of personal income tax was used, which is in effect in developed countries. On October 21, 2020, the State Duma of the Federal Assembly of the Russian Federation adopted in the first reading a bill on the introduction of a 15% personal income tax rate for income over 5 million rubles in a year. However, this decision, firstly, will not significantly increase budget revenues, since the increased rate slightly exceeds the base rate, and secondly, it does not establish the abolition of this tax for the population with low incomes, which usually implies a progressive scale of income tax. In particular, in China, where the maximum tax rate is set at 45%, if a person receives less than 4,000 yuan ($ 620), then he is exempted from paying income tax.

Opportunities for corporate social responsibility decrease to a certain extent with an increase in capital outflow from Russia, which, in turn, depends on the political and economic situation in the country and the world. According to the Central Bank of the Russian Federation, in the 2000s, the largest capital outflow from Russia was observed in 2008 (- $ 133.6 billion) and in 2014 (- $ 154.1 billion). It should be noted that in these years, in comparison with the previous period, the proportion of the population with cash incomes below the subsistence level also increased. In recent years, the minimum indicator of capital outflow from Russia was observed in 2019 (- $ 26.7 billion), with the lowest share of the population with money incomes below the subsistence level over the past five years (12.3%). The Central Bank of the Russian Federation has downgraded the initial forecast for net capital outflow from Russia in 2020 from $ 25 billion to $ 53 billion.The negative consequences of capital outflow include a decrease in tax revenues to the budget of the Russian Federation and a decrease in the possibilities of financing social programs.

In those years when the country's gross domestic product is shrinking and the possibilities for implementing corporate social responsibility programs are decreasing, it is advisable for state bodies to increase the share of financial resources allocated for social purposes from the state budget. So, in 2009, the share of expenditures from the consolidated budget of the Russian Federation on social and cultural events amounted to 52.8% compared to 51.0% in 2008. In 2015, when the most significant happened (after the crisis of 2009,) decrease in real GDP by 3.7%, the share of expenditures from the consolidated budget of the Russian Federation on social and cultural events increased to 57.7% compared to 54.9% in 2014, and from the federal budget - to 36.1% when 32.8% in the previous year.

Corporate social responsibility also implies the responsibility of organizations towards the environment. The UN Conference on Sustainable Development (Rio + 20) was heldin June 2012, at which the demands of corporations and governments to protect the environment and solve environmental problems were expressed. The National Rating Agency compiled a rating of Russian oil and gas companies in 2014, which allowed to obtain objective information on the level of environmental responsibility of organizations in the oil and gas sector. The best was the joint-stock company Surgutneftegas - one of the few oil and gas companies in Russia with the maximum level of information transparency in environmental reporting and publishing data on the costs of environmental protection. The company utilizes associated petroleum gas by generating electricity. The amount of gross economic losses from the irrational use of associated petroleum gas in Russia amounts to 147 billion rubles per year [8]. The rating of openness of Russian oil and gas companies in the field of environmental responsibility, compiled in 2019, ranked Sakhalin Energy (Sakhalin-2) 1st place, Zarubezhneft 2nd place, and Exxon NL (Sakhalin-1) 3rd place. LUKOIL and Surgutneftegaz took 4th and 5th places, as in 2018. Rosneft took 8th place, and Gazprom, which was in 2nd place in 2015, dropped to 6th place in 2018 and 10th in 2019.A similar rating of the transparency of mining and metallurgical companies in the field of environmental responsibility, held among the 40 largest organizations, brought Polyus Gold to 1st place, SDS-coal to 2nd place, Kinross Gold to 3rd place. Of the most famous enterprises, Severstal ranked 6th, ALROSA ranked 8th, Norilsk Nickel ranked 10th, Rusal ranked 17th, EVRAZ ranked 18th, UMMC ranked 20th. Some researchers note that a significant part of companies in the Russian Federation, which have proved themselves in the area of corporate social responsibility to the greatest extent, participating in the implementation of social programs and publishing social reports, carry out their activities with the participation of foreign capital or the state [10].

4 Conclusion

We can draw the following conclusions, based on the research conducted in the work:

- When analyzing corporate social responsibility, it is necessary to take into account its features, which are determined by the degree of organizations independence when implementing their responsibility to society. In this regard, there is a distinction between open and hidden approaches to CSR, the use of which differs in various countries.
- Taking into account its economic and moral-ethical conditions, the CSR concepts of European countries, Japan and China are most suitable for Russia as a European-Asian country.
- Companies in the resource sector play a significant role in modern Russia. They have a higher profitability than manufacturing enterprises. An increased level of social responsibility for these organizations is necessary due to the fact that their activities are associated with the use of national wealth that belongs to all members of society.
- Corporate social responsibility of resource-extracting enterprises to their employees is realized through a higher level of wages in these industries. As for CSR towards the society, it is implemented through the charitable activities of these organizations and the costs of environmental protection.
- Taxation is an important means of increasing corporate social responsibility of organizations, which creates the potential to increase the impact of organizations on the welfare of society and the environment through the growth of state budget revenues.
- CSR opportunities are limited to a certain extent by capital outflow from a specific country, also demonstrated in Russia.
- The structure of expenses of the consolidated and federal budgets of the Russian Federation has a significant impact on the implementation of social programs.
- The economic consequence of coronavirus infectionspreading is the economic crisis, which means the need to increase the corporate social responsibility of organizations and strengthen the social role of the state.

References

1. O.V. Pokramovich, Economic Sciences, **10 (179),** 97 (2019)
2. A.V. Vavilina., T.V. Komarova., I. Velence, R.S. Raikher, Leadership and management**, 6 (4)**, 425 (2019).
3. O.V. Kadnikova, G.R. Lobkov, Leadership and management, **5 (3)**, 105 (2018)
4. E.L. Pavlov, Economic relations, **9 (4)**, 2611 (2019)
5. International standard ISO 26000, http://www.ksovok.com/ (2020)
6. K.V. Volnukhin, Economic sciences,**12 (169)**, 54 (2018)
7. A.A. Morozov, Russian Journal of Entrepreneurship,**19 (10)**, 3119 (2018)
8. V.A. Onuchak, E.N. Levitskaya, Public-private partnership, **3 (3)**, 221 (2016)
9. A.S. Naumov, Economic sciences, **6 (175)**, 68 (2019)
10. I.V. Yakhneeva, R.I. Khansevyarov, A.P. Zhabin, E.V. Volkodavova, Russian Journal of Entrepreneurship, **19 (12)**, 3903 (2018)
11. N.S. Sarakhanova, N.A. Golubkova, Modern management: problems and prospects,**2**, 188 (2016)
12. A.M. Stepanova, Y.N. An, Bulletin of the Omsk Regional Institute, **2(2)**, 154 (2017)
13. D.F. Kuratko, J.S. McMullen, J.S. Hornsby, C. Jackson, Business Horizons, **60 (3)**, 271 (2017)
14. I. Martinez-Conesa, P. Soto-Acosta, M. Palacios-Manzano, Journal of Cleaner Production, **142 (4)**, 2374 (2017)
15. E. Park, K.K. Joon, S.J. Kwon, Journal of Business Research,**76**, 8 (2017).
16. V.V. Ilyashenko, *Macroeconomic and microeconomic aspects of inflation in the transforming economy of Russia* (2006)
17. V.V. Ilyashenko, L.N. Kuklina, Economy of the region, **14 (4)**, 1477 (2018)

Business model of regional institutional design of digital state entrepreneurship: management aspect

Alla Nikolaevna Golovina, and *Roman Yurevich* Levchenko

Ural State University of Economics, st. March 8, 62/45, 620144 Yekaterinburg, Sverdlovsk region, Russia

Abstract. In the article, the authors present the transformation of the concept of "institutional design". Its methodological core is considered in the context of the social and information and communication direction of digital state entrepreneurship. They offer the author's business model of the regional institutional design of digital state entrepreneurship.

1 Introduction

The scientific community has relatively recently begun to use such a concept as institutional design in almost all spheres of public life. Particularly noteworthy are the ideas that design is a capacious philosophical concept, the content of which goes beyond the boundaries of a narrow understanding of it as an art. It covers, but is not limited to it. Thus, the well-known design methodologist K. Jones claims that design should not be mixed with art, since it can be used as a tool for finding an optimal solution, and also used to reformulate the problem itself. Through its inherent constructivism, design allows not only to embody knowledge, information about the subject in the project, but also to understand the meaning of the problem [4].

Ch. Mills believed that any design has a social meaning, which allows it to be considered as a socio-cultural phenomenon of a developing society [10]. Here the ideas of G. Simon and Ch. Mills come down to the fact that today the subject of design extends to the construction of a way and style of life, synthesizing new cultural, moral, social values and is the bearer of social changes.

Summarizing the existing scientific approaches of foreign and Russian researchers to the definition of institutional design, one can conditionally distinguish two directions [3, 9, 5, 17].

The first direction defines design as social construction, creation or transformation of socio-economic reality [8, 13]. The transformations of Peter the Great, Vladimir Lenin, and others can be cited as an example of institutional design. Institutional design can be in narrower formats as well. For example, within a region or a specific organization.

Another direction considers institutional design as a Russian original method for the creation and transformation of social, economic, industrial, and other institutions [1, 2, 11, 16]. Institutional design here acts as one of the ways to change the modern economic situation, since it re-cultivates the existing structure of building a social space, based on the principles of the presence of power relations, self-organization and equality in society [7].

Thus, all the above statements taken together reveal the content of institutional design in modern society either as a socio-cultural phenomenon, or as a philosophy of design, development and formation of the built environment. At the same time, the purpose of the design is to improve the quality of life of people through the formation of a harmonious spatial environment [6].

In this regard, in our opinion, institutional design should be considered as social transformations, a set of institutions created and reorganized by the authorities in order to form a new quality of life in society.

Based on the above definition of institutional design, let us designate its methodological core in the context of digital state entrepreneurship. So, it is based on the provisions of the following regulatory documents.

1. Forecast of scientific and technological development of the Russian Federation for the period up to 2030 (approved by the Government of the Russian Federation on January 3, 2014). It identifies the most important areas for the development of science and technology that provide the country's competitive advantages in areas such as medicine and health care; information and communication technologies; rational nature management; biotechnology; transport and space systems; new materials and nanotechnology; energy saving and energy efficiency. In each direction, problems, threats, as well as development opportunities, promising markets, products and services, as well as areas of scientific research have been highlighted.
2. The strategy of scientific and technological development of the Russian Federation, approved by the Decree of the President of the Russian Federation of December 1, 2016 No. 642 "On the Strategy of Scientific and Technological Development of the Russian Federation". For the next 15 years, the priority areas for the development of the Russian Federation will be the development of advanced digital technologies; robotization of industrial systems; use of new materials; transition to resource-saving and environmentally friendly energy; creation of new sources of energy storage; health preservation technologies; counteraction to technogenic, biogenic, as well as socio-cultural threats; connectivity of the territory due to telecommunication systems, etc.
3. The strategy for the development of the information society in the Russian Federation for 2017 - 2030, approved by the Decree of the President of the Russian Federation of May 9, 2017 No. 203 "On the Strategy for the Development of the Information Society in the Russian Federation for 2017-2030". This strategy is focused on the formation and development of the digital economy, i.e. economic activity of a company, in which digital data is a key factor in production, processing of large volumes and the use of the analysis results of which, in comparison with traditional forms of management, can significantly increase the efficiency of various types of production, technologies, equipment, storage, sale, delivery of goods and services.
4. The program "Digital Economy of the Russian Federation", approved by the order of the Government of the Russian Federation, dated July 28, 2017 No. 1632-p. This program is aimed at stimulating the development and implementation of digital technologies in various sectors of the economy. An important substantive aspect here can be identified as three main areas in which digital state entrepreneurship is manifested: industrial, social, information and communication.

As one of the main directions of digital state entrepreneurship, one can single out the formation of new opportunities for improving the quality of life of the population and developing human potential through the information services that allow you to quickly exchange knowledge, improve labor skills, receive and process large amounts of information, etc.

However, the comparative analysis of countries presented in Figure 1 for the indicator "Internet use by the population" indicates its low value in the Russian Federation. This is

https://doi.org/10.1051/shsconf/20208907003

due to the low user activity of Internet services among the population of retirement age, associated with their mentality and the lack of habitual skills in working with IT technologies [15].

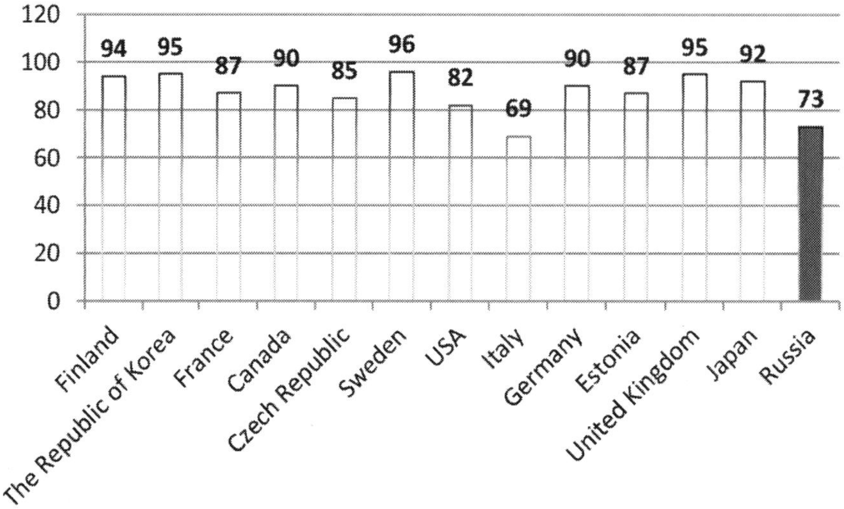

Fig. 1. Internet use by population by country (as a percentage of the total population aged 15–74)

In addition, the economically active population also has weak digital skills in working with information resources. Here the Russian indicator is inferior to the corresponding indicator of Great Britain, Germany, Finland, France, Sweden, Estonia and other European countries. According to the results of opinion polls conducted by the HSE, the main constraints are: lack of need (70.1% of the respondents aged 15-74); lack of skills for working on the Internet (29% of respondents aged 15-74); high connection costs (11.5% of surveyed respondents aged 15-74); lack of technical connectivity (4.1% of respondents aged 15-74); for security and confidentiality reasons (3.2% of respondents aged 15-74) [15].

At the same time, it can be noted that remote learning is gaining momentum in Russia. The annual growth rate has averaged 23% since 2012. Among the main advantages of massive open online courses, students note an increase in the accessibility of education (59% of respondents); the possibility of learning at an individual pace (30% of the surveyed respondents); more up-to-date content of training courses and the possibility of their versatile general development (19–20% of the interviewed respondents); variety of courses offered (17% of respondents) [14]. Also, 25% of the students surveyed believe that such courses can help in employment, and 11% associate training in such courses with improved academic performance. This, in turn, serves as a signal for further demand for this type of educational services. But despite this, online education in the total volume of educational services still takes a very small share - less than 3% [12]. This accounts for the low Russian percentage of Internet use by the population for remote learning (4%) (Figure 2) [15].

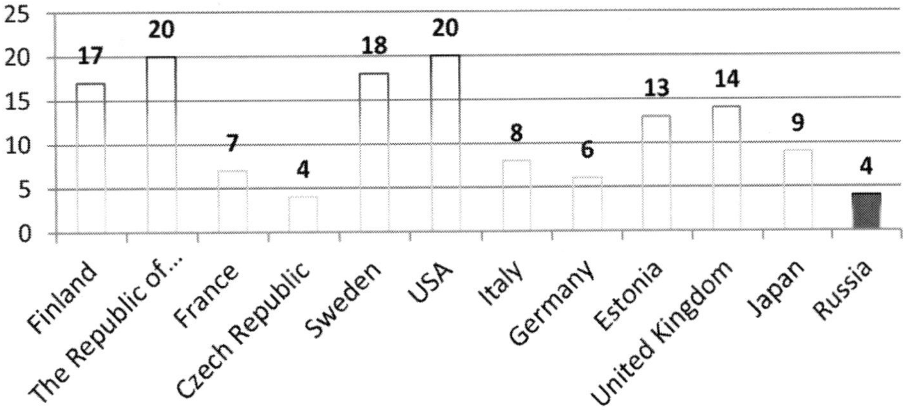

Fig. 2. Internet use by the population for remote learning by country, as a percentage of the population aged 15–74 using the Internet

In general, it can be noted that the population began to use more online financial services and electronic government services. Thus, over the past two years, the number of respondents aged 15-74 using the Internet to carry out financial transactions has grown by more than 10%. At the same time, the Far Eastern Federal District and the Siberian Federal District show the highest growth rates. At the same time, an intercountry analysis of this indicator suggests that Russians are significantly inferior in performing financial transactions via the Internet from Finns (93%), Swedes (90%), Estonians (90%), etc., (Figure 3) [15].

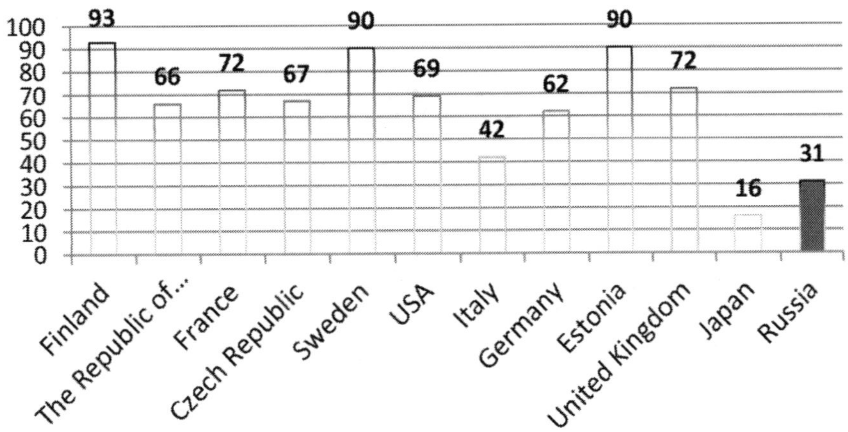

Fig. 3. Use of the Internet by the population for financial transactions, as a percentage of the population aged 15-74 using the Internet

Thus, we find that the social direction of development of digital state entrepreneurship, first of all, should be focused on improving the indicators presented above, reflecting the use of the Internet by the population. In addition, the following Internet services should appear and expand: digital services for real-time monitoring of movements and health status; scientific research based on "big data"; medicine "4P"; an expanded package of electronic public services and automated control of legally significant actions.

Another important area of digital state entrepreneurship concerns the development of the industry related to software production, data processing and transmission, storage, creation of Internet services, and electronic commerce.

According to the Ministry of Digital Development of Communications and Mass Media of the Russian Federation, the development of the information technology sector is advancing at a faster pace, the average salary in it exceeds the average for the economy by 50-60%. This sector of the economy can be considered young in terms of the age composition of its employees, which increases its potential for creativity, the perception of new knowledge, the development of innovative ideas of a revolutionary nature (Figure 4) [15].

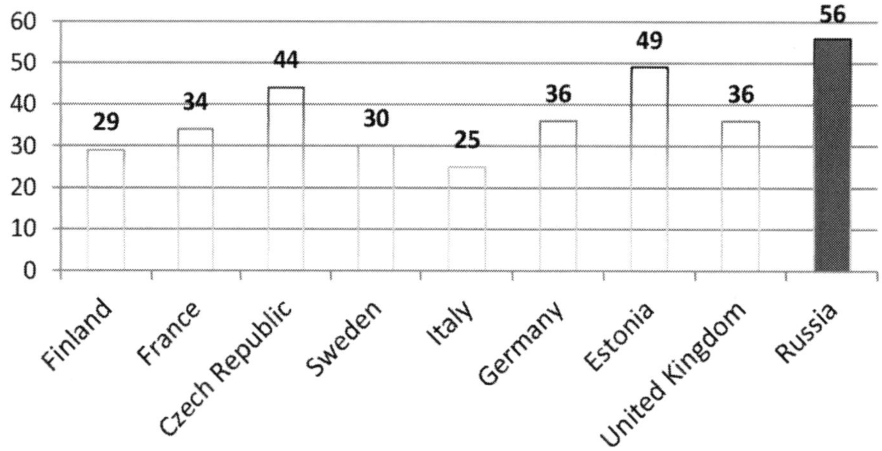

Fig. 4. Share of ICT specialists under 35 years old in the total number of ICT specialists,%

If we consider certain types of software, then according to the Unified Register of Russian Programs for Electronic Computers and Databases (reestr.minsvyaz.ru), it contains information on more than 3500 software products. Russian ICTs take part in the creation of information retrieval systems, social networks, design systems (Compass, Adem), anti-virus programs, ERP systems (Galaktika, 1C).

At the same time, this sector of the economy is at the initial stage of its development, as the share of the ICT sector in the gross value added of the business sector is only 3.4%, while the same indicator for other countries is significantly higher: Republic of Korea - 10.3% ; Sweden - 7.3%; Finland - 6.9%; USA - 6%; Estonia - 6%; Czech Republic - 5.9%; Great Britain - 5.4%; Germany - 5%; France - 4.6%; Canada - 4%.

Considering the position of Russian ICT in the cross-country context on the export of goods and services, it can be noted that they are minimal (Figure 5) [15]. Moreover, a distinctive feature of the Russian Federation is the specialization of the Russian Federation in finding the solutions in the most competitive segments of the world's software (artificial intelligence, big data, computer vision, information security, machine learning).

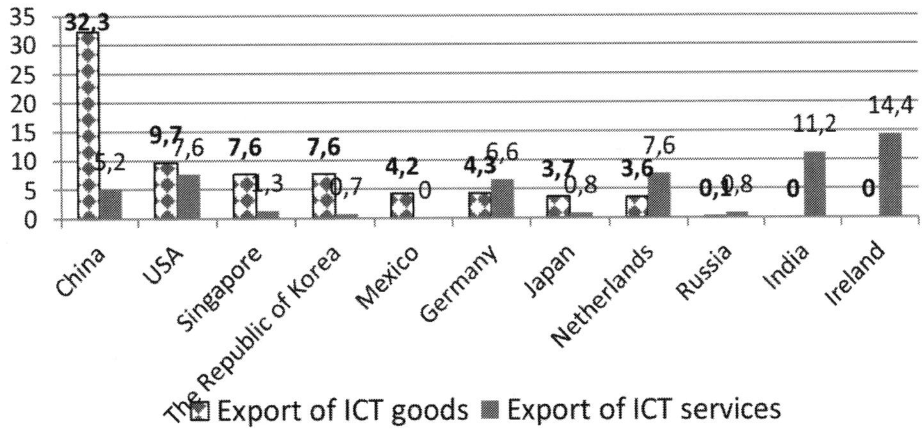

Fig. 5. Exports of ICT goods and services, as a percentage of global exports of goods/services

In addition, the Russian ICT sector remains highly import-dependent, as for every dollar of ICT goods exports accounted for $ 10.11 of imports, and for every dollar of ICT services exports accounted for $ 1.11 of imports of services. Therefore, institutional measures are needed to develop import substitution in this sector.

Another important area of ICT development is the creation of new digital platforms, including new networks, cloud technologies, data centers, applications for processing and providing data. These software products are aimed at forming fundamentally different business models based on completely different principles of interaction between the consumers and manufacturers.

Summarizing the above trends, the authors believe that in order to create a favorable environment for the development of digital state entrepreneurship, it is necessary to develop regional institutions to support the main directions of digitalization of the economy. In our opinion, such a design should be based on the following institutions that are in continuous interaction: Institute of Industry Digitalization; Institute for Digitalization of Society and Institute for ICT Development. Figure 6 shows the author's model of the regional institutional design of digital state entrepreneurship. The model includes three modules:

1) The institutions of the digitalization of the industry;
2) Institutions of digitalization of society;
3) Institutes for the development of information and communication technologies.

Each module has a specific institutional structure and is focused on the specific results. This approach systematizes and integrates all the existing developments in the field of achievements of digital state entrepreneurship.

Institutes	1. Structure	2. Results
1. Institutes of the digitalization of the industry	1.1.1 additive technologies; 1.1.2 expansion of automation and robotics; 1.1.3 formation of the industrial Internet of things; 1.1.4 cloud computing; 1.1.5 use of smart networks; 1.1.6 end-to-end automation, CRM, ERP, PLM applications 1.1.7 implementation of cyber-physical systems; 1.1.8 development of industrial networks.	2.1.1 "smart" services and products; 2.1.2 decentralization of production management; 2.1.3 flexible supply chain; 2.1.4 flexible innovation in the product lifecycle; 2.1.5 flexible production networks; 2.1.6 improving the manageability of technological processes; 2.1.7 mass customization; 2.1.8 reduce costs and time spent on the product lifecycle.
2. Institutes of the digitalization of the society	1.2.1 storage and processing of "big data»; 1.2.2 creation of artificial intelligence; 1.2.3 creating the Internet of things; 1.2.4 virtual reality applications; 1.2.5 cloud computing; 1.2.6 geolocation applications.	2.2.1 participation in online trading; 2.2.2 use of geolocation technologies; 2.2.3 getting new educational services; 2.2.4 new ways and speeds of knowledge sharing; 2.2.5 availability of e-government services; 2.2.6 simplification of financial transactions and non-cash payments; 2.2.7 development of telemedicine and accurate health diagnostics.
3. ICT development institutes	1.3.1 increasing the number of University graduates in technical specialties related to ICT; 1.3.2 creating new platform companies and increasing jobs; 1.3.3 increase investment in fixed assets of ICT enterprises 1.3.4 import substitution and increasing the volume of exports of ICT products and services	2.3.1 standardization of processes; 2.3.2 organization of remote work; 2.3.3 emergence of new professions and the growth of the labor market; 2.3.4 intellectual property protection; 2.3.5 availability of skilled workers; 2.3.6 investment in research; 2.3.7 new business models.

Fig. 6. Structural model of regional institutional design of digital state entrepreneurship

The first module of the proposed model unites industrial digitalization institutions, reveals their structure and expected results of functioning.

The integration of modern information technologies and the production sector of the economy will lead to the emergence of new products, models and business formats, an increase in the quality and efficiency of the production process, and ensuring the stability of its work in any conditions, including during the COVID-19 epidemic.

The second module combines the institutions of digitalization of society. The digitalization of society is currently fundamental in its economic and social modernization. It leads to the acceleration of vital processes, access to services, that part of the population for which they were inaccessible. So, digitalization of public services, education, medicine can improve the life of a significant part of society. The doctor can not only promptly diagnose, but also carry out the operation, regardless of the location of the patient. A resident of any region of the country can get an education, improve their qualifications or get a new profession remotely. Customers have a huge selection of products available in online stores. At the same time, the speed of rendering services and their quality increases.

The third module presents institutions for the development of information and communication technologies. The development of information and communication technologies will also create an information society, improve the quality of life of citizens, develop the economic, socio-political, cultural spheres of society, improve the public administration system, and ensure the competitiveness of products and services in the field of information and telecommunication technologies.

Using the methodology of tensor modeling of complex technical and economic systems (which includes state digital entrepreneurship), we can transform this structural model into a conceptual model of the main states of digital entrepreneurship, and also highlight the presence of invariants. These invariants are: reproductive scientific and technical cycle; types of activity (including by industry); resources consumed and produced (costs - results). Tensor algebra allows us to build general and specific models characterizing individual aspects and situational states of state digital entrepreneurship. In each specific case, you can use a typical mathematical apparatus of nonlinear optimization. The level of detail in the calculations is determined by the subject of management.

The indicated approach allows us to fix three blocks (Figure 6), and within the blocks there are various situational combinations that characterize certain states of digital entrepreneurship. These states characterize digitalization in terms of economic levels (national, regional, local) and types of activity, taking into account the complexity, technology features, industry, personnel training level and society's readiness to perceive digitalization processes.

To determine the most rational digitalization chains at the enterprise-to-enterprise level; territory-to-territory; industry-to-industry should use a non-linear modeling apparatus.

Each of the constituent links of the reproduction cycle of updating digitalization objects{ 1) development of a new idea of digitalization; 2) familiarization; 3) replication; 4) modernization of the digitalization object; 5) exhaustion of potential, transition to a new digital idea }, respectively, we denote A, B, C, D, E, their specific weights, respectively - α, β, γ, δ, ε. Let us introduce a unified notation for the coefficients r_1, r_2, r_3, r_4, r_5. Each of the phases has its own directional vector and its own quantitative characteristics. Then the optimization of digitalization development processes can be expressed by the following mathematical model.

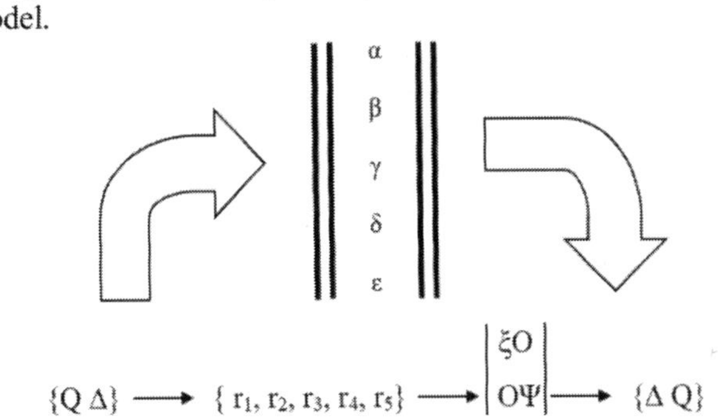

Fig. 7. Model of optimization of digital processes through changing the power of streams

where: Q Δ is the gradient of increasing the power of the flow-process of changing state digital entrepreneurship (incoming flow);

r_1, r_2, r_3, r_4, r_5 – conversion factors for the capacity of state digital entrepreneurship;

α, β, γ, δ, ε - vectors of individual phases of the renewal cycle of state digital entrepreneurship;

ξO is a possible increase in the power of the process flow (changes in state digital entrepreneurship);

OΨ - the actual increase in the power of the flow-process (changes in state digital entrepreneurship);

Δ Q - increase in flow power due to optimization of the interaction of elements (outflow).

The model makes it possible to comprehensively monitor the pros and cons of products (services), regions and industries and determine the most rational basic chains of state digital entrepreneurship, and accordingly develop adequate mechanisms for managing them.

Summing up the above we can draw the following conclusions:

1) the successful development of digitalization in the basic sectors of the economy is the source and consequence of a strategically adjusted scientific and technical policy at the state and regional level, combined with market methods for managing business processes and the development of the direct market for digital services and goods in the territory;

2) the basis for the successful development of digitalization is the unity of theoretical, methodological, procedural approaches, taking into account the transformation of the business model for the development of the country's economy and a specific region;

3) it is necessary to observe the principle of proportionality and evenness in the development of state digital entrepreneurship, given that the stability of the economy is determined, first of all, by the level of development of the industrial base as a whole, and not by its individual, even quite advanced fragments;

4) industrial corporations, being historically more organized and dynamic structures integrated into the economy of the country and the region, are highly sensitive to its general state and development trends, since the latter generate market activity, affect the strength of intersectoral and interregional information links, determine the investment climate, including in digitalization. This makes it possible to use the principle of the leading link in solving common problems of digitalization of the economy and society.

5) Digitalization processes built into multi-level scientific and industrial cooperation contribute to the integration of the economic space and accelerate the business processes of the reproduction cycle;

6) A successful solution to the development of state digital entrepreneurship requires an integrated approach: transformation of the digital law system, ideological and marketing support, systemic management of business processes at the regional level; harmonious integration into the socio-economic and social life of the population.

References

1. T.V. Anakhin, Bulletin of the Buryat State University, 6 (2013)
2. V.V. Volchik, L. A. Shafirov, A.A. Oganesyan, Journal of economic regulation (Issues of economic regulation), **4(4)**, 41 (2013)
3. A.O. Glazacheva, O.E. Perfilova, Social and Humanitarian Knowledge, **12**, 156 (2008)
4. A.O. Glazacheva, Social and humanitarian knowledge, **4**, 327 (2019)
5. M.S. Glatko, Omsk Scientific Bulletin, **3(119)**, 96 (2013)
6. M. S. Kagan, *Philosophy of culture*, 415 (1996)
7. I. A. Krehbel, Philosophy and social dynamics of the XXI century: materials of the II Intern. scientific conf, 156 (2008)

8. M.A. Kulikov, Izvestia of the Russian State Pedagogical University named after A.I. Herzen, **64**, 84 (2008)
9. G.G. Kurierova, MSU Bulletin. Ser. 7. Philosophy, **2**, 74 (1994)
10. V.A. Lyashenko, Humanities. OSU BULLETIN, **9(91)**, 21 (2018)
11. Official website dolgorukov.ru, http://dolgorukov.ru/
12. Official website dolgorukov.ru, https://www.openbusiness.ru/
13. Official website wikipedia.org, http://en.wikipedia.org/
14. Ya. M. Roshchina, S.Yu. Roshchin, V. N. Rudakov, Questions of education, Educational Studies, **1**, 174 (2018)
15. G.I. Abdrakhmanova, K.O. Vishnevsky, L.M. Gokhberg, *Digital economy: 2020*, a brief statistical collection, 112 (2020)
16. R.E. Goodin, The theory of institutional design, **1** (1996)
17. H. Simon, The Sciences of the Artificial. Cambridge, 55 (1969)

Interaction of technology entrepreneurship and higher education as a factor in the innovative development of the region

Elza Mantaeva, *Victoria* Goldenova*, *Dayana* Batyreva, and *Dolgan* Batyrev

Kalmyk State University named after B.B. Gorodovikov, Pushkin Str., 11, 358000 Elista, Russia

Abstract. The purpose of the study is to substantiate the strengthening of the role of scientific and educational institutions as centers for the formation of technological, managerial and innovative competencies in the formation of an innovation ecosystem, as an environment for the functioning of technological entrepreneurship. In the course of the work, it was revealed that the motivation of the entrepreneurial sector in funding scientific research of educational and scientific institutions should come to the fore. It has been substantiated that it is precisely the close interaction of business, educational and government structures that is the key to increasing investment and innovation activity in the region. The theoretical significance of the article is to substantiate the need to strengthen the role of educational institutions as centers for the formation of technological, managerial and innovative competencies in the formation of an innovation ecosystem, as an environment for the functioning of technological entrepreneurship.

1 Introduction

End of XX century characterized by the emergence of a new stage in the development of the leading countries of the world, a feature of which was the transition to the fifth technological order, the distinctive features of which are the active use of information and communication technologies, bio- and nanotechnologies, genetic engineering, renewable energy sources, which is quite naturally reflected in priority areas development of science and technology. Today the world economy is on the verge of the sixth technological order, the outlines of which are beginning to form in developed countries of the world, primarily in the USA, Japan and China, the vector of development and application of science-intensive, or so-called "smart technologies" becomes the system-forming signs of a new technological order [1].

In turn, the transition to a knowledge-based economy aggravates a number of problems of the world economy, among which, first of all, it is necessary to highlight the sustainable development of economic systems, the development and implementation of social innovations, the development and implementation of renewable energy sources, and much more. In this regard, the demand for information and knowledge-rich technologies and goods to meet the growing material and social needs is growing. All this determines the modern vector of development of human civilization in the forecasted future.

* Corresponding author: goldenova-vs@yandex.ru

2 Materials and methods

The main content of the study is an analysis of the activities of the Kalmyk State University, which received the status of a pivotal regional university, on the example of which it is shown what role educational structures can play in the formation of the innovation ecosystem of the region as an environment for the functioning of technological entrepreneurship. It is substantiated that it is the state that must create conditions in which it is entrepreneurship that will become not only the customer of qualified personnel with innovative, managerial and technological competencies, but also the customer of scientific research carried out by educational institutions.

3 (Results and Discussion)

Structural change in the determinants of the external and internal environment of economic systems as a result of scientific and technological progress leads to the dominance of the development of the technosphere, which acts as a set of artificial systems created by man [2]. The result of this development is the emergence of a qualitatively new type of founders of business, with fundamentally new technological, managerial and innovative competencies. It is economic entities with such competencies that become representatives of technological entrepreneurship [3].

One of the reasons for the rather low level of innovative activity of economic entities in the Russian economy is the low level of spread of technological entrepreneurship. At the same time, according to some estimates of researchers, during the first five years, most technology startups go bankrupt, while in 80% of the reasons for failures are problems of marketing and management, as well as capitalization of assets. Proceeding from the fact that it is the institutions of science and education that are the centers for the formation of technological, managerial and innovative competencies, it becomes necessary to clarify their role in the development of technological entrepreneurship.

There is generally no connection in the public mind between successful technological companies and higher education institutions. However, it is universities that are traditionally centers for generating innovations, it is there that the emergence and transfer of new knowledge takes place, and it is in universities that the greatest concentration of young people is observed, presenting demand for new technological, managerial and innovative competencies [4].

It is fair to note that research carried out in universities can be divided into the following types. First, it is fundamental research, which is deep in nature, which is the source of the formation of fundamentally new theories and directions of development of science and technology, and determines the directions of scientific and technological progress. Moreover, this type of research is extremely financially costly, requires unique equipment and highly qualified personnel, but does not bring profit. As a rule, this type of research is within the power of only highly developed countries, which can afford to finance basic science through sufficiently tangible budgetary expenditures. As a rule, large-scale fundamental research results in a number of narrower applied research areas that have broader opportunities for practical application. Thus, there is the formation of applied science, the results of which are easier to commercialize and are designed to create new products or services or create new properties of products and services required by the consumer. Moreover, it is natural that the period of applied development is much shorter and cheaper than the period of the corresponding fundamental research [5]. Thus, it is the fundamental developments that are the basis for applied science, the results of which are reflected in various types of technological entrepreneurship.

Regional economic systems as a consequence of the development of technological entrepreneurship on their territory receive a list of technologies, products and services characterized by a high level of competitiveness in the market. This accelerates the rate of economic growth in the region, changes the structure of the economy and increases the investment attractiveness of the territory. The experience of developed countries confirms that it is the level of development of technological entrepreneurship that is one of the most important indicators of the level of regional development [6]. Consequently, it is necessary to strengthen the role of universities in the development of technological entrepreneurship, and the state should create not only favorable conditions for innovative activities of universities, but also conditions for the commercialization of these innovations [7].

In modern conditions, the creation of a system of flagship universities in Russia presupposes positioning them as the cores of the region's innovation ecosystems [8]. For example, in the Republic of Kalmykia, the only institution of education and science on the territory is becoming one of the most important drivers of the socio-economic development of the region. At the same time, it should be noted that in these conditions regional universities are faced not only with the task of providing high-quality educational services and research activities, but also independently attracting investors to finance these activities. The consequence of this situation is that a regional university has to become not only a university for students, but also an entrepreneurial university in order to obtain an opportunity to finance the research activities. In this regard, it is the development of technological entrepreneurship on the basis of the university that opens up wide opportunities for educational and scientific institutions to attract resources to finance scientific research and more active interaction with the regional community.

Today, after receiving the status of a reference regional university, Kalmyk State University plays a major role in the formation of a regional innovation ecosystem. The higher educational institution of the Republic of Kalmykia is the central research site of the territory and is focused on the formation of priority areas for the socio-economic development of the region.

Kalmyk University has strong friendly relations with the Innovation Support Fund, within the framework of which KalmSU acts as the main platform for competitions within the framework of the "U.M.N.I.K." and "START", the annual forum "Innovative Kalmykia", which are the centers of attraction for the best innovative projects in the south of the country. Also, innovative activity at KalmSU is represented by a network of small innovative enterprises, which are a clear example of the development of technological entrepreneurship on the basis of an educational institution.

It is innovation and entrepreneurial activity that is becoming a promising direction for the development of the university in modern conditions. As the tasks of the implementation of this direction, it should be noted the formation of the innovation ecosystem of the region, stimulation of the development of technological entrepreneurship, support of innovative activities, search and development of talented youth as the basis of the human potential of technological entrepreneurship, formation of technological, managerial and innovative competencies of future entrepreneurs.

The studies carried out indicate that modern Russian share of universities acting as partners in technological entrepreneurship is extremely small. [9]. At the same time, as international experience shows, in highly developed countries, the level of interaction between technological entrepreneurship and higher educational institutions is much higher. The following mutually beneficial reasons for such a high level of cooperation between technology business and universities in world practice can be distinguished:

1) universities are the main subject of the supply of qualified personnel for the labor market, while the peculiarity of the new federal state educational standards is that it is the request of employers that determines the profile of training areas;

2) educational institutions are a source of new scientific research and innovative developments;

3) universities are the basic platform for creating technology startups.

Thus, mutually beneficial cooperation between business and educational structures can bring significant social, cultural and financial results for both parties. Businesses will benefit from the supply of skilled workforce and applied university research. Educational and scientific institutions can act as producers of knowledge commercialized through technology entrepreneurship, which will help build the capacity of resources for further research funding. Thus, it is precisely the close successful interaction of business and higher educational institutions that will accelerate the socio-economic development of the territory. World experience testifies that in modern conditions it is becoming an important factor of success to build up regional interaction between education and business.

However, it is the trust between potential partners that becomes the main problem in the development of cooperation between universities and business. This is clearly seen in the modern relations between educational and business structures. The asymmetry of information about research conducted within the higher educational institutions leads to a mismatch between supply and demand in the market for technological innovations, which then results in a low level of commercialization of scientific and technical developments and a low level of patenting [10].

An insufficiently high level of cooperation between business and educational institutions, which is reflected in the insufficient level of development of technological entrepreneurship, can be called as a factor that inhibits development on the technological development of Russia. The reluctance of business structures to invest in research in universities leads to a slowdown in research growth and an increase in demand for public investment. At the same time, the effect of state funding for scientific research is usually lower than the effect of commercial funding. There are many examples in the world when there were the startups that emerged on the basis of an educational institution that, over time, were transformed into successful large businesses. It is the high-quality cooperation of business and educational structures in modern conditions that becomes the key to the socio-economic development of the territory. At the same time, it is the enterprises that are actively involved in cooperation with educational institutions that are the first to gain access to technologies for the production of unique intellectual products and services.

4 Conclusions

The effective promotion of innovation and the achievement of technological leadership can only be built on the basis of close cooperation between educational and business structures. The state should play a key role in this process, but funding of scientific activities from the state budget alone is not enough. The state must create conditions in which it is entrepreneurship that will become not only the customer of qualified personnel with innovative, managerial and technological competencies, but also the customer of scientific research carried out by educational institutions. Therefore, the motivation of business structures in investments in scientific research of educational institutions comes to the fore. Here the state can use such instruments of state incentives as state guarantees and orders, as well as measures to improve the social status of research activities of universities with the aim of innovative development of territories.

As for the regions, here the regional innovation policy should come to the fore, the purpose of which should be the formation of an effective innovation ecosystem that will promote the spread and successful functioning of technological entrepreneurship. In particular, the Kalmyk State University today can become the platform on which the interaction of the scientific and entrepreneurial potentials of the region will take place so that they produce unique intellectual goods and services and contribute to an increase in the general level of well-being of the region.

In this regard, the main goal of the regional policy to stimulate innovation is to create favorable conditions for the development of technological entrepreneurship on the basis of the university and through the private investment. The positive consequences of such a policy will be an increase in the competitiveness of local goods and services, an increase in the number of jobs created, an increase in the investment attractiveness of the territory and, as a consequence, the solution of many socio-economic problems of regional development. Fundamental is the creation of the necessary institutional environment for the implementation of innovative activities, which implies the creation of an appropriate regulatory legal framework, the development of targeted programs, the creation of a system of grant support and much more. The main tasks of regional target programs should include stimulating interaction between subjects of supply and demand for research activities, dissemination of best practices for creating innovative products, and developing a general strategy for innovative processes in the region.

At the same time, with a lack of its own resources for regional and federal funding of scientific research, the state must create conditions under which the entrepreneurial structures would benefit from participation in the innovation processes taking place in the universities. It is precisely the close interaction of business, educational and government structures that will be the key to increasing investment and innovation activity in the region. Another priority area for the development of technological entrepreneurship at the regional level is the creation and development of a technological platform, in which the efforts of entrepreneurial, educational and government structures will be consolidated. It is possible to implement this direction within the framework of the cluster approach by creating and developing a technological cluster, the directions of activity of which will determine the vectors of development of priority sectors and complexes of the region's economy. Kalmyk University in the status of a reference regional university can become a key participant in the technological cluster of the Republic of Kalmykia, solving the problems of reproducing the intellectual potential of the region and its effective use in innovation. The result of the creation of a technological cluster can be the formation of an effective innovation ecosystem in the region, an increase in the investment attractiveness of the territory, a decrease in educational migration, an increase in regional and local budget revenues, and much more.

References

1. E. N. Kablov, Science and life, **4**, 2 (2010)
2. C.M. Nacu, S. Avasilcri, Advanced Materials Research, **837**, 639 (2014)
3. H. Montiel-Campos, Y.M. Palma-Chorres, Journal of Technology Management & Innovation, **11(3)**, 77 (2016)
4. T. Bailetti, Technology Innovation Management Review, **2(2)**, 5 (2012)
5. D.E. Tolmachev, K.V. Chukavina, Regional Economy, **16(2)**, 420 (2020)
6. I.N. Kichigin, Social and Human Sciences: Theory and Practice, **1(2)**, 298 (2018)
7. E. Badzicska, Entrepreneurial Business and Economics Review, **4(3)**, 57 (2016)
8. I. V. Korchagin, Bulletin of the Research Center of Corporate Law, Management and Venture Investment of Syktyvkar State University, **2**, 96 (2019)
9. A.V. Vershitsky, E.R. Vershitskaya, Service in Russia and abroad, **13(2)**, 162 (2019)
10. J. Alvedalen, R. Boschma, European Planning Studies, **25(6)**, 887 (2017)

Government, business and NPOs: formation of a new model of interaction and assessment of its effectiveness

Alla Aleksandrovna Bylinskaya, and *Sergey Aleksandrovich* Sudin

Lobachevsky State University of Nizhni Novgorod, 23, Gagarin avenue, Nizhni Novgorod, Russia

Abstract. The relevance of the research topic is due to the denationalization of the social services market and, as a result, the growing importance of the third sector in their provision; the development of civil society, the construction of social problems, in the solution of which non-profit organizations play an important role. A comparative analysis of assessments of the emerging model of interaction between civil society and the state will allow us to determine its current state, show strengths and weaknesses, the most pressing problems and development trends. A large role in the study of the issues raised belongs to the study and critical understanding of the leading practices of building interaction between the state and civil society, which will not only adapt the most appropriate practices, but also avoid possible problems. The proposed research will help to establish feedback between the state, business and supported NPOs and develop recommendations for optimizing this interaction. The aim of the study is to comprehensively evaluate the effectiveness of the emerging model of interaction between the state and NPOs as new actors in the country's economic and socio-cultural policy and forecast the prospects for its development. The study is based on the study of the perceptions of leaders and key actors of NPOs about the prospects for the development of their organizations in the context of the emerging model of interaction with business and the state. The authors put forward a hypothesis about the need to study the assessments of the existing model of interaction between the state and NPOs by representatives of executive authorities responsible for the implementation of relevant programs, direct managers and key actors of NPOs, as well as representatives of target groups of their activities in the regions to determine the problems and prospects of interaction between the state, business and NPOs.

1 Introduction

The transition to the polysubject market of social services, initiated by 442-ФЗ "On the Basics of Social Services for Citizens of the Russian Federation", put an end to the state monopoly on their provision and led to the emergence of not only registers of social service providers, but also numerous socially oriented NPOs that spread their activity in all kinds of areas of social support and assistance. The social construction of problems, attention to the needs of deprived social groups, the implementation of the principle "nothing for us without us" when planning measures for their social protection has become a positive example of adapting international experience and increasing trust in non-state actors of social protection and assistance. This policy also became a response to the de facto ban on financing of civic activism of social orientation by foreign funds, which for a long time provided the opportunity for systematic work for leaders of the main areas of activity. Over the three years that have passed since the launch of the grant distribution system in its modern form, there has been a need for its assessment both on the part of NPOs, participating or not participating in the application process and implementation of initiatives, and on the part of government agencies. In 2016, under the direcrion of S.A. Sudin, a large research was carried out to study the interaction of NPOs and the state, during which problems were identified that hinder their constructive interaction: bureaucratization and control as a result of a lack of trust, strict regulation of the types of services, focus on image projects in the distribution of funds, high entry barriers for obtaining funding through the provision of services and many others. The system of presidential grants has become a fundamentally new step in supporting not only civil society as such, but also those social groups and individuals to whom its activity is directed. A comprehensive assessment of the effectiveness of socially oriented NPOs, an analysis of the range of thematic areas as an indicator of the most pressing social problems, as well as the prospects of the system itself is the subject of the proposed study.

2 Materials and methods

The research questions presented in this study are characterized by relative novelty and are being implemented within the framework of related thematic areas. First of all, NPOs and their activities are considered as a component of a civil society that is in the stage of formation and, therefore, experiencing specific difficulties associated with trust from state institutions, the formation of civic consciousness, stimulation of population activity, and the possibility of using foreign models of interaction.

The issues of the formation of civil society are in the focus of M.K. Gorshkov [1], [2], Z.T. Golenkova [3], I.A. Khalii [4]. The issues of forming interaction and monitoring the state of civil society are presented in the works of the staff of the Center for Civil Society Research and the Non-Profit Sector of the Higher School of Economics: L.I. Yakobson, I.V. Mersiyanova, V.B. Benevolensky, M.N. Gromova et al. [5], [6], [7], [8], [9]. Another area is the study of volunteering as a manifestation of civic engagement and the basis for the formation of non-profit structures. This aspect is at the center of research of the employees of the Ural Federal University. The first President of Russia B.N. Yeltsin: G.E. Zborovsky [10], [11], M.V. Pevnoi [12], A.A. Kuzminchuk [13].

The works of foreign authors reflect the issues of relationship in the development of civil society and the non-profit sector [14], cooperation between the state and the non-profit sector [15], the impact of democratization on the development of NPOs [16], network interaction of NPOs [17], financial management of NPOs and diversification of sources their income [18], the problem of changing the scale of the spread of NPOs and indicators of their development [19].

The scientific novelty of the research consists of the goal and objectives of the research and consists in an attempt to create a system for assessing state support for NPOs through grants, funding through the provision of services, as well as by comparing with the practices of international experience. The problem under study is to assess the effectiveness of the existing system of interaction between business, government and NPOs and compare with the results of the previous research. The social-constructionist approach presupposes the need for further construction of social problems in the sphere of NPO activity, the structural and functional one - a description of the place of NPOs in the civil society system and an analysis of their functions; systemic - analysis of the activities of NPOs in the system of social relations in the close relationship of structures, institutions and processes of modern Russia.

When performing the work, approaches and methods were used to solve theoretical and practical research problems. The theoretical basis will be made up of the system, structural and functional and problemological approaches taken together, which make it possible to form the conceptual apparatus of the research and interpret the basic concepts in a social and scientific context.

The following issues are used as empirical methods aimed at collecting and processing information:

1. In-depth interviews with the representatives of the executive authorities responsible for interaction with civil society institutions, leaders and key actors of NPOs, as well as representatives of target groups to which their activities are directed. The results of in-depth interviews will form the basis for the formation of tools for mass surveys.

2. Questionnaires allowing to give a quantitative and qualitative assessment of the effectiveness of NPOs from the point of view of their clients, to highlight the forms of their participation in the life of society. In addition to traditional technologies for collecting personal data, it is planned to use paperless technologies, which will facilitate the process of collecting data, increase the number of respondents to provide the highest quality content.

3 Content analysis of open sources devoted to the activity of NPOs, assessing their resource capacity and effectiveness in various areas of social functioning.

3 Results and discussion

The results of the research are consistent with the tasks. First of all, a quantitative and qualitative analysis of the existing thematic areas of activity of NPOs receiving support from the state was carried out. This is considered as one of the main indicators of the recognition of their relevance for the economic socio-cultural development of the country. The study of assessments of the existing model of interaction between the state and NPOs in the future will become a reliable feedback tool and provide valuable analytical material. In the course of the study, assessments were collected from employees of executive authorities, from leaders and key actors of NPOs, and from representatives of target groups to whom their activities are directed. The data obtained make it possible to assess in dynamics the effectiveness of public policy measures in relation to civil society based on a comparison of new data with the results of previous studies.

The cross-cultural nature of the study also involves the study of the latest trends in the dynamics of interaction between the state and civil society abroad, as well as a critical analysis of existing practices and an assessment of the prospects for their implementation in Russia.

The involvement of the leading actors of NPOs in the application system and the implementation of project initiatives contributes to the formation of not only assessments of the current state of the interaction system, but also scenarios for the further development of relations between the state and NPOs. Of particular interest for future research is the assessment of the prospects for the development of social entrepreneurship, which we understand as the next step in the formation and development of NPOs.

The research work was carried out in two main stages.

The first stage is theoretical and methodological one. At this stage, the development of the methodological base of the study was carried out, the analysis of the domestic and foreign discourse of the problems of interaction between civil society and the state, the existing socio-cultural context, the development and testing of research tools and the solution of organizational and technical issues was carried out.

Within the framework of this stage, the following procedures have been implemented:

1. Development and approval of the theoretical basis of the study and its methodological foundations; preliminary analysis of the object, putting forward working hypotheses. In the course of implementing this step, a comprehensive research program was created and publications were prepared on the theoretical conceptualization and empirical operationalization of various models of interaction between the state and civil society institutions.

2. Conducting a series of pilot studies to test questionnaires, creating data entry matrices in SPSS. Data analysis and formulation of the final versions of the questionnaires. The result of this step is a data collection tool for all groups of respondents.

3. Preparation of publications on the theoretical part of the work. Within the framework of this stage, the publication of a collection of scientific works "Civil society in Russia: modern challenges" [20], including works on the problems of interaction between the state, business and civil society institutions, was carried out.

The second stage of the work is the empirical and analytical stage. Within the framework of this stage, the planned research procedures were implemented, the main array of data was obtained, their processing and meaningful analysis were carried out.

During the implementation of this stage:

1. A sample was formed for all groups of respondents.

2. Research has been carried out at all empirical sites. The bases of matrices in the sav format were formed, their processing and meaningful analysis were carried out. During the implementation of this stage, databases on questionnaires and protocols of in-depth interviews were obtained.

It is planned to build frequency and linear distributions by research parameters, cross tables, multivariate data analysis, identification of functional dependencies of research parameters, etc. preparation of publications based on the results of an empirical study of models of family practices in a civil society.

The implementation of the study made it possible to obtain the following results:

1. Development and approval of the theoretical basis of the study and its methodological foundations; preliminary analysis of the object, putting forward working hypotheses. During the implementation of this step, a comprehensive research program was created and publications were prepared on the theoretical conceptualization and empirical operationalization of various models of interaction between the state and civil society institutions.

2. A series of pilot studies have been carried out to test the questionnaires, create data entry matrices in SPSS. The analysis of the data and the formulation of the final versions of the questionnaires were carried out. The result of this step is the generated data collection toolkit for all groups of respondents.

S.A. Sudin conducted a study "The state as a partner of Russian NPOs", during which the heads and key actors of non-profit socio-medical organizations were interviewed by the method of expert interviews (N = 47). As a result of the study carried out during the period of radical transformation of the model of interaction between NPOs and the state, the following problems were identified:

1. Lack of effective mechanisms of state subsidies for public organizations, implemented through contractual relations;

2. The actual lack of interest on the part of state authorities in enhancing interaction with public organizations;

3. Weak feedback mechanisms in the chain of interaction between the state and NPOs, which do not allow for a comprehensive assessment of the effectiveness of both the interaction itself and the activities of public organizations. A narrow range of indicators and

data, represented by the data contained in the financial statements and indicators of the budgetary efficiency of NPOs.

As a result, since public authorities are not sufficiently well aware of the activities of public organizations, this does not contribute to increasing confidence and expanding the range of areas of interaction between the state and NPOs.

The transformation of state policy towards NPOs, expressed in the creation of registers of providers of social services and, later, the activation of grant programs, should have changed the attitude towards state policy in this area. A comprehensive assessment of the emerging system of interaction can become a source of important information for its further planning.

With the support of the ISEPI Foundation, the project "Research on partnership between civil society institutions, government and business for the development of dialogue, strengthening civil society and dissemination of best practices in this area" was implemented. The implementation of the project included a study of the main practices of social partnership and the identification of problems standing in the way of the development of social partnership. The study was conducted in Moscow, St. Petersburg, Nizhny Novgorod, Pskov, Samara regions and in the Perm Territory. An expert survey of 120 people (20 in each of the regions) was carried out, 6 focus groups were conducted, with a total of 62 participants. A total of 182 experts took part in the project. Table 1 presents data on the composition and structure of the expert group.

Table 1. Composition and structure of the expert group

Region	Men/ women	Government departments	Business	NPOs	Experts/ universities	Media
Moscow	8/12	4	5	11	0	0
Saint-Petersburg	12/8	4	5	7	4	0
Nizhny Novgorod	13/7	3	7	9	1	0
Perm region	7/13	4	6	7	2	1
Pskov	13/7	5	5	5	3	2
Samara	7/13	5	6	6	3	0
Total	60/60	25	34	45	13	3

The age structure of the research participants includes respondents and experts aged 20 to 70 years, of which the age group 20-30 years old was 26 people, the group 31-40 years old was 31 people, the group 41-50 years old was 31 people, the group 51-60 years was 21 people, the group of 61-70 years was 5 people.

According to the results of the study, it was revealed that among the experts there is no unambiguous interpretation of both the very concept of "Social partnership" and the basic forms of its implementation.

Based on the results of the expert study, the basic forms of social partnership implementation include:

1. formation of collegial advisory bodies at the federal, regional and municipal levels, mainly in the form of public councils, in order to expand the possibilities of interaction between the state, business and NPOs and to provide conditions for the inclusion of business in solving issues of the development of public institutions;

2. financing of the activities of NPOs on a competitive basis, involving the participation of interested partners from both government bodies and business structures in order to increase the efficiency of allocation of both state and charitable funds and the effectiveness of joint projects;

3. charity as the simplest and most important form of interaction between different sectors.

In accordance with the resource base, based on the results of the study, a number of forms of social partnership were named, such as the implementation of joint projects, volunteer actions, charitable actions, the provision of the infrastructure by the business for the implementation of projects within the framework of social partnership and material assistance provided by the business of the NPO.

According to the results of the study, it was found that grants are of particular importance for financing social partnership projects, according to experts. It is they who act as direct investments that provide both the solution of problems and stimulate the institutional development of social partnership. According to experts, the main questions that business and NPOs have to the authorities are related to the organization of effective interaction.

According to the results of the study, based on expert assessments, it was revealed as follows:

1. Lack of interest and initiative among officials responsible for the implementation of projects within the framework of social partnership. Since responsible officials are accountable to the state, they are unable to ensure effective representation of the interests of society. Thus, bridging the gap in the "power-society" chain becomes a big problem.

2. The existence of a demand for social partnership is a consequence of the insufficiently effective work of the state and social services.

3. The inertial nature of interaction between government and society. In most cases, interaction is carried out in the format of social protection that is familiar to everyone. Despite the fact that this form of interaction is passive and a transition to active forms of interaction in this area is required, the inertial effect is triggered and this transition does not occur.

4. Despite the awareness of the importance of moving away from the primitive distribution of funds to the unprotected strata of the population to the development of active forms of interaction with active strata of society interested in working on solving common problems, this problem has not yet been overcome.

In the course of the study, the main criteria and requirements for participants in social partnership were tested. The structure of the criteria is shown in Table 2.

Table 2. Criteria and requirements for partners

Group of criteria	Components of the criteria group
Requirements for the competences	if potential partners have the necessary competencies clarity and consistency in decision-making and interaction within the framework of social partnership diligence, punctuality and responsibility in relation to the undertaken obligations compliance with legal regulations, requirements and criteria
Moral and ethical requirements	honesty and decency willingness to take the path of making reasonable compromise decisions ability to suppress "corporate selfishness" ban on solving your own/personal problems through a joint project moral stability
Technological requirements and criteria	effective financial control mechanisms transparency environmental friendliness and respect for the environment ability to implement the effective projects activity and willingness to work together

Table 2. Continued

Communication requirements	trust in partners mutual interest in potential participants in a joint project ability to carry out the search for and finding compromise solutions and desire
Image criteria	whether potential partners have a positive reputation, whether potential participants in the interaction have the recommendations, negative attitude of participants towards corruption, rejection of various corruption schemes stable positions of potential partners in the market whether potential partners have a portfolio of projects

According to the results of the study, the following were named as undesirable partners: illegal business, interaction with which was marked as inadmissible within the framework of social partnership, classifying it as criminal; types of industries whose products are capable of harming health; organizations engaged in political activities; organizations that appear on various lists of unfair partners.

On the basis of the study carried out by the authors, the key factors at various levels that hinder the implementation of social partnership in Russia were identified. The results are shown in Table 3.

Table 3. Factors hindering the implementation of social partnership

Levels	**Obstacles to the implementation of social partnership**
Level of organization	-weak institutional support for the activities of NPOs, -insufficiently clear understanding of the interests of partners, -incompleteness of information about the activities of NPOs both from business and from the state, - unsatisfactory quality of management, - lack of a clearly formulated mission of the NPOs
Region level	-indifferent attitude to serious problems, -reduction of funding for NPOs' projects due to the crisis, -low level of public solidarity, - lack of formation and lack of consideration of public opinion in matters of social partnership, - business misunderstanding of the meaning of interaction with NPOs, - rejection of business and NPOs by the state as equal partners, and unwillingness to jointly solve common problems, - lack of government interest in involving strong NPOs and businesses in solving major social problems, - administrative barriers and attempts at administrative pressure on both business and non-profit organizations, - lack of effective mechanisms for a reliable comprehensive assessment of the professional activities of NPOs

Table 3. Continued

National level	- gaps in the regulatory framework governing the activities of NPOs, - low efficiency of the state in terms of performing a number of functions, combined with high bureaucratization and insufficient customer focus, - lack of flexible forms of interaction, attempts to establish a rigid framework by the state, lack of interest among the officials in working in social projects, -administrative pressure from the state, expressed in attempts to exclude business from the number of partners and organize work only with NPOs within the framework of social partnership projects, - lack of effective mechanisms for a reliable comprehensive assessment of the professional activities of NPOs.

4 Conclusions

1. Prospects for the implementation of social partnership at the present stage are primarily associated with the need to create conditions conducive to increasing the interest of business in interaction with NPOs.

2. Strengthening of institutional support for the activities of NPOs is required.

3. Formation of civil society is possible only with the formation of effective mechanisms of social partnership, requiring the active participation of the non-governmental sector in making managerial decisions.

4. Social partnership is a powerful stabilizer of not only the political, but also the socio-economic situation in the country. As a powerful tool for supporting social innovation and civic initiatives, effective social partnership creates conditions for stability in the future.

5. The current stage is characterized not only by an increase in the understanding of the very idea of social responsibility and interest in the topic of social partnership, but also by the intensification of involvement in social projects, especially among young people. In Russia, there is a gradual departure from social support and a transition to social partnership. A kind of new "trend" is the intensification of partnership interaction between business, the state and NPOs, the involvement of various structures in projects.

5. To increase the efficiency of social partnership, it is necessary to solve the problem of expanding to increase the motivated circle of participants in the process of partnership interaction by creating appropriate motivational mechanisms. In this regard, the questions of search, support for leadership, and education of leaders acquire particular relevance. It also forms new development tracks, as to become a leader is not only fashionable, but interesting as well. Thus, new horizons for development, both personal and eventually federal, are opening up.

5. Today, work is needed in a number of areas to solve the problem of enhancing interaction between the state, business and NPOs through social partnership. The main tasks to be solved:

-creation of favorable conditions for business development;

- creation of effective mechanisms to promote the development of small business as a potential subject of state orders by increasing the competence and professionalism of this participant in social partnership projects;

-expansion of the range of areas of work of the main sites providing interaction between the state, business and NPOs;

-increasing transparency and accountability in the implementation of grants;

- solving the problems of popularizing social partnership, its informational support, the formation of management skills in the non-state sector and fundraising;
- creation of a system for training leaders, which provides an effective mechanism for involving children and youth in social partnership.

References

1. Mikhail K. Gorshkov, Sociological studies, **12** (2016)
2. Mikhail K. Gorshkov, Sociological studies, **1** (2017)
3. Z.T. Golenkova, Sociological studies, **3** (1997)
4. I.A Khaliy, Polis. Political research, **3** (2012)
5. Irina V. Mersianova, Vladimir B. Benevolenski, Public Administration Issues, **4** (2016)
6. V.B. Benevolensky, Civil Society in Russia and Abroad, **1** (2015)
7. I.V. Mersiyanova, I.E. Korneeva, Monitoring Public Opinion: Economic and Social Change, **2** (2017)
8. Irina V. Mersianova, Vladimir B. Benevolenski, Public Administration Issues, **2** (2017)
9. M.N. Gromova, I.V. Mersiyanova, Civil Society in Russia and Abroad, **1** (2016)
10. G.E. Zborovsky, Bulletin of the Perm National Research Polytechnic University, Socio-economic sciences, **3** (2017)
11. G.E. Zborovsky, M.V. Pevnaya, A.A. Vedernikov, Bulletin of the Perm National Research Polytechnic University. Socio-economic sciences, **4** (2018)
12. M.V. Pevnaya *Volunteering Management: International Experience and Local Practices* (2016)
13. A.A. Kuzminchuk, M.V. Pevnaya, E.R. Timirshina, Bulletin of the Surgut State Pedagogical University, 2 (2018)
14. J. Casey, *The Nonprofit World: Civil Society and the Rise of the Nonprofit Sector*, (2015)
15. L.M. Salamon, S. Toepler, Journal of Voluntary and Nonprofit Organizations, **6** (2015)
16. S. Appe, Voluntas, **1** (2016)
17. T. Brandsen, M. Oude Vrielink, T. Schillemans, E. Van Hout, Nonprofit Organisations, Democratisation and New Forms of Accountability: A Preliminary Evaluation. In A. Ball& Osborne (Eds.), Social Accounting and Public Management London: Routledge, (2011)
18. J. Lu, J Shon, P. Zhang, Nonprofit & voluntary sector quarterly, **1** (2020)
19. A. Pennerstorfer, A.C. Rutherford, Nonprofit & voluntary sector quarterly, **2** (2019)
20. *Civil Society in Russia: Challenges of the Modernity*, Collection of scientific papers, Ulyanovsk State Technical University (2016)
21. Collection of materials on the results of the project, *Creation of a resource center for socially oriented non-profit organizations of the Pskov region* (2015)

Optimization of management control costs in the context of transaction expenses (in terms of an educational organization)

M.V. Chuvashlova[1], *M.N.* Pavlenkov[2], and *O.F.* Ermishina[1,*]

[1]Ulyanovsk State University, Lva Tolstogo street, 42, 432970 Ulyanovsk, Russian Federation
[2]Lobachevsky National Research Nizhny Novgorod State University, Gagarin Ave., 23, 603950 Nizhny Novgorod, Russian Federation

Abstract. The article examines the importance of management control for optimizing the transaction expenses in an educational organization. The author highlighted the opinions of the economists from various theoretical positions. Examples of changes in the optimal control costs, the occurrence of losses from the organization of ineffective control nave been considered, possible ways of solving the problems that have arisen have been given. The extensive material has been structured, which made it possible to highlight the relevance of the study of transaction expenses. According to the authors, the minimum amount of control costs is determined at the point of equality of explicit and implicit transaction costs, that is, the losses of the organization and the costs of executing contracts. At the same time, the optimal amount of control costs will be in the area of relevance of the marginal costs of control and the marginal benefits from the implementation of control measures. Deviations from this zone in any direction are undesirable for the organization.

1 Introduction

An integrated interdisciplinary approach to the management of organizations from the standpoint of the theory of management control examines the processes and mechanisms used by organizations in order to influence the behavior of subjects within the organization and, through this, facilitate the implementation of various kinds of tasks that arise everywhere in the organization.

The results of the study of the economic literature indicate that this subject area was justified from various theoretical positions, including the concepts and principles from such disciplines as systems theory, institutional economics, organization theory and psychology, which helps us to better understand control as a phenomenon and as a process. Nevertheless, the question of the use of certain tools or methods of control corresponding to emerging management situations, as well as the justification of their effectiveness, including for an educational organization, remains open for scientific discussion.

* Corresponding author: ermofl9@mail.ru

As a rule, in educational organizations the internal control system is carried out by a specially created department. The specialists in this structure have a certain level of qualification to know all the "pitfalls" for effective management. However, at the moment, the internal control system and optimization of costs for it requires further study.

2 Materials and methods

Organizational control systems serve several purposes. Jensen and Mecklin (1992) [1] argue that a control system provides incentives by measuring and evaluating performance, and by using reward and punishment instruments. The authors agree that along with the system that delimits the rights to make the decisions, the control system of the organization compensates for the absence of an "invisible hand" coordinating the activities in the market. Speckle (2004) [2] takes a similar broad view of control systems for governance. He reduces the main objectives of management control systems to knowledge gathering, coordination, and incentives through rewards and punishments.

Various combinations of control instruments are used to achieve these goals. [3]. For example, the formation of knowledge that remains within the organization and is available for use in decision-making can be accomplished by delegating tasks and responsibilities to those who have this knowledge. In addition, financial rewards provide incentives and can induce purposeful behavior.

The authors believe that the effectiveness of control tools in terms of managing the transformation costs of any organization lies in the plane of management accounting, from budgeting to analysis of variances, and is not the subject of this article.

But the effectiveness of the control tools in terms of managing transaction expenses is the subject of scientific discussion of modern management. However, not all authors clearly share the difference in approaches to control in the theory of transformation and transaction expenses. So, G. Demsetz criticizes the theory of the firm solely from the standpoint of transaction expenses, sees its weakness in the absence of accounting for production costs, which leads to neglect of differences between firms if they are not related to the control function [4].

Let us turn again to the theory of transaction costs, which has already become mainstream in the history of research on management control [5]. Despite the fact that in most of the previous works, the main attention was focused on the problems of control in specific situations, or control tools, the benefits of using the theory of transaction expenses are far from being limited to a narrow focus - in fact, this theory allows you to lay a reliable basis for building a more general and meaningful theoretical basis for the subsequent study of the structures of management control.

The theory of transaction expenses allows you to take the first steps towards substantiating such an important concept as optimal control costs. Calculation of the optimal amount of expenses is necessary for making many management decisions in any organization, including an educational one. First of all, calculating the optimal amount of control costs helps to plan these costs, include them in the organization's budget. Secondly, knowing the actual value and structure of such costs, their best option, it is possible to develop measures to reduce control costs, thereby creating new motivational incentives for the controllers themselves. And thirdly, the calculation of the optimal amount of control costs will be the first and most important step towards the study of efficiency [6].

The optimal control costs we consider as the best state of the management control system "in relation to its goals under the given external and internal conditions" [7]. This interpretation of the concept of economic optimum involves the study of management processes and factors that determine this optimum. What influences the optimal amount of control costs? To do this, it is necessary to understand the structure of control costs and identify the factors that affect them. Control costs or control expenses are broadly a part of transaction expenses. They are associated with the process of concluding and implementing the internal contracts of the organization. May include both explicit and implicit costs. These include: costs of maintaining a controlling service, costs of ensuring control in other departments and services of the organization, costs of monitoring the implementation of internal contracts, costs of minimizing opportunistic behavior within the organization and other types of the expenses, including losses as a result of improper performance contracts [8], which will be discussed separately.

The relationship between the magnitude of explicit transaction expenses and the magnitude of control costs is obvious. Control costs are an integral part of transaction expenses. With an increase in transaction expenses, there is a corresponding increase in control costs. Let's reflect this proportional dependence on graph 1.

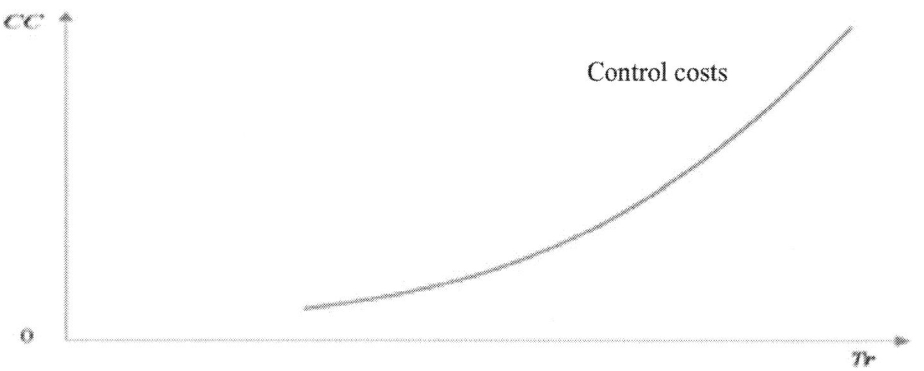

Fig. 1. Schedule of control expenses. *Compiled by the authors.*

The question arises, is the graph describing the dependence of control costs on the value of transaction costs linear? Most likely not, since control costs do not directly depend on the level of business activity or opportunistic behavior of controllers, or on any other quantifiable factors. Undoubtedly, only their stable proportional dependence to the value of explicit transaction expenses. If there is an increase in transaction costs caused by the conclusion of new contracts, entry into new markets, production and sale of new types of products, all this will undoubtedly require new control costs, which will be reflected in an increase in the corresponding costs. Let us assume that point Tr1 in the presented graph corresponds to the current level of transaction expenses and the corresponding level of expenses for control CC1. In the case of an increase in transaction expenses (movement to the right along the 0 Tr axis), other things being equal, there will be an increase in control costs (movement up along the 0 CC axis).

Consider what other factors besides the magnitude of transaction expenses affect the total costs of control. This is, of course, an applicable control model that may or may not be effective in an organization. A more suitable control model will allow maintaining the same amount of control costs with an increase in transaction expenses, the shift of the curve "control costs" to the right (Figure 2.).

Other factors that influence the value of control costs in addition to transaction expenses can be: opportunistic behavior (shift of the curve to the left), developed corporate culture (shift of the curve to the right), and other factors. A shift in the control cost curve to the right entails a decrease at the same value of transaction expenses.

In addition to explicit transaction expenses, which include the costs of executing internal and external contracts, the amount of control costs is also affected by implicit transaction expenses, which include various kinds of losses.

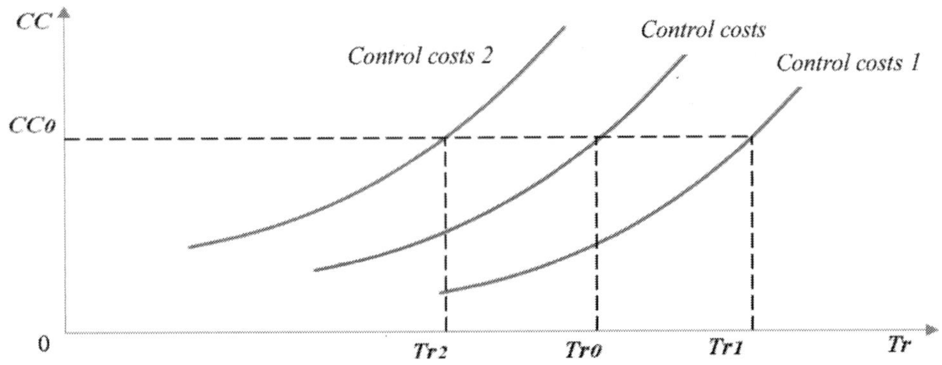

Fig. 2. A shift in the control cost curve under the influence of various factors. *Compiled by the authors.*

These include losses from shortages, theft and damage to property, shortages from depreciation of inventories, losses from overdue debts, losses from non-fulfillment of the terms of contracts, and others. A certain relationship can also be traced between the amount of losses and the amount of control costs. The less an organization spends on control, the higher its potential losses. Let's reflect this dependence on graph 3.

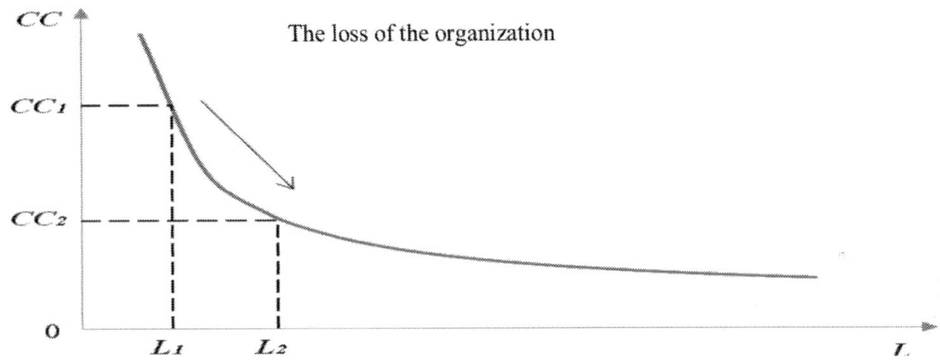

Fig. 3. Graph of the organization's losses from ineffective control. *Compiled by the authors.*

As can be seen from graph 3, if the control costs are insufficient (CC1), then various kinds of losses to the organization will be great, since proper control over compliance with internal and external contracts is not ensured. As the number of control activities increases, control costs begin to increase, which reduces the implicit costs, namely, losses and expenses of the organization. Thus, the dependence is inversely proportional.

Consider what factors, in addition to control costs, can affect the amount of losses to the organization? These are primarily factors that ensure the effectiveness of the control system. A shift in the "loss of organization" curve to the right indicates ineffective control instruments, since at the same value of transaction costs, losses to an organization will be higher.

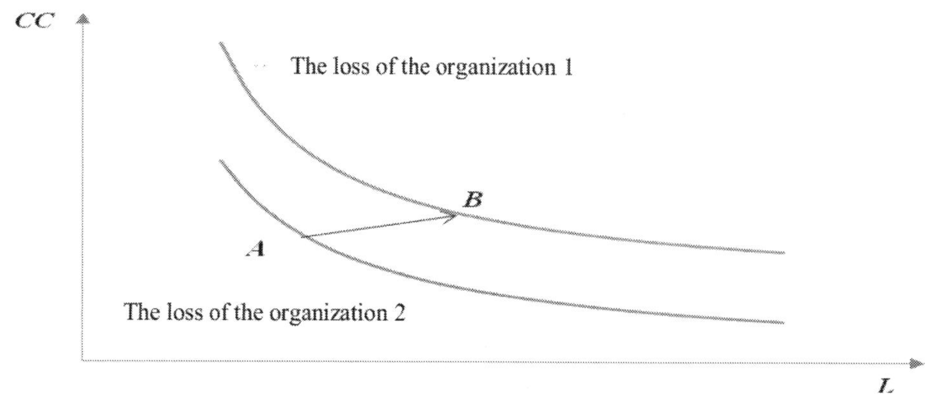

Fig. 4. A shift in the organization's loss schedule from ineffective control under the influence of various factors. *Compiled by the authors.*

Thus, having substantiated the main factors that ensure the determination of the optimal amount of control costs, as part of the amount of losses and the amount of transaction expenses, we can conclude that the optimal amount of control costs can be in the zone of optimum losses and the transaction expenses themselves. Let's show the result of the study in graph 5.

Z curve, which summarizes the total expenses of the organization for the provision of control measures and compensation for damages and losses, has a U-shape. As can be seen from the graph, in order for the total control costs to be minimal (Cmin), the costs of ensuring the management control system must be comparable to the damages and losses from ineffective control.

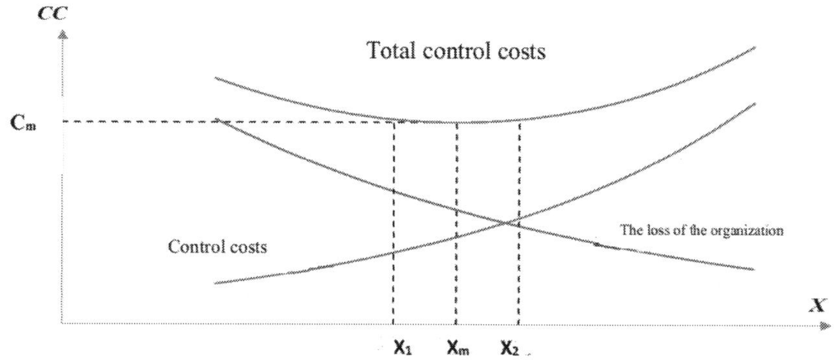

Fig.5. Schedule of total control costs. *Compiled by the authors.*

This will happen in an area where the marginal cost of control will not exceed the marginal benefit of the control. Thus, it is possible to identify a certain area within which the total costs are close to their absolute minimum. **[X_1;X_2]**The minimum amount of control costs is possible at the point of equality of explicit and implicit transaction costs as well, that is, the losses of the organization and the costs of executing contracts. But this condition is not at all obligatory for every organization. At the same time, the optimal amount of control costs will be in the area of relevance of the marginal costs of control and the marginal benefits from the implementation of control measures. Deviations from this zone in any direction are undesirable.

3 Results and discussion

Let us calculate the optimal amount of control costs using the example of FSBI HE Ulyanovsk State University using actual data. For this purpose, we will take for the analysis the reporting data for 2010-2019. (Table 1).

Table 1. Reporting data of FSBI HE "Ulyanovsk State University" for 2010-2019.

Years	Losses from shortages, theft and damage to property, thousand rubles	Fines and penalties, thousand rubles	The amount of losses from the execution of internal and external contracts, thousand rubles	Expenses for the execution of internal contracts (adminis-trative expenses), thousand rubles	Expenses for the execution of external contracts (commercial expenses), thousand rubles	The amount of expenses for the execution of internal and external contracts, thousand rubles	Estimated value of control costs, thousand rubles
2010	0	13	13	4399	11870	16269	44
2011	0	24	24	5622	16838	22460	56
2012	487	765	1252	15185	14928	30113	152
2013	601	147	748	17695	16086	33781	177
2014	0	390	390	20380	18285	38665	204
2015	10602	0	10602	23319	15923	39242	233
2016	8389	0	8389	25511	11997	37508	255
2017	8771	208	8979	26636	7785	34421	266
2018	8439	1002	9441	35488	15395	50883	355
2019	235	1061	1296	33584	27168	60752	336

To plot the control costs, we will take the amount of costs for the execution of external contracts and the amount of costs for the execution of internal contracts as the amount of transaction expenses. For the explicit costs of internal contracts, we will take the general business expenses of the organization or, as they are called in the statement of financial results, the administrative expenses. For the costs of external contracts, we will take the commercial expenses of the organization, since it is the commercial expenses that include the explicit costs of securing the external contracts. As an estimated value of control costs, we will take 2% (this value was formed based on the results of a survey of employees of internal control services of educational organizations) of transaction expenses according to a questionnaire survey of departments and services of FSBI HE Ulyanovsk State University, the data will be reflected in Figure 6.

The relationship between transaction expenses and control costs is significant, with reliability above 99%. The trend is stable at the 76% level, it can be written as a dependence

$$CC = 29,577 * \exp(4,867E - 05 * Tr) \quad (1)$$

To plot the organization's losses from ineffective control, we also take the data on the estimated value of control costs.

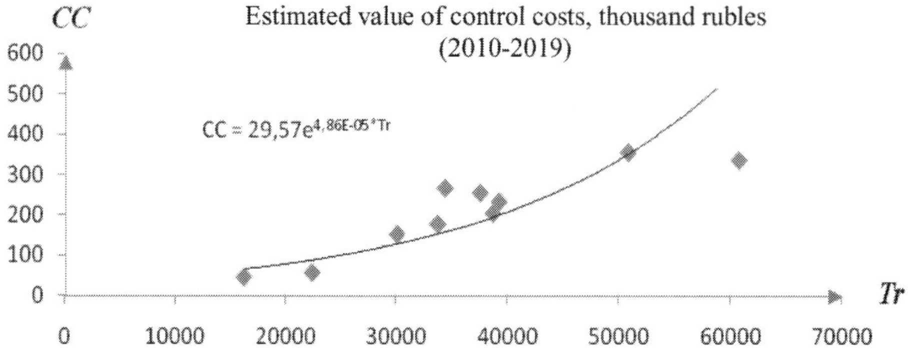

Fig. 6. Schedule of total control costs by divisions of FSBI HE Ulyanovsk State University. *Compiled by the authors.*

The amount of losses (implicit transaction expenses) will be calculated as the sum of losses of the organization, fines, penalties paid, thereby including in the calculations both losses from the execution of internal contracts, and losses from the execution of external contracts. These calculations can be rather arbitrary, but these data in our case are the only available information from official reporting, applicable in the study of transaction expenses. Primary analysis of the data characterizing the magnitude of losses showed that they are unstable, which is confirmed by their visual observation (Fig. 7).

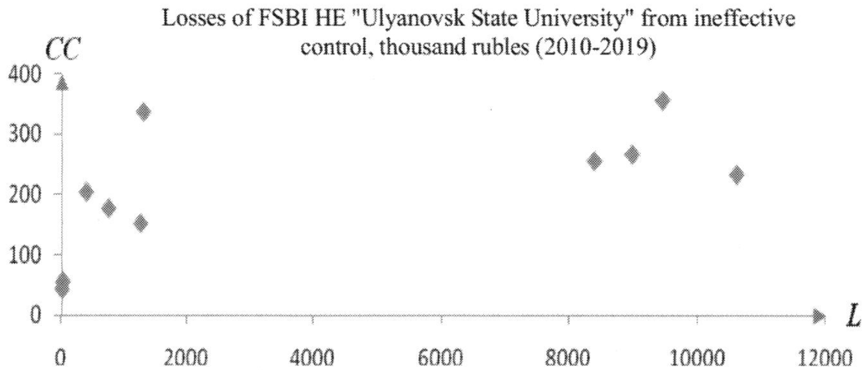

Fig. 7. Losses of FSBI HE "Ulyanovsk State University" from ineffective control. *Compiled by the authors.*

At the next stage of the analysis, the data for the period 2010-2014 were excluded, since for this period the statistical data are small in their absolute value, which can give rise to an error in their assessment, in turn, the data of 2018 should be discarded as uncharacteristic for FSBI HE "Ulyanovsk State University". The indicators of 2018 should not be considered in general statistics, since this period is characterized by an atypical situation with a significant share of losses with a sufficiently high value of control costs. This is due to the fact that during the formation of financial statements in 2018, it was decided to carry out one-time operations for the impairment of non-current assets (equipment for scientific, technical and educational and laboratory facilities).

This was situationally reflected in the net loss and expenses at the end of the reporting period. Let us construct a graph of losses (Figure 8) of FSBI HE Ulyanovsk State University from ineffective control.

The relationship between the amount of losses to the organization and control costs is significant with a reliability of 98%. The trend is stable at the level of 95%, it can be written as a dependence

$$CC = 395,955 * \exp(-3{,}703E - 05 * L) \tag{2}$$

Now it is possible to draw up a graph of the total control costs according to the data of FSBI HE "Ulyanovsk State University" (Figure 9). In this case, the point (C_I, X_I)=(239,9; 25770,7) is the point of minimum total control costs. This means that for a given educational organization the optimal value of control costs is 239.9 thousand rubles, which will ensure the minimization of the organization's losses and its explicit transaction expenses.

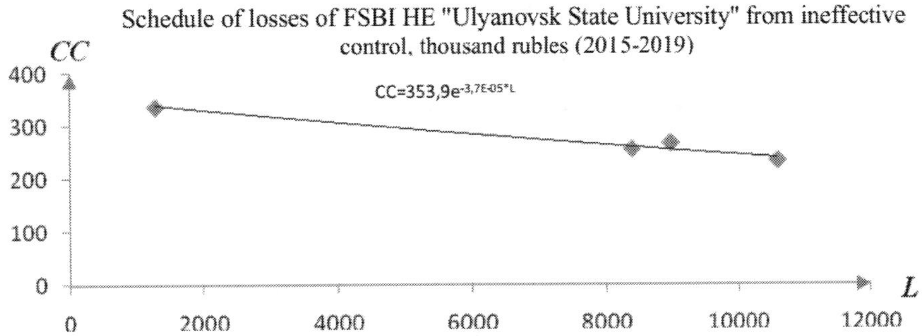

Fig. 8. Schedule of losses of FSBI HE "Ulyanovsk State University" from ineffective control. *Compiled by the authors.*

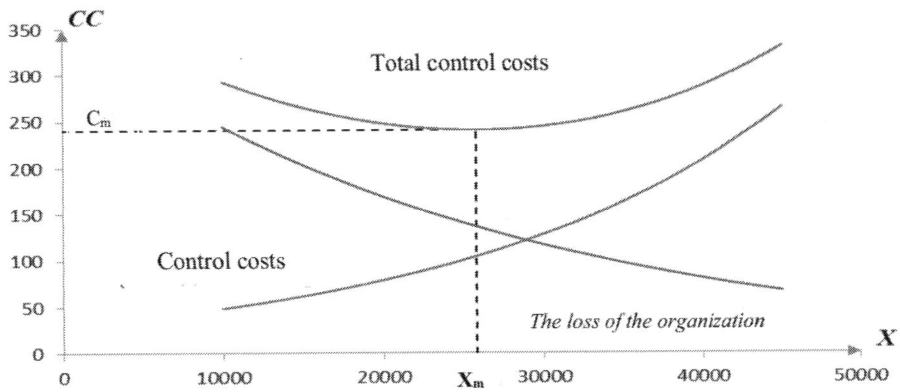

Fig. 8. Schedule of total control costs of FSBI HE Ulyanovsk State University. *Compiled by the authors.*

4 Conclusion

This study can be expanded to include a third dimension and a third parameter that affects the effectiveness of control - the degree to which the organization has achieved its goals. It is possible to assess the influence of this factor and substantiate what amount of control costs will be optimal, taking into account the implementation of KPIs, by resorting to expert judgment. As a result of a survey of employees of the internal control service and heads of departments of the FSBI HE "Ulyanovsk State University", it was revealed at what amount of expenses for control the achievement of KPI is maximum. It can be assumed that there are areas of insufficient control, the achievement of KPI is minimal, optimal control - KPI is the most achievable and excessive control, KPI is again decreasing.

Traditionally, transaction expenses include the cost of completing business transactions, including the monetary estimate of the time spent finding a business partner, negotiating, concluding a contract, and ensuring the appropriate performance of the contract. The main idea of using the theory of transaction expenses in solving management control problems is, in our opinion, that the achievement of an organization's goals depends on the participation and personal contribution of a large number of people and their interaction, and the structures of management control act as a means for solving coordination problems. adaptation, motivation and control of performance by concluding an agreement while monitoring the personal contribution of each participant.

References

1. M.C. Jensen, and W.H. Meckling, Contract economics, 251 (1992).
2. R.F. Speklé, *Configurations of control: a transaction cost approach (working paper)* (2004).
3. K.A. Merchant, Sloan Management Review, 43 (1982)
4. 4.H. Demsetz, Quarterly Journal of Economics, **1**, 243 (2000)
5. Spekle, 2001, Colbert and Spicer, 1995; Van der Meer-Kooistra and Vosselman, 2000; Spicer and Ballew, 1983; Swieringa and Waterhouse, 1982; Tiessen and Water-house, 1983.
6. M. Chuvashlova, European Research Studies Journal, **3**, 675 (2018)
7. Economic dictionary, http://abc.informbureau.com/
8. R. Coase, R. *Firm, market and law* (2007)

Role of industrial cooperation and inter-cluster interaction in the development of localization of production in shipbuilding

Alexey Serbulov, and *Marina* Shaliapina*

Immanuel Kant Baltic Federal University, Gorkogo str., 23, 236000, Kaliningrad, Russia

Abstract. An effective tool that allows you to dampen economic losses from the Western sanctions imposed on our country is the development of international cooperation and the associated processes of localization of production. However, along with the positive effect, there are a number of administrative and organizational problems that impede the expected results from both foreign partners and domestic manufacturers. On the example of the shipbuilding industry, the problems of further development of inter-cluster interaction are shown both at the national and international levels. It is shown how participation in the formation of the Panel of Equipment Suppliers, which is an extensive computer system for managing procurement activities and material and technical supply of shipyards, will expand the opportunities for foreign companies to participate in the implementation of large-scale government orders.

1 Introduction

The complexity of the current foreign policy situation in which Russia finds itself requires an urgent solution to the most urgent task facing the country's economy, formulated as "import substitution and expanding the range of the products".

Given the fact that our economy is still attractive for foreign direct investment, the state economic policy pays serious attention to the development of local production, considering it as a way to solve the above problem.

The Government of the Russian Federation creates a legal environment favorable for investors, counting on the creation with their help of competitive industries in strategically important industries for the country, which will lead to a decrease in Russia's dependence on imported products [1].

Federal Law No. 488-ФЗ dated December 31, 2014 "On Industrial Policy in the Russian Federation" provides for a system of state support measures aimed at the formation of a competitive industry corresponding to the innovative type of development of the domestic economy. Following the aforementioned law, a number of regulations have been adopted regarding the localization of industry, taking into account: a specific industry; specifics of financing an investment project (in the case of using public funds); the specifics of the process of selling the finished products, if the buyer is the state or the

* Corresponding author: m-shaliapina@mail.ru

companies with the state participation, or implementation in the B2B market is envisaged.

In recent years, domestic practice has accumulated significant experience in the creation of innovative industrial clusters by well-known foreign companies together with Russian manufacturers [2; 3]. Of course, the leaders in this matter are German, Chinese, Korean and Japanese automakers. But there are positive examples in other sectors of the economy. For example, in the Lipetsk region, with the active participation of one of the largest manufacturers of household appliances in Europe, the Italian machine-building company Indesit Company (bought by the American Whirlpool Corporation in 2014), an innovative territorial cluster of complex household appliances (refrigerators, freezers, washing machines) was created. Today the cluster cooperates with more than 250 foreign suppliers of spare parts. Therefore, the prospects for the development of the anchor enterprise JSC "Indesit International" are associated with the localization of production within the national territory.

In partnership with the Lazio Regional Health Cluster (Italy), C.H.I.CO. (Cluster of Health, Innovation and Community) a pharmaceutical cluster has been created in the Kaluga region, the main task of which is to conduct preclinical and clinical research, development, synthesis and introduction into production of pharmaceutical substances and radiopharmaceuticals, industrial production of finished medicines and pharmaceutical substances. Today it can be stated that 70% of the products manufactured by the enterprises of this domestic cluster are innovative.

However, the presence of a number of specific problems arising in the process of cluster development: growing competition within the cluster, lack of resources for the implementation of certain projects, etc. forces clusters to look for new ways to grow.

For example, the organization of inter-cluster interaction, and at the highest level, which makes it possible to successfully combine the interests of all parties (participants in an integrated structure, regions, industries, the state).

Developing a system of inter-cluster interaction, clusters support entrepreneurship within their competence, based on the principles of business cooperation, which favorably affects the activities of all participants and allows them to predict steady economic growth [4].

The past decade has been characterized by the development of international industrial cooperation processes. To further build the potential of economic instruments for ensuring national competitiveness, localization of industrial production within its own territory is necessary. Localization allows solving a number of problems that arise at the present stage of the economic development of society as a whole and its productive forces. Among them are the following:

1. Preservation of the country's economic sovereignty.
2. Overcoming export and import restrictions.
3. Unification of Russian requirements and approaches to localization in accordance with existing international practice.
4. Development of our own productive forces through the development of advanced technology and technological practice, acquisition of new personnel competencies, activation of the resource and energy factor, etc.

A partner company that locates production on Russian territory also pursues certain goals that do not contradict the goals of the Russian participants, the host region and the state as a whole. Among them, improving market positions through guaranteed sales of products in Russia and the member states of the Eurasian Economic Union (EAEU), reducing costs due to the cheapness of labor, geographical proximity to sources of raw materials and energy resources, reducing the adverse impact on the ecosystem of their country.

Let's consider the localization of production within the framework of inter-cluster interaction using the example of shipbuilding clusters.

Shipbuilding is one of the leading sectors of the economy with high scientific, technical and production potential, influencing the development of technologies in related industries.

The existing production capacities and resources of the Russian shipbuilding industry allow increasing the volume of civil and defense products.

According to some estimate given in [5], the shipbuilding industry includes about 600 enterprises, including 180 shipyards, more than 90 design and research organizations, 310 enterprises that produce ship components for various purposes, the cost of which depending on the type of ships and samples of marine equipment, can be more than 50%.

To increase the competitiveness of the industry, the Government of the Russian Federation created a number of preferences for both export-oriented industries and the domestic market, aimed at increasing the motivation for locating and expanding production in Russia (Table 1).

Table 1. Measures of state support in the shipbuilding industry implemented by the Ministry of Industry and Trade of Russia, contributing to the effective promotion of the policy of import substitution and localization

Measures to support the shipbuilding industry	Budget allocations, mld. Rub.				Number of subsidized vessels, units
	2020	2021	2022	2023	
Subsidies for reimbursement of part of the cost of interest on loans for the purchase of civil ships and lease payments under lease agreements	3.8	3.8	3.8	-	120
Ship recycling grant (Resolution of the Government of the Russian Federation No. 502 dated April 27, 2017)	0.5	0.5	0.5	-	10
Leasing program for sea and river civil vessels	-	The Ministry of Industry and Trade of Russia plans to prepare an application for additional financing in the amount of 105 billion rubles.			82
Encouraging the construction of small and medium-sized fishing vessels at domestic shipyards (within the framework of the decree of the government of the Russian Federation No. 1917 dated December 27, 2019)	0.393	0.360	0.360	-	-
Granting subsidies from the federal budget to Russian organizations to financially support part of the costs associated with the construction of large-tonnage vessels (within the framework of the RF government decree No. 1584 dated December 4, 2019)	4.915	7.2	9.9	-	18

*Compiled by the authors from the source [6]

At the level of the Government of the Russian Federation, such measures are considered to stimulate the entry of foreign companies into a joint business on terms of localization, such as the possibility of VAT refunds upon subsequent re-export of products. In addition, upon signing a special investment contract (SPIC), a foreign investor can claim to simplify the administrative procedures for the development of its activities, obtain the status of a Russian manufacturer, and, therefore, access to government contracts.

Shipbuilding and ship repair enterprises have a rather pronounced specificity in the organization of production and economic activities.

Firstly, this production is highly specialized, which affects the nature of the resources used.

Secondly, it is a high-tech production with a high level of application of innovative technologies. In shipbuilding, 3D modeling, specialized software for cutting, cutting and processing workpieces with CNC machines, and robotic welding are widely used. Fireproof aluminized coatings are more and more actively used, which guarantee corrosion resistance and significantly reduce the cycle of painting.

And, finally, thirdly, this is a geographically concentrated production, which is concentrated in large maritime centers with direct access to sea areas (Arkhangelsk, Astrakhan, Kaliningrad, Murmansk, St. Petersburg) and having a long-standing shipbuilding tradition.

The above key characteristics make it possible to classify the shipbuilding industry as a strategically important sector of the economy, which implies that it is under the unconditional care of the state.

However, it is precisely this development strategy that has led to the fact that almost all the leading enterprises of Russian shipbuilding and ship repair today are concentrated as part of a state-monopoly structure - United Shipbuilding Company JSC (USC) with 100% of shares backed by federal ownership. This industrial giant has an annual production turnover of over 300 billion rubles and a number of employees of over 90 thousand people at enterprises located in various regions of the country.

The main share of the use of the resources of the USC enterprises is associated with the fulfillment of state orders, namely, orders for the construction of ships for the Navy, marine equipment for the development of the Arctic shelf and orders for foreign trade cooperation. All of the above are the federal projects.

However, lately there have been more and more opinions [7] that the desire for excessive centralization forms a number of negative characteristics of the economic environment. In particular, internal competition in the industry is eliminated, the quality of management decreases, and interagency barriers are created.

An example of this is the conflict of interests of shipbuilding and ship repair enterprises, which are simultaneously part of the USC and the composition of the ubiquitous industrial shipbuilding clusters. And it is connected with the fact that, in parallel with the state order at the sectoral and regional level, a stable interest of investors has formed in the implementation of commercial projects related to the renewal and modernization of the fleet. In particular, this is due to an increase in demand for fishing vessels, cruise passenger ships, ferries, pleasure yachts, as well as vessels of the research fleet.

The complexity of the current foreign policy situation in which Russia finds itself requires the prompt solution of the most urgent problem facing the shipbuilding industry, formulated as “import substitution and expanding the range of products” [8].

However, the high price, low quality, problems associated with repair and maintenance have led to the low competitiveness of domestic products, which results in a large share of the cost of this type of foreign-made product in the structure of the cost of ship component equipment (for civil shipbuilding - up to 85%, for military - up to 60%) [9].

The current situation explains the significant dependence of the domestic shipbuilding industry on foreign supplies, which, along with the sanctions policy of the individual states, the global economic crisis entailing fluctuations in exchange rates, poses a threat to the entire industry, creating the risks of increasing costs and disruption of construction time, and in some cases, endangering the very possibility of building certain types of products in Russia.

Since the formation and development of intersectoral clusters in Russia is declared as one of the most important directions of the industrial development strategy (for the period up to 2035), the cluster approach can be considered as a new round of localization of both foreign and Russian companies in the regions of the Russian Federation, within which the formation of pool of domestic suppliers around an enterprise with foreign investment [10].

2 Materials and methods

The study of the processes of localization of production in industry is based on the study of data from sectoral statistical reporting provided by the Ministry of Industry and Trade of the Russian Federation, data from official statistics bodies. A significant amount of information for further analysis is contained in regulatory documents (Federal laws, resolutions of the Government of the Russian Federation, local regulations). In particular, the decrees of the Government of the Russian Federation on the program of import substitution in the shipbuilding industry in 2015 and 2019 are of great practical value for understanding the strategy of state influence on the processes of import substitution.

In the process of studying this issue, the materials of scientific reviews of the European Cluster Observatory "Cluster Collaboration and Business Support Tools to Facilitate Entrepreneurship, Crosssectoral Collaboration and Growth" (2014), the working group on shipbuilding of the Secretariat of the Organization for Economic Cooperation and Development "Shipbuilding Market Developments" (2018).

When conducting the research, the authors used general logical methods of the theoretical level (analysis, synthesis, generalization, analogy), as well as the empirical level (observation, description, comparison).

3 Results and discussion

The use of production localization within the framework of inter-cluster interaction with the help of a pool of domestic and foreign manufacturers of spare parts, raw materials, equipment, directly contributes to import substitution, as well as ensuring a complete production chain within the cluster.

Based on the materials of the study of the processes of interaction of shipbuilding clusters in the North-West region [4], 6 enlarged groups of resources were identified that are used in the implementation of large inter-cluster projects and require resource interaction between its participants.

The economic characteristics of each resource can be presented as follows (Table 2).

Table 2. Characteristics of the resource potential in the field of inter-cluster interaction

Resource potential type	Resource potential characteristics
Production potential	Demonstrates the provision of industry enterprises with basic production assets and production infrastructure.
Human resources	It is a set of demographic and social characteristics of the economically active population working at industry enterprises in the region.
Innovation potential	It is characterized by a set of all types of resources and conditions for the creation and practical development of the results of scientific research and development.
Investment potential	Determined by the volume of investments aimed at modernizing and developing the industry in the region.

Table 2. Continued

Financial potential	It is a collection of financial resources that can be attracted for the implementation of inter-cluster projects.
Information potential	It is represented by a set of information resources that provide connections between the participants in inter-cluster interaction.

The owners of all the listed resources are the business entities. However, the decision to transfer these resources on a reimbursable basis to other participants in cluster projects is made by a specially created structure - the Inter-Cluster Interaction Control Center (ICICC), which acts as a coordinator providing the process of interaction between clusters based on the information management platform, one of the functions of which is to optimize the procurement activities. [11]

It should be noted that the principle of retribution for various types of resource potential is implemented in different ways. In this sense, it is possible to divide the resources transferred within the framework of the inter-cluster exchange to be returned into the following groups:

1. Resources completely consumed in the process of use (raw materials, materials, quickly worn out tools, equipment, intellectual property, etc.). They are returned not in material, but in monetary form in full amount of their commercial value.
2. Resources returned with a loss of value (immovable production facilities, equipment, etc.). Returned in tangible form with payment of the cost of depreciation and the commercial benefit of the resource owner.
3. Resources returned without loss of production capabilities with payment of the cost of use (remuneration of workers of other enterprises seconded to carry out work on the implementation of inter-cluster projects).
4. Irrevocably transferable resources with full payment of the cost or on the principles of mutually beneficial exchange (information, design and technological documentation, intellectual creations, etc.).

In the process of designing and building a ship, a three-way interaction is formed - customers, manufacturing plants, equipment suppliers, where each of the parties has its own interests. To harmonize the interests of all the participants in shipbuilding projects, a regulatory framework is needed that regulates the relationship between the parties, which would determine interaction with resource suppliers, relations within the corporation, as well as between partners within the framework of inter-cluster interaction.

Such a platform for managing MTS processes in long, constantly repetitive business processes is the "Panel of suppliers in civil shipbuilding", which is a management system for procurement activities and material and technical supply of shipyards of the United Shipbuilding Corporation group of companies. Developed since 2017, this system is designed to ensure the formation and constant updating of an array of information about the suppliers and the nomenclature of their products in an automated mode.

The modern practice of material and technical support within the framework of the activities of state corporations suggests that there should not be only one supplier for any item of equipment [12]. This is prohibited by Russian competition law.

4 Conclusions

Inclusion in the shipbuilding clusters of the "pilot pool of suppliers" localized on the territory of the cluster in such areas as: main mechanisms ("Wartsila", "MAN", "Yanmar", "Schottel", "Steerprop", "ABB"); marine equipment ("Fluidmecanica", "Vulkan Nova", "Adria Winch"); ship navigation equipment ("Furuno", "Polar", "OSK Technologies");

equipment for cargo ("Liebherr", "Sormec", "Palfinger", "Gurdesan"), as well as the attraction of clusters from related industries (machine tools, metallurgy, etc.), will provide the full production cycle with the necessary resources, and therefore will provide an opportunity to localize the production of equipment and component parts most in demand in shipbuilding.

For the timely provision of the processes with the necessary equipment, components, tooling, etc., it is necessary to provide for a pre-order period for negotiations with the customers. This provision is especially important when organizing and carrying out ship repair and emergency operations, where the possibilities of preliminary planning and ordering of equipment necessary for replacement or restoration are very limited.

Following the classification of the resources proposed by us above, the suppliers of the first and, partially, the second group of the resources will be included in the developed "Supplier Panel". At the same time, not only the requirements for the product should be formed, but also the requirements for the supplier in terms of ensuring the delivery time, the quality of equipment and its service.

To develop the processes of inter-cluster interaction and create conditions for the localization of production, it will be advisable to create a distributed database (DDB) of equipment suppliers in all existing and newly created territorial clusters of shipbuilding and ship repair.

The inclusion of the potential partners in the Supplier Panel will stimulate not only domestic, but also foreign manufacturers to fulfill the government's requirements for expanding localization in the shipbuilding industry.

References

1. V. A. Tsvetkov, M. N. Dudin, A. A. Yurieva, Economy of Region, 691 (2020)
2. A. Andrésen, E. Laaksonen, H. Mäkinen, Smart Comp Research Report (2012)
3. Laaksonen, E. Mäkinen, H. Journal of East-West Business, 95 (2013)
4. M.A. Shalyapina, Kazan. econom. bulletin, 618 (2018)
5. V. Khanichev, D. Stoyanov, Sh. Petrosyan, Business magazine Neftegaz.RU, 23 (2019)
6. N. Valuisky, S. Vinarova, G. Yegorov, PortNews IAA, 3 (2020)
7. V.N. Zazimko, Expert Union, 12 (2019)
8. A.M. Butov, *Shipbuilding products market 2018*, 125 (2018)
9. Strategy for the development of the shipbuilding industry for the period up to 2035 No. 2553-p dated October 28, 2019 ConsultantPlus (2019)
10. O.P. Ivanova, The region's economy, 1215 (2018)
11. M.A. Shalyapina, Actual problems of economic development and management: collection of scientific papers, 425 (2018)
12. V.S. Ismagilova, A.R. Ismagilova, Economic management: methods, models, technologies: proceedings of the XIX International scientific conference, 157 (2019)

Transformation of the company's HR management systems based on digital literacy measurement

Elena Galenko*, *Elena* Shumakova, and *Natalia* Ovcharenko

Far Eastern Federal University, School of Economic and Management, FEFU Campus, 10 Ajax Bay, Russky Island, 690922 Vladivostok, Russia

Abstract. Corporate culture in each company has its own particularities, which allows to assess the efficiency and sustainability of a given company, design its development strategy as well as the capacity to attain its strategic goals. The current study, based on the model of O'Reilly, Chatman and Caldwell, allowed to define the key problem, that is the absence of digital literacy on the part of the personnel of the studied hotel chain. The identified problem may become an impediment towards formation of the digital culture in the context of digital transformation

1 Introduction

In the current conditions the development of Russian corporate sector, including hospitality sector, requires enormous investments. The members of the competitive battle (that is happening among the countries), while struggling to attract foreign investment, pay major attention to corporate management. In this respect the question of globalization and digital transformation in the hospitality sector, which profoundly influences hospitality sector, becomes a hot topic. Therefore, the analysis of current tendencies and problems of corporate management of the hospitality industry companies in the context of digital transformation, possibility of application of best practices in the domestic practices become particularly relevant.

The corporate management of hospitality industry companies is looked as a tool that is used in the process of management aimed to keep an adequate balance between the shareholders (owners), the needs of board directors, management of the company, which provides the structure for the setting out of the for the company as well as achievement of these goals and control over the functioning of the company [1]. It is necessary to note that corporate culture takes a special place among the above-mentioned elements of corporate management. The corporate culture may be defined as the most important internal reserve of a company which allows to ensure its sustainability and provide an impetus for its development and transition to new levels in the context of digital transformation of business development in the hospitality industry [2]. Corporate culture is a key element of the success of any company, but for hospitality industry this approach is particularly relevant, since the positive experience of its employees is vitally important in satisfying the needs of

* Corresponding author: galenko_tgey@email.ru

its clients. Corporate culture influences every aspect of the business: from staff retention to the strategy and its realization.

However, notwithstanding corporate culture's unique strategic advantage the following questions arise: Are hospitality companies ready for efficient functioning in the context of current digital transformation? How new technologies influence the relation between price and quality? What role corporate culture plays in relation to digital transformations of hospitality companies?

In the view of the above-mentioned, the aim of the present article is to define the particularities of corporate culture in the system of company's corporate management in the context of digital transformation using the example of hotel chain 'AZIMUT.'

2 Materials and methods

The development of corporations of the world-known hotel chains is the result of concentration and internalization of the capital as well as monopolization of the hospitality services' market. Hotel company functions as voluntary association of individuals and legal persons created by its founders for carrying out of a specific activity, achieving common goals and realization of specific tasks. Company acquire special organizational form which is based on its legitimate functioning, voluntary (not public) membership, self-management and hierarchical structure as well as independent adaptation by its founders of the charter, strategy and business objectives [3].

There are thousands of world-wide hotel corporations such as Marriott International, USA (5,952 hotels, 1,164,668 suites), Hilton Worldwide, USA (4,875 hotels, 796,440 suites), IHG (InterContinental Hotels Group), UK (4,875 hotels, 796,440 suites), Wyndham Hotel Group, США (8,035 hotels, 697,607 suites), Shanghai Jin Jiang International Hotel Group Co., China (5,977 hotels, 602,350 suites), AccorHotels, France (4,144 hotels, 583 161 suites), Melia Hotels International, Spain (376 hotels, 96,355 suites), APA Group, Japan (361 hotels, 56,734 suites), Nordic Choice Hotels, Norway (188 hotels, 32,342 suites), Grupo de Turismo, Куба (83 hotels, 2,163 suites), Azimut Hotels Россия (25 hotels, 8,124 suites) and many others [4]. Their effective corporate management is based on the organization of good management of the company, which functioning provides its owners with income in the current period, the companies adapting adequately to the market environment to achieve sustainable development in the long-term perspective. All of the companies mentioned above function in the sphere of computer software, use latest developments in the sphere of global digitalization and development of modern information society. The use of digital technologies, which is introduced both in providing hotel services aimed at its clients and in its internal processes, made the companies competitive and opened new possibilities for business [5].

Digital economy is regarded in the governmental program 'Development of digital economy in Russia until 2035' as 'the totality of social relations formed during the use of digital technology, digital infrastructure and services, technologies of analysis of large sets of data and forecasting with the aim to optimize the production, distribution, exchange, consumption and increasing the level of social-economic development of the government' [6]. This definition emphasizes that the priority in identification of factors of influence on the formation of digital economy is given to digital technology and analysis of large datasets. Decision of the Supreme Eurasian Economic Council entitled 'On the main trajectories of realization of digital agenda of Eurasian economic union until 2025' digital economy is defined as 'part of the economy in which the processes of production, distribution, exchange and consumption have gone through digital transformations with the use of information and communication technologies' [7]. Digital economy in a hotel chain may be seen as innovation in the form of new developments of a hotel service. For example, modern hotel locks appear not only as the security elements that effectively resolve the problem of control of access to interior rooms of the hotel, but also as having some specific function. Paper cards, mobile phones, iris recognition, fingerprints have become an alternative to digital locks. Clients' identification with magnetic and chip card has become widespread. Distant support of software programs, which guarantees operational communication with respect to a given inquiry, is also of great significance. PMS, CRS, CIS, CRM are among some of the specialized hotel systems. In hotel industry the concept of 'intelligent' knowledge comprises of the following solutions: possibility to manage the real estates, increase comfort, information exchange as well as to prolong the guaranteed period of the functioning of electronic devices. However, digitalization requires not only technology, but also corporate culture. Corporate culture which is seen as the most important internal reserve of an organization allowing ensure its sustainability and give an impetus towards the development and transition to qualitatively new levels of business' organization taking into account ongoing adaptations and changes.

Gold K. in its research sees corporate culture as unique characteristics of the perceived features of an organization, something that distinguishes it from its competitors [8]. Schein E. in his studies defined corporate culture as beliefs, behavioral norms, attitudes and values which are those unwritten rules that define how people in this organization must work and act. The author sees this phenomenon as the 'complex of basic assumptions invented, detected or developed by the group in order to learn how to deal with problems of external adaptation and internal integration' [9]. Corporate culture includes different characteristics and elements, including internal environment of the company, its mission, management style, ethics, expectations and aims. Depending on the management style some enterprises may form team culture in which employees take part in the development of the company at all levels, while other companies have culture which values formal, traditional or hierarchical management. Personnel working in companies developing the culture with traditional management style knows perfectly well its job duties, while there may be no possibility for career growth. In this respect it is important to understand that the culture needs to be managed, it needs to be improved and developed in accordance with changes that are happening both internally and externally. Successful companies invest substantial amount of time, money and efforts in its cultures because they have experienced its benefits. They know that corporate culture is a strategic asset, it improves the level of guest service, the number of their hotel visits, thus increasing the company's net profit [10]. The main functions of corporate culture of the company are image formation, maintaining the values, support of new employees in socialization, the establishment of rules and values of behavior, formation of the sense of belonging to common cause on the part of the employees. Thus, the main function of corporate culture is the creation of unique business environment which becomes the part of competitive advantage and provides value to consumer.

Different approaches may be used in order to analyze the company's existing culture For instance, the model of corporate culture developed by Deal T.E. and Kennedy A.A., which encompasses the history of the company, key values and rituals, heroes and myths, symbols and artifacts. The model of O'Reilly, Chatman and Caldwell requires a particular attention. This model includes seven dimensions: innovation and readiness to take risks; stability; respect for people; results orientation; attention to details; team orientation; aggressiveness [11].

The set of the above-mentioned parameters may be used in practice in order to analyze current corporate culture of hotel chain and to define its particularities with the use of content-analysis of the media, the study of visitors' reviews of the chain, interviews of top-managers and managers of the companies, websites of the companies.

3 Results and discussion

Each organization has its history and set of rules and prescriptions that are constantly changing and improving while forming the unified team.

The history of international hotel chain 'AZIMUT Hotels' begins in 2004 with the acquisition of the hotel in Samara and from this time it becomes one of the first rapidly developing Russian companies managing the hotel chain under the same name and under the management of Mr. Alexander Klyachin. During the following two years the brand 'AZIMUT Hotels' is formed and in 2006 it already unites 8 hotels within the Russian territory: in Samara, Ufa, Kostroma, Saints-Petersburg, Vladivostok, Astrakhan and Murmansk. All the hotels introduced uniform standards of service and hospitality and common corporate culture. In 2008 'AZIMUT Hotels' became the first Russian brand that has entered the European market. The chain 'AZIMUT Hotels' comprised of hotels in the biggest cities in Germany (Berlin, Dresden, Cologne, München, Nurnberg, Erding). In 2012 the chain 'AZIMUT Hotels' unified 23 hotels of the 'three-four stars' category in Russia, Germany and Austria. The total number of bedrooms was 5,500 suites, including more than 3,500 in Russia. In 2016 the chain 'AZIMUT Hotels' acquired 4-star hotel in Austria. Delta Hotel Vienna became the eighth hotel of the company in Europe. The remaining European hotels of AZIMUT Hotels are located in Germany – Berlin, Dresden, Cologne, München, Nurnberg and Erding. In 2018 'Azimut Hotels' became the most actively developing brand in the Russian hotel market addint to its portfolio 1,300 suites more: hotels in Kemerovo, Pereyaslavl, Nebug, Mirny, Suzdal, Penza, Nalchik have joined the chain. Today AZIMUT Hotels is the world-known hotel chain combining under its brand the property in the cities of Russia, Germany, Austria and Israel, being the first operator that managed to launch the Russian hotel brand to the international level. The developing brand in the business hotels segment AZIMUT Hotels involves 40 hotels in 32 cities in Russia, Europe, Middle East, the total number of bedrooms being more than 7,000 of rooms. AZIMUT Hotels manages hotels owned under the long-term lease agreement.

Attention to details – this position of the corporate culture has been well described in the interview with Mr. Walter Neumann. Mr. Neumann has clearly defined the company's approach and its development strategy. Attention to such details as new design concept and SMART organization of space was aimed at business and has three principles: functionality, processability and conciseness. The company adapted the list of services in accordance with the principal needs of business tourists of new generation and put accent on the combination of comfort, technological solutions and possibilities for work 24/7.

In order to maintain its stability AZIMUT Hotels chain undertakes regular education of its employees and advanced training by putting them to other hotels of the chain. The chain's portfolio includes more than 30 developed programs. There is a special program related to the development of the personnel at three levels. The first level for managers of services, the second level is for operational director and the third level – for general manager. It helps to put stress on the process and performance, the obligations of the employees are clearly defined, advancement in the career follows defined trajectories.

Respect for people can be seen through the 'AZIMUT Bonus' loyalty program which offers unique privileges during the stay in the hotels. For the study of the aggressiveness the authors of this article analyzed 25,660 reviews in 35 hotels presented in 'tripadvisor.ru' website. Not all the hotels rite responses to the reviews that are made by its visitors. In general, there are no responses to the negative service reviews. Therefore, it may be stated that the personnel works within the framework of traditional culture where the work is based within the tasks and barriers, realization and support of current standing and acceptance of barriers.

Result-based performance of the personal was calculated based on the average review throughout the whole chain of AZIMUT Hotels (tripadvisor.ru) and is presented in Figure1.

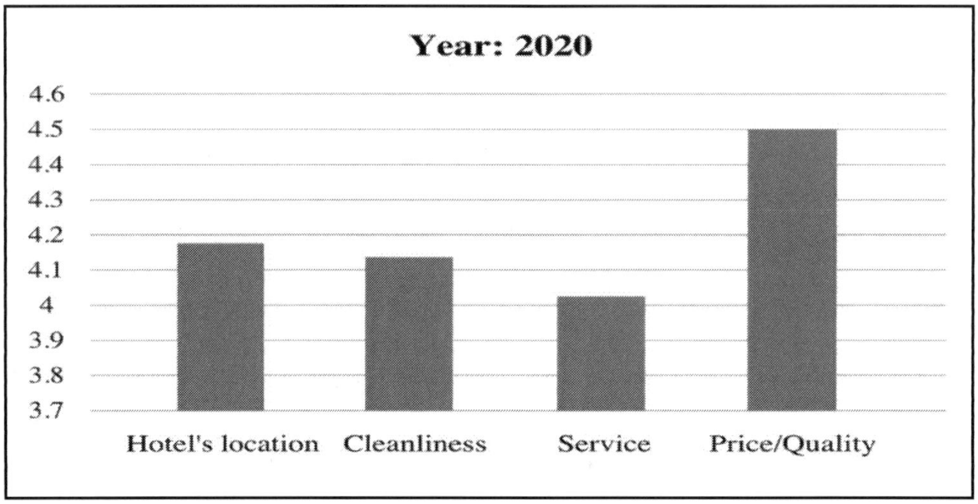

Fig. 1. Result orientation of the personnel

The average review that was calculated based on the reviews to all of the hotels of 'AZIMUT Hotels' chain (tripadvisor.ru) has produced the following results: visitors evaluated the location of hotel – 4.19, cleanliness – 4.14, service – 4.01, price/quality – 3.9. The mentioned indicators of the hotels of the chain are oriented at the sales/offer, but not at the client.

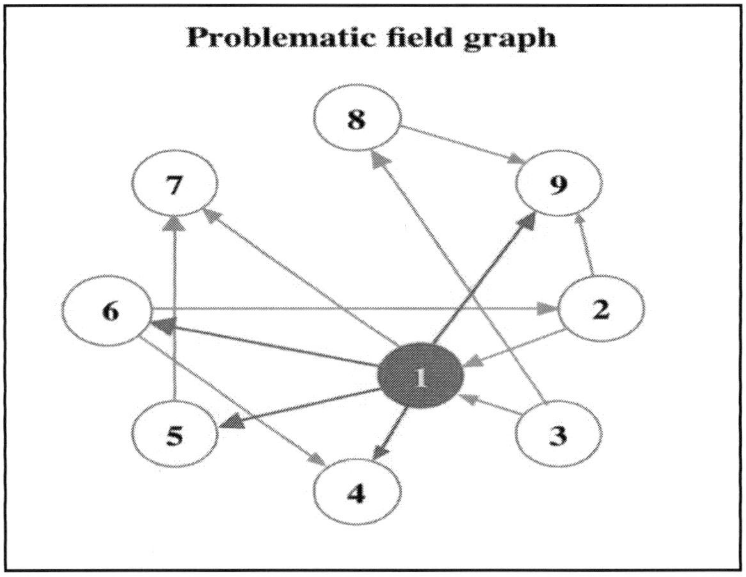

Fig. 2. Analysis of the problematic field.

Innovativeness and readiness to risk was analyzed based on the following identified problems: 1) absence of digital literacy on the part of the personnel; 2) hotels of the chain are oriented towards sale/offer, but not the client; 3) orientation at the process and task performance, obligations of the employees are strictly defined; 4) realization, maintenance of the current state of events and acceptance of barriers; 5) experience and stability are taken into consideration; 6) work within the framework of tasks and barriers; 7) career advancement within the pre-defined trajectories; 8) ignorance of the digital culture.

Within the accepted assumptions, the key problems are the following: 1) producing influence on the major part of other problems; 2) producing influence on more than four problems. Relation between the problems are shown in the graph of the problematic field where the key problem is 1. Absence of the digital literacy on the part of the personnel are dependent problems 4, 5, 6, 9.

4 Conclusions

Corporate culture of AZIMUT hotel chain was studied. Based on this study and within the framework of O'Reilly, Chatman and Caldwell model the following particularities were identified:

- hotel chain AZIMUT is one of the actively developing Russian brand in the business-hotel segment;
- corporate culture of the hotel chain is developing in the traditional format without consideration of digital transformation;
- the elements of digital corporate culture to various extent are present in AZIMUT hotel chain, but the main problem remains the absence of digital literacy on the part of the personnel.

The identified problem within the context of digital transformation may become an impediment on the way of formation of digital culture. Zavyalova E.B., Kondratev V.B. (eds). Korporativnoe upravlenie kak faktor povysheniya investitsionnoi privlekatel'nosti [Corporate Management as a Factor of Increasing Investment Attractiveness]. Moscow, MGIMO-Universitet Publ., 2014. 311 p.

References

1. E.B. Zav'yalova, V.B. Kondrat'ev et al., *Corporate Management as a Factor of Increasing Investment Attractiveness*, 240 (2014)
2. I.L. Shevchenko, Bulletin of the South Ural State University. Ser. Economics and Management, Corporate Culture as a Mechanism of Corporate Governance in Russian Companies, **10**, 116 (2016)
3. O.N. Kolomyts, A.S. Boyko, R.A. Kulikov, Scientific Bulletin of the southern Institute of management, **3**, 57 (2019)
4. TOP hotel companies in the worldHotelier.PRO (2017), https://hotelier.pro/
5. P. Parviainen, J. Kääriäinen, M. & Tihinen, S. Teppola, International Journal of Information Systems and Project Management, 5, 63 (2017)
6. Development of digital economy in Russia until 2035, http://innclub.info/wp-content/
7. Decision of the Supreme Eurasian Economic Council on the 'Main Lines of Realization of the Digital Agenda of the Eurasian Economic Union until 2025' No. 12 dated 11 November 2017, http://www.eurasiancommission.org/
8. K. Gold, *Managing for Success: A Comparison of the Private and Public Sectors* (1982)
9. E. Schein, *Organizational Culture and Leadership* (1985)
10. P. Kansakar, A. Munir, IEEE Consumer Electronics Magazine. Shabani, Technology in Hospitality Industry: Prospects and Challenges, **8 (3)**, 60(2019)
11. V. Zhravleva, *Corporate Culture - the mirror of the head?* (2014)

12. Interview with Mr. Maksim Brodovsky, General Director of AZIMUT Hotels, https://www.vedomosti.ru/
13. Interview with Ms. Tatyana Popova, the General Director of 'AZIMUT Hotel Vladivostok,' https://www.vedomosti.ru/

AUTHOR INDEX

AUTHOR INDEX